Structure and
Bonding in
Noncrystalline
Solids

Structure and Bonding in Noncrystalline Solids

Edited by

George E. Walrafen
Howard University
Washington, D.C.

and

Akos G. Revesz
Revesz Associates
Bethesda, Maryland

Plenum Press • New York and London

Library of Congress Cataloging in Publication Data

International Symposium on Structure and Bonding in Noncrystalline Solids (1983:
Reston, Va.)
  Structure and bonding in noncrystalline solids.

  "Proceedings of the International Symposium on Structure and Bonding in Non-
crystalline Solids, held May 23–26, 1983, in Reston, Virginia"—Verso t.p.
  Includes chapters submitted independently of the conference.
  Includes bibliographical references and index.
  1. Chemical structure—Congresses. 2. Chemical bonds—Congresses. 3. Amorphous substances—Congresses. I. Walrafen, George E. II. Revesz, Akos G. III. Title.
QD461.I64 1983   530.4'1   86-22544

ISBN-13: 978-1-4615-9479-6   e-ISBN-13: 978-1-4615-9477-2
DOI: 10.1007/978-1-4615-9477-2

Based on the proceedings of the International Symposium on Structure and Bonding in
Noncrystalline Solids, held May 23–26, 1983 in Reston, Virginia

© 1986 Plenum Press, New York
Softcover reprint of the hardcover 1st edition 1986

A Division of Plenum Publishing Corporation
233 Spring Street, New York, N.Y. 10013

FOREWORD

Noncrystalline (NC) solids, as is well known, lack the long-range order of crystals. Accordingly, they exhibit scattering, e.g., x-ray, electron, and neutron, but not the diffraction patterns characteristic of crystals. The intensity distributions from NC solids are usually transformed into radial distribution functions (RDF), but the interpretation of the RDF's is not unique. The lack of long-range order, and the non-uniqueness of the structural interpretation, have constituted the main obstacles to the usual solid state physics approach which has been so successful in dealing with crystals. As a corrolary, questions of local order and structure have frequently been de-emphasized.

This monograph contains a collection of chapters many of which emphasize local-structure and chemical bonding as opposed to long-range order. Most of the chapters were chosen from talks given at the international symposium, "Structure and Bonding in Noncrystalline Solids," held in Reston, Virginia in May of 1983. Other chapters, however, were simply submitted independently of the Reston conference. Thus, this book is not a proceedings of that conference, nor was it ever intended to be.

The chapters presented here range from theory of bonding and structure, to structurally oriented measurements of various kinds, e.g., ESR, Raman, to more applied chapters. Our goal was to produce a monograph that enhances understanding of the structures of NC solids.

In preparing this book we wish particularly to acknowledge the able assistance of Mrs. Rebecca L. McVicker who typed and assembled the camera-ready copy. We are also grateful to Professor P. N. Krishnan for preparing the index and for general assistance.

George E. Walrafen
Akos G. Revesz
Washington, DC

CONTENTS

APPLICABILITY OF THE MOLECULAR DYNAMICS TECHNIQUE TO SIMULATE THE

VITREOUS SILICA SURFACE

Stephen H. Garofalini

Rutgers University, College of Engineering, Department
of Ceramics, Brett and Bowser Roads, Busch Campus
P. O. Box 909, Piscataway, New Jersey 08854

INTRODUCTION

Vitreous silica (V-SiO$_2$) has been of technological importance
for many years in the ceramics and glass industries and, more
recently, in catalysis, microelectronics, fiber optics and nuclear
radiation waste containment. Because of its importance in a number
of diverse areas, v-SiO$_2$ has been the subject of a large number of
experimental and theoretical studies aimed at determining the properties and structure of this material. In recent years, the
molecular dynamics (MD) computer simulation technique has been used
to determine the structural and dynamic properties of v-SiO$_2$ and
silicate glasses at the atomic level [1-6]. In most of these MD
simulations the modified Born-Mayer-Huggins equation has been used
as the form of the pairwise interatomic potential function, although
other potentials have been used [6,7]. The Born-Mayer-Huggins potential, being most suited for ionic systems, has also been used in
simulations of alkali halides and BeF$_2$ [8-11]. Although v-SiO$_2$ is
about 50% covalent, the modified Born-Mayer-Huggins equation can be
considered as an effective potential in stimulations of vitreous silica which reproduces a number of structural and dynamic features with
surprisingly reasonable accuracy. In particular, the simulations
reproduce the tetrahedral arrangement of oxygen atoms around a
silicon, with predominantly corner sharing between tetrahedra, generate a reasonably accurate radial distribution function, with an
appropriate SiO bond length of 1.62 Å, reproduce the observed Si-
O-Si bond angle distribution from 120° to 180°, and generate the
main features of the frequency spectrum of v-SiO$_2$. In addition, the
simulations indicate the presence of overcoordinated Si and O species as defects which can influence transport properties (especially
the pentacoordinate cation defect [12]) although the concentration

1

of such species in the simulations are much higher than might occur
in real glasses because of the very high quench rates inherent in
the simulations.

The application of v-SiO$_2$ as catalyst supports or passivation
layers in microelectronics or as the main constituent in nuclear
radiation waste containment glasses has created an increased
interest in the surface properties of v-SiO$_2$ and silicates. The use
of the common surface analytical techniques such as Auger electron
spectroscopy or low energy electron diffraction are difficult or
impossible to apply to studies of the v-SiO$_2$ surface because of the
insulative and amorphous nature of the material. A technique such
as ion scattering spectroscopy (ISS) provides a useful method for
determining the composition of the uppermost atomic layer of v-SiO$_2$
or silicate glasses [13,14], but does not provide information
concerning surface structure in these amorphous systems. A
technique which could potentially provide information about the
surface structure, surface EXAFS (SEXAFS), has not yet been applied
to studying these surfaces. Although IR and Raman spectroscopy have
provided much useful information concerning the v-SiO$_2$ surface [15-
18], these, as well as the other abovementioned experimental
techniques, only provide average properties from large numbers of
atoms over fairly long times and detailed atomic level behavior must
be inferred. The importance of understanding such atomic level
behavior lies in the fact that average properties are based on a
large number of individual events and understanding individual
events could provide answers which had previously been masked.
Understanding detailed events could provide answers to problems
which currently have different models to explain the same phenomenon
(e.g., leaching of sodium silicate glasses [19,20]) or provide additional information previously not considered. For instance, the
mechanism of stress corrosion cracking of glasses in water
has recently been described as a concerted nucleophilic-
electrophilic reaction between a water molecule and a strained
siloxane (Si-O-Si) bond at the crack tip [21], where the strained
bond results from an opening of the Si-O-Si bond angle at the tip.
However, quantum mechanical studies [22,23] and molecular dynamics
simulations [24] show that an increase in bond angle creates only a
minor increase in bond energy and bond length, whereas a decrease in
bond angle causes much greater bond energy and bond length increases. Consideration of reactions at compressed bond angles would
therefore be worthwhile. (Compression at the glass surface does not
show enhanced stress corrosion cracking, but the effect of macroscopic stress, compressive or tensile, on the local atomic level
bond angle distribution at the surface is not well understood.)
There are many other areas where knowledge of precise atomic level
behavior would be beneficial.

The application of the MD simulation technique in studies of
the v-SiO$_2$ surface could provide useful information concerning surface behavior at the atomic level. Although the MD technique has

been used successfully to simulate bulk v-SiO$_2$ and silicate glasses, incomplete knowledge about the forces between atoms at the surface versus those between atoms in the bulk creates an uncertainty about the extension of the technique to simulate the surface behavior of these glasses.   The work presented here describes the first attempt to determine the adequacy of using the MD technique to simulate the surface of v-SiO$_2$.   Comparisons between the properties generated in the simulations with known or expected properties derived from experiments are made.   Because of the tetrahedral network structure of v-SiO$_2$, most oxygen atoms act as the 'bridge' between two tetra- hedra and are called bridging oxygen (BO).   The Si-bridging oxygen bond length is $\sim$ 1.62 Å (see Figure 1a).   This structure also creates the Si-O-Si bond angle, which, while having distinct values in the crystalline forms of silica, has a wide distribution between 120° to 180° in v-SiO$_2$, with an average near 150° [25].   In addition to bridging oxygen, oxygen linked to only one Si exist and are called non-bridging oxygen (NBO) (see Figure 1b).   The pristine v-SiO$_2$ surface has oxygen as the outermost species rather than silicon [13,16,18,26], with both BO and NBO present (although a surface NBO would be hydrated in real systems and would be called a silanol (Si-OH) group).   Also, because of the lack of neighbors on one side of the surface oxygen, surface BO may be expected to form strained siloxane bonds [16,27].

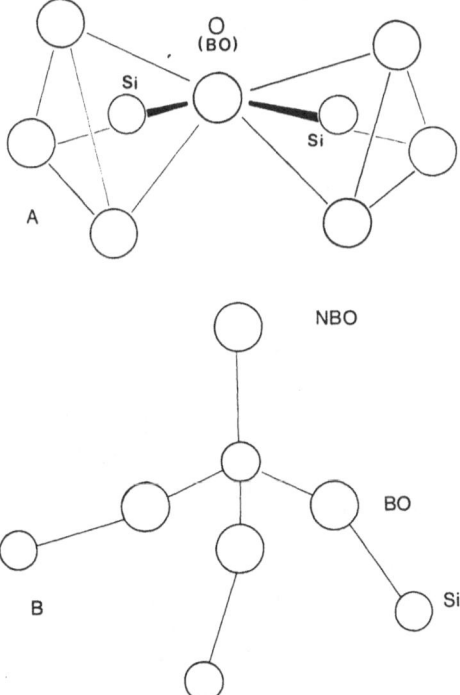

Fig. 1   (A) Two tetrahedra connected by a bridging oxygen (BO).
         Other oxygens attached to other silicons which are not shown.
         (B) Nonbridging oxygen (NBO) attached to only one silicon.

COMPUTATIONAL PROCEDURE

    MD computer simulations involve solving Newton's equations of
motion for a system of particles (atoms) interacting via an assumed
interatomic potential. A net force on each atom due to all the
other atoms is calculated and each atom of the system is moved
according to that force. Reiteration provides for additional
movements of the atoms, with the subsequent time evolution of the
system. A number of algorithms are available for solving Newton's
equations. In the simulations reported here, a fifth order
Nordsieck algorithm [28] using a predictor-corrector scheme is
employed. In the prediction part of the scheme, this method
predicts the next position of each atom using the position and its
first five time derivatives and moves the atom according to the time
step t. The predicted values of all of the time derivatives are
similarly calculated. The forces on each atom are then calculated
according to the predicted configuration and each atom's position
and time derivatives are corrected according to the new force in
order to account for the changing configuration of the system. New
predictions, configurations, and corrections are determined
reiteratively, creating a time evolution of the system for a time
equal to the number of reiterations, or moves, times the time step.
The algorithm used here is very stable and allows for constant
energy runs after an initial velocity rescaling to the desired
temperature for the first several hundred moves of the run. Average
temperatures invariably show variations of approximately 1% over
several thousand move runs and energy drifts are less than 1 part
in 10000.

    The modified Born-Mayer-Huggins potential, used as the
effective potential in these simulations of the v-SiO$_2$ surface,
gives the potential between atoms i and j, separated by distance
r_{ij}, as

$$\phi_{ij} = A_{ij}\exp\ (-r_{ij}/\rho_{ij}) + \frac{z_i z_j e^2}{r_{ij}}\ \text{erfc}\ (r_{ij}/\beta_{ij})$$

The parameters used here are given in Table I. Although the
coulomb term is modified by the first term of the Ewald sum, with
β_{ij} being the product of the length of the box containing the
system of atoms in three dimensional periodic boundary conditions
and a parameter which enables rapid convergence of this term,
neglect of the complete Ewald sum makes this function empirical,
or effective. In this case, rather than considering β_{ij} as
originally proposed, one may view β_{ij} and the complementary error
function as a means of creating effective ionic charges on the
species of interest. Although fully ionic charges are used for
Si and 0, +4 and -2 respectively, the modification reduces these
charges as a function of separation distance. Variations in β_{ij}
can therefore create variations in the effective charges between pairs.

<u>Table I</u>

$$A_{Si-O} \quad = 2.9620 \times 10^{-9} \text{ ergs}$$

$$A_{O-O} \quad = 0.7254 \times 10^{-9} \text{ ergs}$$

$$A_{Si-Si} \quad = 1.8770 \times 10^{-9} \text{ ergs}$$

$$\varepsilon_{ij} \quad = 2.9 \times 10^{-9} \text{ cm}$$

$$Z_{Si} \quad = +4$$

$$Z_{O} \quad = -2$$

$$\beta_{ij} \quad = 2.538 \times 10^{-8} \text{ cm at } 300^{\circ}K$$

The formation of the v-SiO$_2$ surface occurred as follows (see Figure 2). Bulk v-SiO$_2$ was simulated using 288 atoms in a box with equal X and Y dimensions and three dimensional periodic boundary conditions. The system was initially melted at $9000^{\circ}K$ (see Figure 2a) and quenched to lower temperatures by rescaling velocities to

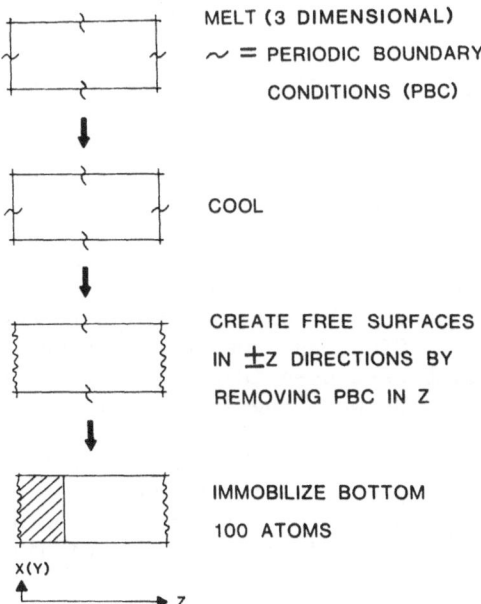

Fig. 2 Schematic showing the procedure used to create the v-SiO$_2$ surface.

the desired temperature every tenth move for the first 500 moves,
followed by several thousand moves in a constant energy run. At
1500°K, free surfaces in the Z directions were made by removing the
periodic boundary conditions in the Z direction but keeping them in
X and Y. The system was subsequently quenched to 300°K and allowed
to equilibrate thermally. The bottommost 100 atoms were immobi-
lized and only the upper 188 were mobile for the remaining 20,000
moves (20 psec). The final 2 psec were used for data collection.
Variations on this procedure included immobilizing the bottommost
atoms at 1500°K, then quenching to 300°K, or not forming the surface
until 300°K, with or without immediate immobilization of the bottommost atoms. In all cases the results are very similar to those
reported here. The length of the sides of the box parallel to the
surface at 300°K was 14.27 A (with volumes rescaled at the higher
temperatures).

All data analysis was averaged over the last 2 ps of the
simulation. Analysis of the surface includes density profiles of
individual species as a function of distance perpendicular to the
surface, pair distribution functions calculated as the number of
atoms at a distance between r and r + Δr from a central atom,
averaged over all central atoms located within specific regions of
the system, and the Si-O-Si bond angle distribution calculated for
all two fold oxygen (BO) within the selected regions. Local structural differences between atoms in the interior and at the glass
surface could be ascertained by evaluating the above mentioned
properties.

RESULTS AND DISCUSSION

Density profiles of the Si and O atoms are shown in Fig. 3
and include only the mobile atoms. The immobile atom-mobile atom
interface is at the left, the free surface is at the right. The
abscissa indicates the depth below the outermost position of the
atoms, which is arbitrarily set as the zero point. The profiles
show that the oxygen are the outermost species, indicating that the
simulations reproduce the required predominance of oxygen at the
surface instead of silicon. Other runs using different starting
configurations and quench procedures invariably show oxygen as the
outermost species in v-SiO$_2$, as shown here. Labels 'A' and 'B'
indicate regions which were used in subsequent analysis to determine
structural differences between the interior (A) and the surface (B).

The Si-O-Si bond angle distribution for twofold oxygen located
within the interior is shown in Fig. 4A and for twofold oxygen
located at the surface in Fig. 4B. Twofold oxygen are those
oxygen which have only two silicon neighbors within 2.0 A during the
run (the normal Si-O bond length is 1.62 A). This precludes
incorporation of threefold oxygen in the distribution which would
have smaller bond angles, although it does not prevent inclusion of

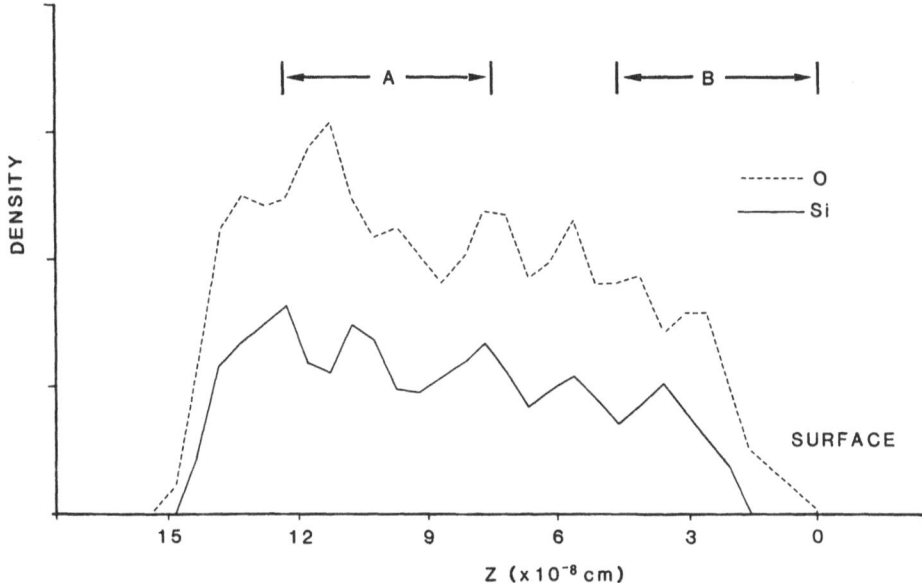

Fig. 3   Density profiles of mobile oxygen atoms (dashed line) and
         silicon atoms (solid line) as a function of distance per-
         pendicular to the surface. Immobile atoms-mobile atoms
         interface at left; surface at right. A and B denote regions
         of system used for subsequent analysis.

a twofold oxygen which also has a third Si neighbor near to, but
greater than, 2.0 Å. Comparison between Figure 4A and Figure 4B
shows that an increase in the smaller bond angles occurs at the
surface and may indicate the presence of strained bonds.

     More important information can be found in the pair distri-
bution functions around central atoms located in the interior
and in the surface, as shown in Figures 5A and 5B, respectively.
The peak positions in the PDF of the interior (Figure 5A) occur at
distances which are similar to those found in MD simulations of bulk
v-SiO$_2$ [1-3]. However, a broadening of the Si-O and O-O peaks in
comparison to bulk v-SiO$_2$ occurs due to slightly higher intensities
at the larger distance sides of each peak. Direct evaluation of the
coordination number of the atoms located in the interior region
indicated a high concentration of overcoordinated species (fivefold
Si and threefold O) which would cause this broadening. Additional
simulations using different starting configurations and quenches
gave similarly high concentrations of defect species in the subsur-
face regions. Such defects can also cause a slight increase in the
distance of the peak maximum in the Si-O peak as can be seen in the
width of the top of the Si-O peak in Figure 5A. Much lower concentrations of overcoordinated defect species are found at the surface
and this is reflected in the lower intensity of the high distance
side of the Si-O and O-O peaks for the surface species (Figure 5B)

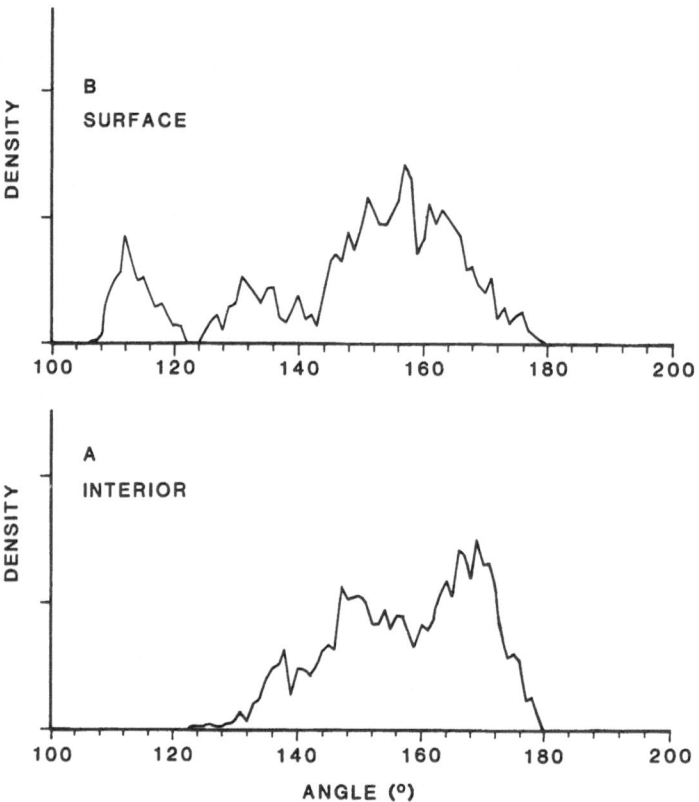

Fig. 4 Si-O-Si bond angle distribution twofold oxygen located
        in (A) the interior (see region A in Figure 3); (B) the
        surface (see region B in Figure 3).

in comparison to the interior (Figure 5A). Also, the Si-BO bond
length at the surface is about 0.01 Å-0.02 Å greater than the 1.62 Å
value found for Si-BO bond lengths in bulk v-SiO$_2$ [2-4], with the
longer bond lengths occurring for bonds closer to the surface. Such
bond length increases between surface silicon and bridging oxygen
consistently appear in all of our simulations. In addition, simulations in our lab using different β_{ij} parameters in the coulomb
portion of the potential than those reported in Table I produced
glasses with little or no overcoordinated species, yet still showed
the Si-BO bond length increase at the surface in comparison to
either the interior of those glasses or to normal bulk v-SiO$_2$.
Although such bond length increases have not been established experimentally, these increases may not be just an artifact of the
simulations caused by the use of the pairwise potential and the
removal of one-half of the system in creating the surface (a surface
expansion effect), as will be shown below. Nonetheless, experiments
should be considered to establish this observation.

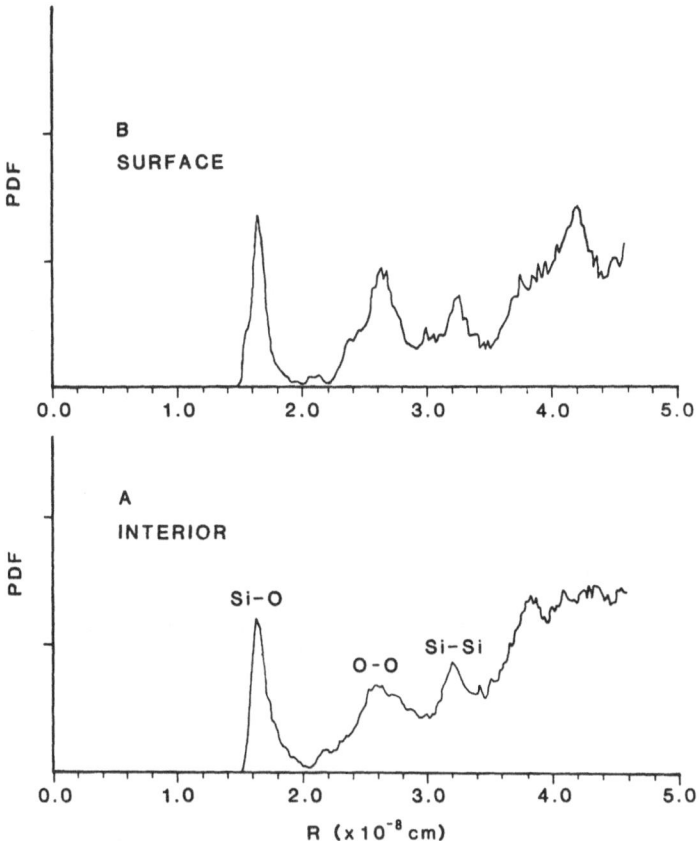

Fig. 5 Pair distribution function for central atoms located in (A) interior, and (B) surface.

The shoulder occurring at the short distance side of the Si-O peak at the surface, Figure 5B, is not observable in experimental RDF's of bulk v-SiO_2 and only becomes discernable when the RDF of just the surface region is made. Detailed analysis of the surface species indicated that five NBO were present. A PDF generated by jsut these NBO was made and is shown in Figure 6. As can be seen, the Si-NBO bond length occurs at the same distance as the shoulder in the Si-O peak in Figure 5B. This distance, 1.54 Å is 0.08 Å less than the normal Si-O bond length of 1.62 Å. Calculations [29] based on vibrational changes indicate that the Si-NBO bond length should be approximately 0.08 Å less than the Si-BO bond length. Additional simulations have given similar decrease in Si-NBO bond length. The relaxation of the Si-NBO bond length to shorter distances is very significant in that it indicates the adequacy of this MD technique to simulate an important structural feature in the v-SiO_2 surface. Also, if the Si-BO bond length increase at the surface, as mentioned above, was only due to creation of the surface and removal

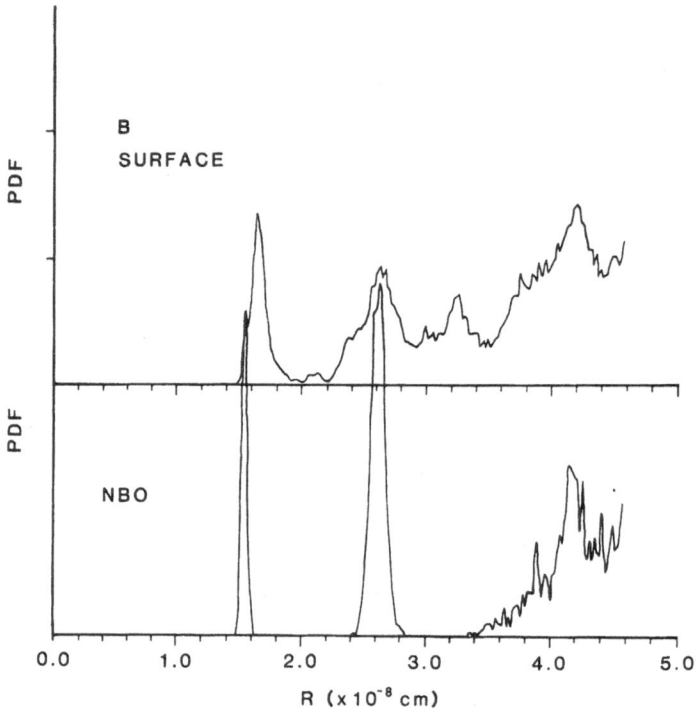

Fig. 6 Comparison between the pair distribution function averaged
        over all central atoms located in the surface region
        (Surface) and the pair distribution function around non-
        bridging oxygen (NBO).

of atoms in the +Z direction (the surface expansion effect), then
one might expect a similar expansion in the Si-NBO bond length. The
decrease in the Si-NBO bond length helps alleviate some of the
concern about the surface Si-BO bond length expansion, although it
does not completely validate the observed expansion. Corroboration
of Si-BO expansion at the surface by experimental techniques would
be beneficial. IR data from high surface area silica, porous Vycor
($\sim$ 250 m^2/gram), showed a decrease in vibrational frequencies at the
three main peaks ($\sim$ 400 cm^{-1}, $\sim$ 800 cm^{-1}, and $\sim$ 1200 cm^{-1}) for the
Vycor in comparison to bulk vitreous silica [30]. An increase in
the Si-BO bond length of about 0.01 Å-0.02 Å would cause a decrease
in the vibrational frequencies and may be the cause of the decreases
seen in the IR spectra. However, the shifts seen experimentally
could also be viewed as changes in intensity of already existing
vibrational modes and may not be due to bond length changes. Experiments specifically aimed at elucidating Si-BO bond lengths at
the silica surface are required.

An interesting feature seen in the simulation results presented here as well as in additional simulations of the v-SiO_2 surface concerns the nearly consistent presence of NBO at the surface. The density of NBO present is usually around 2.5-3.0 per 100 $\overset{\circ}{A}^2$. Considering that such NBO in a "real" glass would form surface hydroxyls due to the presence of water, a value near 3.0 silanols per 100 $\overset{\circ}{A}^2$ would occur on these simulated glass surfaces. This is only slightly lower than experimentally determined values near 4.6 (experimental values are highly dependent upon surface history) [27]. If the simulated surface contains strained siloxane bonds, and such strained bonds hydrolyzed in the presence of water, then the concentration of silanol groups on the simulated glass surface would be well within the experimentally determined values.

CONCLUSION

The results of the work presented here indicate that the molecular dynamics computer simulation technique can be used to reproduce the main structural details of the v-SiO_2 surface. The simulations generate the experimentally observed predominance of oxygen at the outer surface, indicate the presence of strained siloxane bonds, and generate non-bridging oxygen defects. Using a simple pairwise potential in this study, the simulations indicate that a slight bond length increase occurs between Si and bridging oxygen within 2 $\overset{\circ}{A}$ to 3 $\overset{\circ}{A}$ of the surface and a bond length decrease occurs between Si and non-bridging oxygen defects.

Additional studies using the technique presented here have been used to evaluate the surfaces of alkali silicate glasses and the structures of metal catalysts supported on v-SiO_2. Results are again in good agreement with known experimental data. Continued application of this method of investigating v-SiO_2 and silicate glass surfaces at the atomic level seem appropriate given the results presented herein. Interpretation of atomistic behavior observed in such simulations of glass surfaces should provide new insight into the mechanisms of various surface reactions.

REFERENCES

1. L. V. Woodcock, C. A. Angell, and P. Cheeseman, J. Chem. Phys. 65:1565 (1976).
2. T. F. Soules, J. Chem. Phys. 71:4570 (1979).
3. T. F. Soules, J. Non-Crystalline Solids 49:29 (1982).
4. S. H. Garofalini, J. Chem. Phys. 76:3139 (1982).
5. S. H. Garofalini, J. Chem. Phys. 78:2069 (1982).
6. S. A. Mitra, Philos. Mag. B 45:529 (1982).
7. T. Halicioglu, personnal communication.
8. L. V. Woodcock, Advances in Molten State Chemistry, Vol. 3, Plenum, New York, (1975).
9. W. R. Busing, J. Chem. Phys. 57:3008 (1971).

10. S. A. Brawer and M. J. Weber, J. Chem. Phys. 75:3522 (1981).

11. M. J. L. Sangster and M. Dixon, Advances in Physics 25:247
        (1976).

12. S. A. Brawer, Phys. Rev. Lett. 46:778 (1981).

13. R. C. McCune, Anal. Chem. 51:1249 (1980).

14. J. F. Kelso, C. G. Patano, and S. H. Garofalini, Surf. Science
        134:L543 (1983).

15. R. S. McDonald, J. Phys. Chem. 62:1168 (1958).

16. M. L. Hair, Infrared Spectroscopy in Surface Chemistry,
        Dekker, New York (1967).

17. J. B. Peri, J. Phys. Chem. 70:2937 (1966).

18. R. B. Laughlin, J. D. Joannopoulos, C. A. Murray, K. J. Harnett,
        and T. J. Greytak, Phys. Rev. Lett. 40:461 (1978).

19. B. M. J. Smets and M. G. W. Tholen, J. Am. Ceram. Soc. 67:281
        (1984).

20. R. H. Doremus, J. Non-Cryst. Solids, 55:142 (1983).

21. T. A. Michalski and S. W. Freiman, Nature, 295:511 (1982).

22. H. H. Dunken, Treatise on Materials Science and Technology,
        Vol. 22, Academic Press, New York (1982).

23. A. G. Revesz and G. V. Gibbs, The Physics of MOS Insulators,
        Pergamon Press, New York (1980).

24. S. H. Garofalini, to be published.

25. A. W. Wright and R. Sinclair, Physics of Silicon Oxide and Its
        Interfaces, Pergamon Press, Oxford (1979).

26. R. B. Laughlin and J. D. Joannopoulous, Phys. Rev. B. 17:4922
        (1979).

27. R. K. Iler, The Chemistry of Silica, Wiley, New York (1979).

28. A. Nordsieck, Math. Comput. 16:22 (1962).

29. G. Lucovsky, Philos. Mag. B. 39:513 (1979).

30. C. A. Murray and T. J. Greytak, Phys. Rev. B, 20:3368 (1979).

STATISTICAL MECHANICS OF CHEMICAL DISORDER: APPLICATIONS TO SIMPLE GLASSES

R. Araujo

Research and Development Division
Corning Glass Works
Corning, New York 14831

ABSTRACT

A statistical mechanical method for calculating the thermo-dynamic properties of model systems in which the structural units need not all be bonded to the same number of neighbors is presented. Comparison of the calculated miscibility dome with that observed for alkali borates suggests that there is a strong association between two non-bridging oxygen atoms and two alkali ions at temperatures at which phase separation is observed.

A semi-quantitative description of the suppression of phase separation in alkaline earth silicates by alumina is presented. The effect of temperature on the fraction of boron atoms in tetrahedral coordination in alkali borate glasses is predicted. An asymmetrical miscibility dome for binary borosilicate glasses is also predicted.

INTRODUCTION

The structure of a crystalline material is described by the specification of the positions of all the atoms in the system relative to some coordinate system. The task is simplified considerably by the existence of translational symmetry which reduces the problem to the manageable one of merely specifying the position of the atoms in one unit cell. Since glasses do not have long range order, no practical method exists for specifying the positions of all the atoms. Consequently, a discussion of the structure of a glass must be less precise and less complete than one of a crystal. Nevertheless, a meaningful and useful discussion of the structure of a glass is sometimes possible because of the short range order observed.

13

Sometimes a small group of atoms in a given configuration is so stable energetically that it retains its integrity over wide ranges of temperatures and it can be considered to be a fundamental building unit of the material in which it occurs.

In fused silica, for example, the silicon atoms are always bonded to four oxygen atoms in such a way that the oxygen atoms are at the corners of an imaginary tetrahedron. Since each oxygen is found at corners of two neighboring tetrahedra, it is useful to consider the fundamental building block to be a silicon atom bonded to four half atoms of oxygen (denoted in Figure 1 as S_4).

In alkali silicate glasses, the oxygen from the alkaline oxide introduces non-bridging oxides into the coordination shell of the silicon. A group of atoms, such as that denoted as S_3 in Figure 1, conceivably may be an important structural unit in such a glass.

The present paper describes a statistical mechanical technique that can sometimes be used for calculating the properties of a model glass comprised of a small number of assumed building units which interact only with adjacent units as they would in covalently bonded systems. Comparison of the calculated properties of the model with observed properties of real glasses may help to establish if building units similar to those postulated are important entities in the real glasses.

One of the examples of the application of this technique to be included in this paper will be the calculation of the miscibility dome of a model system comprised exclusively of the building units illustrated in Figure 1 and the comparison of the calculated dome with that observed for alkali silicate glasses. Serious discrepancies between the results make it unlikely that a mixture of S_3 and S_4 units is an adequate model of alkali silicate glasses. An alternative choice of structural units will be seen to lead to far better agreement with experiment.

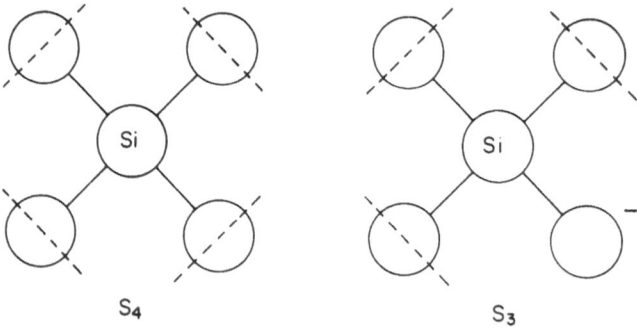

S_4 S_3

Fig. 1   Assumed building units. The slash through the oxygen atom
indicates the concept of half an oxygen atom. The symbols
S_i refer to a silicon atom bonded to i bridging oxygens.

THE PARTITION FUNCTION

   If N atoms are confined to some volume V and if one knows all
the interactions between the atoms, one can, in principle, by solving the Schroedinger equation, determine all possible states of the
system and the energy associated with each state. Knowing all the
energies, one can write a simple but useful function of the energies
and the temperature

$$Q = \sum_i e^{-Ei/kt} \qquad (1)$$

This function is called the partition function or sum over states
(the index i runs over all possible states of the system). All the
thermodynamic properties of the system can be obtained in terms of
the partition function and its derivatives with respect to temperature. The problem of extracting the thermodynamic properties from
the partition function is made somewhat simpler by grouping all the
states characterized by the same value of the energy and by writing
the partition function as a sum over energy levels.

$$Q = \sum_j \Omega j \ e^{-Ej/kT} \qquad (2)$$

where Ωj is the number of different states having the same energy
Ej. The reason this procedure is helpful is that in the determination of the thermodynamic properties of the system, one can replace
the partition function when written as in equation (2) by its
largest term. A system in thermal equilibrium is overwhelmingly
likely to be in one of the states included in this largest term.
Thus the problem is reduced to the problem of finding the largest
term.

   Since one cannot solve the Schroedinger equation for the
general many-bodied problem, one often seeks approximate values
for the Ej in the partition function by assuming that the total
energy can be expressed as the sum of energies corresponding to
simple models. In the present case, only the energy of bonding
between adjacent structural units will be considered. Therefore,
any possible energy of the system can be expressed as the sum of
bonding energies between pairs of structural units, i.e.,

$$E = \sum_{i>j} \sum_j \varepsilon_{ij} \ N_{ij} \qquad (3)$$

where ε_{ij} is the interaction energy between a structural unit
(hereafter called atom) of type i and one of type j and N_{ij} is the
number of bonds formed between an atom of type i and neighboring
atoms of type j.

The essential prlblem of statistical mechanics is to find the number of states associated with any particular value of the energy E. This is more readily done in terms of the number of atoms N_i and N_j than in terms of the number of bonds N_{ij}. Therefore, an expression for N_{ij} in terms of N_i and N_j is sought so that the energy can also be expressed in terms of the number of atoms. If all the atoms are bonded to the same number of neighboring atoms and bonds are formed randomly (i.e., the energy of bonding plays no role), the relationship is especially simple.

$$N_{ij} = \frac{N_i N_j}{\sum\limits_k N_k} \tag{4}$$

Araujo[2] showed that if all the atoms do not form the same number of bonds, equation (4) is replaced by

$$N_{ij} = \frac{b_i b_j N_i N_j}{\sum\limits_k b_k N_k} \tag{5}$$

where b_i is the number of bonds that must be formed by an atom of type i.

Naturally, when the ε_{ij}'s are not all equal, equation (5) must be modified to reflect the effects of energetics. This is done by including a set of appropriate constants f_{ij} in the above relationship.

$$N_{ij} = \frac{f_{ij} b_i b_j N_i N_j}{\sum\limits_k f_{ik} b_k N_k} \tag{6}$$

Only the ratios of the f_{ij}'s to the f_{ii}'s are important so the latter constants are all set equal to unity. The requirement that $N_{ij} = N_{ji}$ introduces a relationship between f_{ij} and f_{ji}. The method of determining the numberical values of the f_{ij}'s will be clarified below.

In order to calculate the approximate number of states associated with any allowed value of the energy, it is sufficient to calculate the number of states associated with a corresponding set of N_{ij}'s (or alternatively, f_{ij}'s). See Appendix. The number of states corresponding to a particular set of N_{ij}'s is easily calculated by considering the bonds between atoms to form a three-dimensional network and by calculating the number of permutations of atoms among the sites corresponding to possible connections of the network. Araujo[3] showed that the logarithm of the number of states is given by

$$\ln \Omega = - \sum_i b_i N_i \sum_j \frac{f_{ij} b_j N_j}{Z_i} \ln \frac{f_{ij} b_j N_j}{Z_i} \qquad (7)$$

where $Z_i = \sum_j f_{ij} b_j N_j$

It should be noted that Eq. (7) yields the entropy of mixing
bonds. The high temperature limit of this expression in which all
the f_{ij}'s are equal to unity is larger than the ideal entropy of
mixing. The reason that the number of states is overcounted is
that every atom which serves to define the symmetry of a central
atom is itself counted as a central atom. Equation (7) is corrected by dividing by the average number of bonds per atom. Thus

$$\ln \Omega_{corrected} = \sum_i b_i N_i \sum_j \frac{f_{ij} b_j N_j}{Z_i} \ln \frac{f_{ij} b_j N_j}{Z_i} / (\sum_i b_i N_i / N) \quad (8)$$

where $N = \sum_i N_i$.

Since both Ω and E can be expressed in terms of the f_{ij}'s,
one need only to find the set of f_{ij}'s which gives the largest
possible value of $\Omega \, e^{-E/kT}$ in order to calculate the equilibrium
properties of the system.

APPLICATION TO PHASE SEPARATION IN ALKALI AND ALKALINE EARTH
SILICATES

When one adds an alkali oxide to silica, nonbridging oxygens
are formed. Since the nonbridging oxygen is negatively charged,
one can assume that an alkali ion is held in the immediate vicinity
by coulombic forces. The model to be studied is one in which the
addition of ome mole of alkali oxide converts two moles of S_4 to
two moles of S_3. This is tantamount to assuming that at most, one
nonbridging oxygen is bonded to any one silicon atom. Obviously,
this model cannot be applied when the alkali content exceeds 1/3 and
it may be a poor approximation when the alkali is very near that
level.

The consequences of assuming that the interaction of an
alkali ion nonbridging oxygen pair with a second such pair is
negligible will be examined first. Values of the ε_{ij}'s are
assumed and free energies are calculated and plotted as a function
of N_3 at each of several temperatures. See Figure 2. Phase
separation is indicated by a negative second derivative. The
extent of phase separation is between the two compositions having
a common tangent line. If the energy of a bond between a silicon
atom and a bridging oxygen is unchanged by the existence of a
nearby nonbridging oxygen atom, the system will show no phase
separation. In contrast to this case, a strong tendency to phase

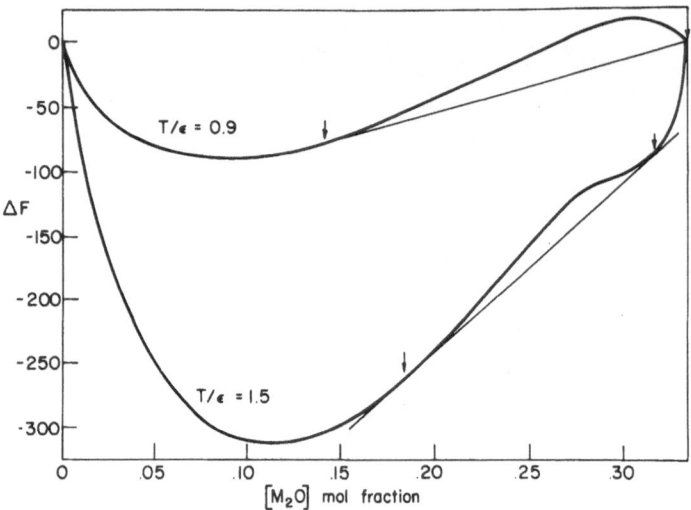

Fig. 2   Calculated free energy relative to that of pure S_4 and
         S_3 vs M (the alkali content).

separate is found if the energy of the bonds between silicon atoms
and bridging oxygen atoms is influenced in such a way that
$2\varepsilon_{34} - (\varepsilon_{44} + \varepsilon_{33}) > 0$. In such a case, f_{33} and f_{44} can each be set
equal to unity and the value of f_{34} (equal to or less than unity)
which leads to the largest term in the partition function is
sought. The asymmetry of the calculated miscibility dome seen
in Figure 3 is easily explained. The requirement that $N_{34} = N_{43}$
leads to the constraint

$$f_{43} = \frac{4f_{34}N_4}{4f_{34}n_4 + 3(1-f_{34})N_3} \tag{9}$$

Although f_{43} and f_{34} approach the same value (unity) as the
temperature approaches infinity, at low temperatures the constraint
expre-sed by equation (9) introduces a degree of asymmetry into
the entropy. Physically this means that at low temperature the
S_4 atoms are less likely to be in an unmixed phase than are S_3
units because, due to the larger number of neighbors of the
former atom, more entropy is produced by adding an S_4 to the
mixed phase than by adding an S_3.

Figure 4 abstraced from Charles [4] indicates that the
asymmetry of the experimentally determined miscibility dome
differs markedly from that shown in Figure 3. It should be
emphasized that different choices for the values of the ε_{ij}'s
do not alter the asymmetry of the miscibility dome; they merely

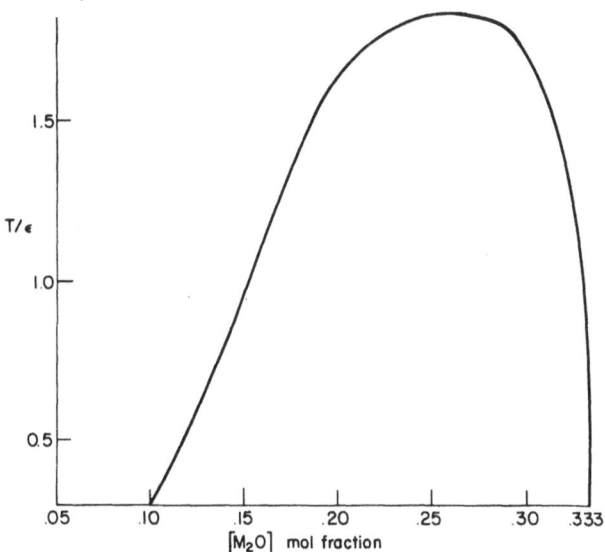

Fig. 3   Calculated miscibility gap for mixtures of S_3 and S_4.

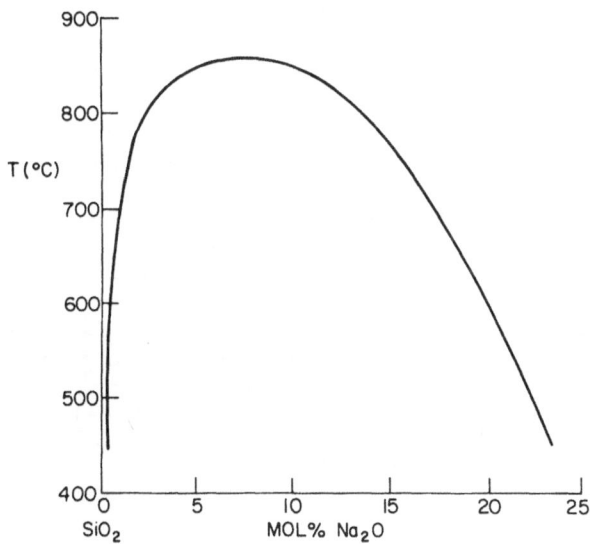

Fig. 4   Observed miscibility dome for mixtures of SiO_2 and Na_2O.

serve to change the predicted upper consolute temperature. Clearly
then, the model is inadequate.

If the non-bridging oxygens always occur in pairs, one
can consider the two coupled S_3 units to function as a single

S_6 unit. (See Figure 5). In such a case the equations for
conservation of mass become

$$N_6 = M \tag{10a}$$

$$N_4 = 1 - 3M \tag{10b}$$

where M is the mole fraction of the alkali. The asymmetry in the
relationship of f_{64} to f_{46}, causes the asymmetry in the predicted
miscibility dome shown in Figure 6. The asymmetry of the dome is
qualitatively similar to that of sodium silicate. It seems
likely, therefore, that in alkali silicates, at the temperatures
at which immiscibility occurs, the non-bridging oxygens have a
strong tendency to occur in pairs as shown in Figure 5.

The reason that the asymmetry seen in Figure 3 is the
reverse of that seen in Figure 6 must be emphasized. In the
latter case any incremental S_6 unit contributes more to the
entropy of the system by being in the mixed phase than does an
additional S_4, while in the earlier case, the S_4 contributes more
to the entropy than does the S_3.

The likelihood of nonbridging oxygens occuring in pairs
is at least as great in the alkaline earth silicates as it is
in the alkali silicates because the coulombic energy is minimized
when the two nonbridging oxygens are both very near the divalent
ion but are separated by it. Therefore, the present model can be
expected to be a reasonable representation of those systems also.
Indeed the miscibility domes observed in barium silicates [5] and
calcium silicates [4] are similar to those shown in Figures 4 and
6. See Figure 7 for an example.

APPLICATION TO ALUMINA IN ALKALINE EARTH SILICATE GLASSES

For the purposes of this investigation, it is assumed that
two aluminum atoms are bridged through oxygen to a single silicon

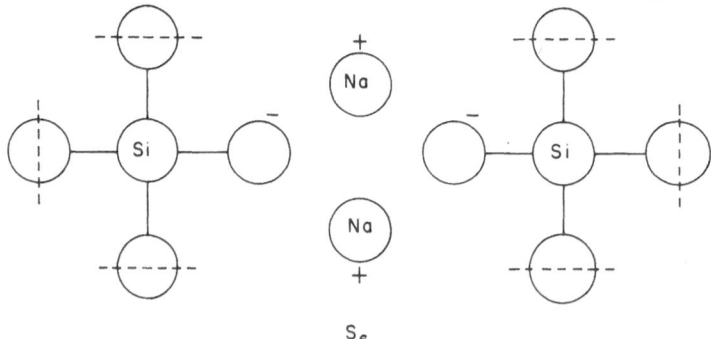

Fig. 5 S_6 building unit comprised of two strongly coupled S_3 units.

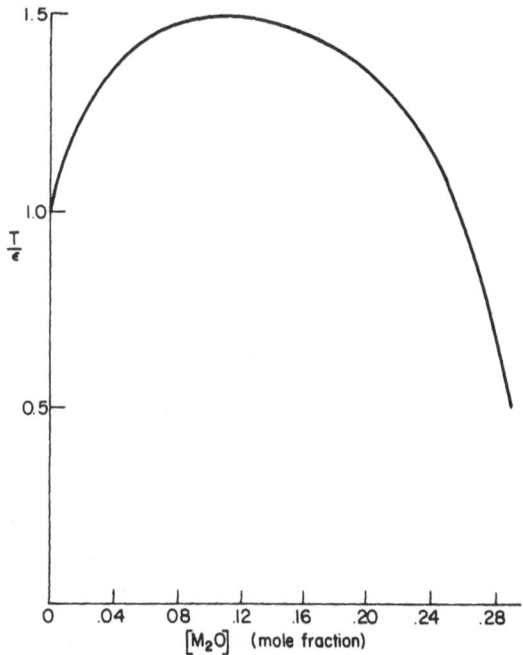

Fig. 6   Calculated miscibility dome for mixtures of S_4 and S_6.

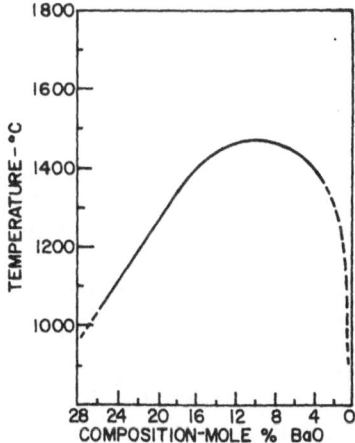

Fig. 7   Observed miscibility dome for barium silicate glasses.

atom as illustrated in Figure 8. Such a structure is postulated
because it does not violate the rule more or less obeyed in silicate
glasses that tetrahedrally coordinated aluminum atoms are never
bridged to each other. Furthermore, it has the configuration
which gives rise to the lowest coulombic energy between the
divalent modifier cation and the two negatively charged aluminum

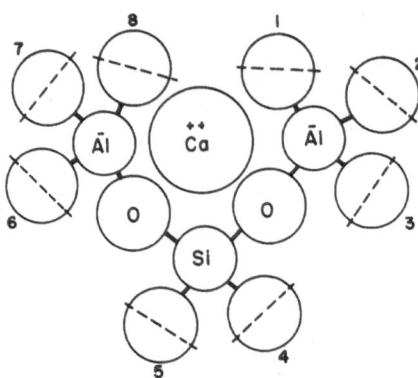

Fig. 8 S_8 building unit comprised of two aluminum atoms bridged
          through oxygen to a single silicon atom.

atoms.   It is further assumed that this structural unit (S_8) can
be bonded to either an S_4 or an S_6 or another S_8 (providing no
two aluminum atoms are bridged by an oxygen atom) with an equal
energy of interaction.

In essence then, the contribution made by aluminum only to
the entropy of the system is being considered.   Unconstrained
permutations of neighbors about the eight sites of an S_8 unit
leads to a higher entropy than the corresponding permutations about
either an S_6 or an S_4.   The rule against tetrahedral aluminum atoms
being bridged to each other, however, diminishes the extra entropy
that would otherwise result from adding alumina to an alkaline
earth silicate glass.   The net result is not immediately obvious.

As in the previous section f_{44} and f_{66} are taken to be unity
and the value of f_{46} is taken to be the one that gives the largest
term in the partition function.   Since S_8 units can bond to each
other only if two aluminum atoms are not bridged through oxygen,
f_{88} is taken to be 7/16 $(1-(\frac{3}{4})^2)$ at finite temperatures but,
of course, is taken as unity at infinite temperature.   All other
f_{18}'s and f_{81}'s are taken as unity.

Again, the requirement that $N_{ij} = N_{ji}$ (the number of bonds
between the ith and the jth species) imposes a constraint on the
f_{ij}'s which is given in Equation (11).

$$f_{64} = \frac{(6N_6 + 8N_8)\, f_{46}}{6f_{46}N_6 + 8N_8 + 4(1-f_{46})N_4} \, . \tag{11}$$

Figure 9 illustrates the free energy as a function of the
concentration of alkaline earth oxide for glasses with and without
alumina.   Figure 10 shows the calculated miscibility domes.   The
results are in qualitative agreement with the work of Seward et al [6]

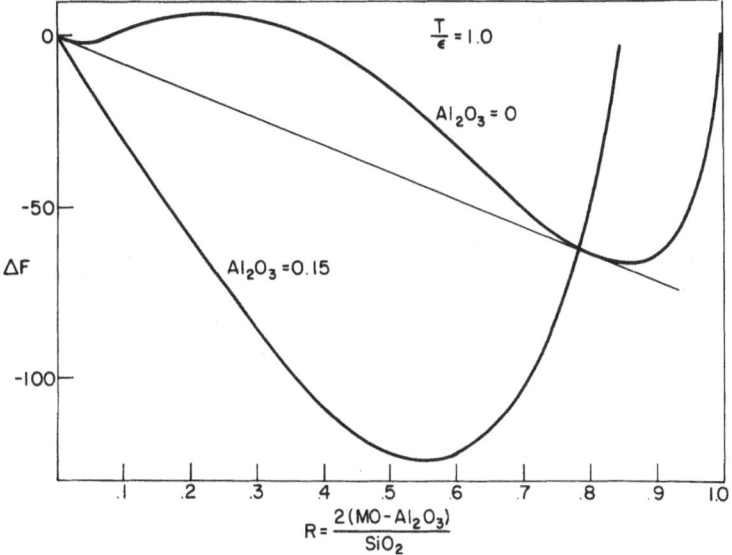

Fig. 9   Calculated free energy vs composition.

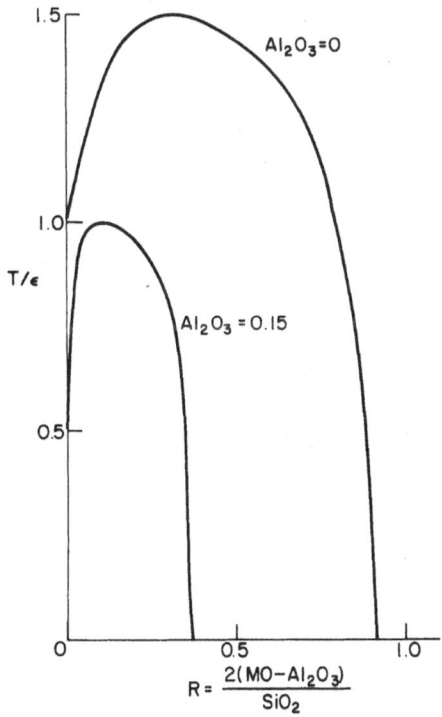

Fig. 10   Calculated miscibility dome for glasses with and without alumina.

which indicates that alumina decreases the upper consolute temperature, narrows the dome, and shifts its center to lower values
of [MO]-[Al$_2$O$_3$]. Because these workers did not extend their
studies to glasses containing more than 3% alumina, no quantitative comparison can be made.

APPLICATION TO ALKALI-BORATE GLASSES

     Boron atoms are trigonally coordinated by the oxygen atoms to
which they are bonded in boric-oxide glasses. When an alkali oxide
is added to boric-oxide glass, some of the boron atoms become tetrahedrally coordinated by oxygen atoms. In some cases, a boron atom
remains trigonally coordinated but one of the oxygens to which it
is bonded becomes a nonbridging oxygen. Jellison, Feller, and Bray [7]
used nuclear magnetic resonance spectroscopy to study alkali-borate
glasses and reported results which were independent of the choice of
alkali. They observed that N_4 (the fraction of boron atoms in tetrahedral coordination) increased with the addition of alkali to a
value slightly higher than 0.4 and then diminished slowly upon
further addition of alkali.

     If Abe's rule [8], that two tetrahedral boron atoms cannot be
bridged to each other through an oxygen atom, is obeyed, then

$$4N_4 \leq 3N_3 + 2N_2 \tag{12}$$

where N_3 is the function of boron atoms bonded to three bridging
oxygen atoms and N_2 is the fraction of boron atoms which are bonded
to two bridging oxygen atoms and one nonbridging oxygen atom. The
simple assumption that at low temperatures N_4 always assumes the
largest possible value consistent with Equation (12) leads without
further assumptions or approximations to Figure 11 which is
indistinguishable from the result of Jellison, Feller, and Bray.

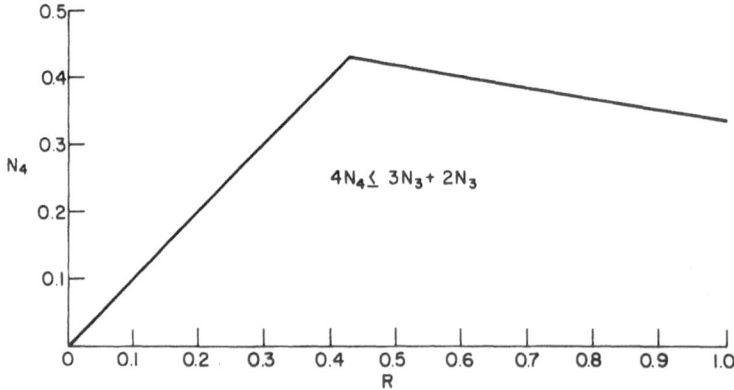

Fig. 11 N_4 bs R low-temperature limit required by Abe's Rule.

If Abe's rule is the basis for the behavior of alkali borate glasses observed in the low temperature limit, then the technique discussed above can be used to predict the behavior at higher temperatures. In order to do so, two points must be clarified. The first point is that since nonbridging oxygens are not observed in alkali borate glasses having low levels of alkali, it is plausible to assume that the formation of such a structural unit must be associated with an increase in the energy of the system, ε_2.

The second point is that it is necessary to extend this technique to systems not having fixed compositions before it can be applied to borate glasses since in such glasses the composition in terms of structural units is not fixed even though the chemical stiochiometry is fixed. In such glasses, the sum of N_4 and N_2 is fixed by the chemical stiochiometry but N_2 and N_4 individually are not. Hill [9] shows that for such a system

$$M\mu_b = -kT \ln \sum_{N_a=0}^{M} e^{\frac{N_a(\mu_a-\mu_b)}{kT}} Q(N_a, M, T) \tag{13}$$

where

$$N_a + N_b = M,$$

$$\mu_a = (\frac{\partial G}{\partial N_a})_{N_b}$$

and

$$G = -kT \ln Q + PV.$$

It is easily shown that the Helmholtz Free Energy (ignoring PV) must be

$$F = -kT \ln Q(N_a{}^*, M, T) + \varepsilon_a N_a{}^* \tag{14}$$

where $N_a{}^*$ is the value of N_a that leads to the largest term in Equation (13) and $\varepsilon_a N_a$ is the amount by which the lowest possible energy of the system is raised when N_a Atoms of type b are replaced by N_a atoms of type a. It is obvious that $\mu_a-\mu_b$ in Equation (13) can be replaced by ε_a thus considerably simplifying the search for the largest term. The canonical partition function, of course, can be expressed as

$$\ln Q = \ln \Omega - E/T \tag{15}$$

and $\ln \Omega$ and E are obtained from Equations (8) and (3), respectively.

A positive value of ε_2 is assumed and Abe's Rule is incorporated by setting ε_{44} larger than ε_2.   The results shown in Figure 12 are independent of the choice of ε_{44} so long as it is larger than ε_2 by a factor of five or greater.   All the remaining ε_{ij}'s are set equal to zero.   Parenthetically it is noted that one can determine whether a given set of assumptions leads to phase separation by plotting the free energy obtained from Equation (14) as a function of composition.   The present assumptions do not lead to phase separation.

Alternative assumptions can be utilized in a completely analogous manner.   Any specially stabilized group such as the diborate group or the boroxyl ring is easily taken into account in Equation (13) by identifying a stabilization energy ε_i with $(\mu_a - \mu_b)$.

OTHER APPLICATIONS

Since a binary borosilicate glass boron must be bonded to three bridging oxygen atoms and silicon must be bonded to four bridging oxygen atoms, it is overwhelmingly likely that one can predict the shape of the miscibility dome by merely replotting Figure 3 with the desired composition variable as the abscissa. The dome must rise very abruptly on the high boron side and much more gradually on the high silicon side.   This is probably the explanation for the asymmetry of the miscibility dome observed when small quantities of barium oxide are included in the borosilicates.[10] The symmetric miscibility dome calculated by Charles and Wagstaff [11] almost certainly reflects the inappropriateness of assuming an ideal

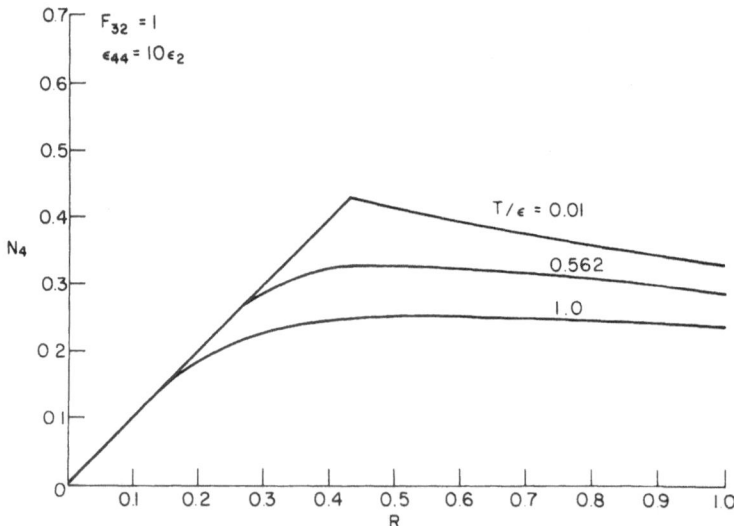

Fig. 12 N_4 vs R as a function of temperature predicted on the basis of Abe's Rule.

entropy of mixing. Note that even the high temperature limit of
Equation (8) is not the ideal entropy of mixing unless all the b_i's
are the same. The maximum value of the entropy is not changed but
the use of Equation (8) introduces a slight asymmetry into a plot
of entropy vs composition.

Since a symmetry miscibility dome implies that the b_i's are
the same, the very nearly symmetrical dome observed for titania and
silica [12] implies that the titania is largely four coordinated in
silica glasses. Likewise, one can fairly safely conclude from the
symmetric miscibility dome observed for stannic oxide and titania
oxide [13] that both metals have the same coordination number.

The strengths of the statistical mechanical calculation described herein lies in the fact that it can be applied to systems in
which not all the structural units need be bonded to the same number
of neighbors. Indeed, inspection of the high temperature limit of
Equation (8) may be sufficient to reach conclusions about the
symmetry of possible miscibility domes for simple systems.

Direct prediction of miscibility domes is possible only for
pseudo-binary systems because only one ε characterizes such systems
and the symmetry of the miscibility dome is not effected by the
value of that quantity. In multicompound systems, however, the
ratios of the various ε_{ij}'s must be considered to be adjustable
parameters and thus the predictive powers of the method are limited.
Nevertheless, for many simple systems, the method is capable of
extracting the thermodynamic consequences of postulating structural
units.

APPENDIX

For a given set of N_i's and ε_{ij}'s, specification of a set
of N_{ij}'s uniquely determines the energy. The converse is not
necessarily true except for binary systems or systems in which all
the ε_{ij}'s but one have the same value. Therefore, the number of
states associated with a given set of N_{ij}'s may be smaller than
the total number of states associated with the corresponding energy.
Nevertheless, selection of the largest term in Equation (A1) is
essentially equivalent to picking the largest term in Equation (2).

$$Q = \sum_{\{N_{ij}\text{'s}\}} \Omega(\{N_{ij}\})e^{-E(\{N_{ij}\})/kT} \qquad (A1)$$

where $\{N_{ij}\}$ indicates the set of N_{ij}'s.

Suppose that there are η (Avogadro's number) terms in Equation
(A1) which correspond to the same energy as that in the largest

term. Then the entropy of the systems lie within the limits given
by Equation (A2).

$$k \, \ln \Omega(\{N_{ij}\}*) \le S \le k \; \ln[\eta\Omega(\{N_{ij}\}*)] = k \, \ln \Omega(\{N_{ij}\}*)$$

$$+ k \, \ln \eta$$

where $N_{ij}*$ is the set of N_{ij}'s which leads to the largest term in
Equation (A1). Since the entropy of a macroscopic system is of
the order of $k\eta$, $k \, \ln \eta$ is a negligible correction. At any rate,
in all the examples discussed in the present report, specification
of E uniquely determines a set of N_{ij}'s.

REFERENCES

1. G. O. Jones, Glass (Methuen, 1956), 14.
2. R. J. Araujo, J. Non-Crystalline Solids, 58:201 (1983).
3. R. J. Araujo, J. Non-Crystalline Solids, 55:257 (1983).
4. R. J. Charles, Ceramic Bulletin 52:673 (1973).
5. T. P. Seward, D. R. Uhlmann, and D. Turnbull, J. Am. Ceram. Soc.
        51:278 (1968).
6. T. P. Seward, F. N. Molea, D. R. Uhlmann, Conference on Small
        Angle X-ray Scattering in Glasses and High Temperature
        Materials, Sept. 1969, University of Missouri-Rolla.
7. G. E. Jellison, S. A. Feller, and P. J. Bray, Phys. Chem.
        Glasses 19:52 (1978).
8. T. Abe, J. Am. Ceram. Soc. 35:384 (1952).
9. T. Hill, Statistical Mechanics, McGraw-Hill, 293 (1956).
10. E. M. Levin, C. R. Robbins, and H. F. McMurdie, Phase Diagrams
        for Ceramists, American Ceramics Society, Fig. 558 (1964).
11. R. J. Charles and F. E. Wagstaff, G.E. Report 67-C-244 (1967).
12. op, cit. Fig. 113.
13. ibid. Fig. 366.

BOND EQUILIBRIUM THEORY AND ITS IMPLICATIONS FOR THE STRUCTURE OF GLASSES

M. Cutler

Physics Department
Oregon State University
Corvallis, OR 97330

INTRODUCTION

The difficulties in experimentally determining the structure of glasses are well known.[1] This has stimulated theoretical investigations which have been undertaken from many points of view. Many of these are based on the minimization of the energy, subject to particular constraints. In the case of covalently bonded glasses, satisfaction of the bonding requirements of the atoms is a primary consideration. At the same time, statistical considerations play a role because of the disorder. Although glasses are not equilibrium systems, one can hope to gain important insights from an understanding of the equilibrium structure of covalent alloys. This is particularly true for glasses formed by cooling a melt, since the final metastable structure is frozen in from an equilibrium structure. A statistical mechanical bond equilibrium theory (BET), which was developed to describe covalently bonded liquid alloys[2,3,4,5], offers a possibility for gaining understanding of glasses from this point of view.

The original purpose of BET, which was derived in the context of twofold bonded alloys such as Se-Te, was to describe the equilibrium of small concentrations of bond defect atoms, and to understand the equilibrium configurations of impurity atoms.[2] The theory showed that n-fold (nF) bonded atomic constituents whose connectivity n differ from 2, and consequently modify the chain structure by introducing chain ends or chain branches, have an additional "polymer factor" p_n in the theoretical expression for the concentration. Aside from p_n, it is identical to the expression derivable from the Law of Mass Action (LMA). The expressions for the polymer factors depend only on the total concentrations c_n of n-F atoms for the different values of n. The nature of this dependence is to cause p_n to be constant when the average number of chain branches per molecule $\lambda \ll 1$. But when $\lambda \gg 1$, the dependence causes p_1 to become large, and p_3 to become small. The result is an interaction between the formation of 1F constituents and 3F constituents which assures that $c_1 > c_3$. This interaction is easy to understand. The polymer factor p_n represents the influence of the configuration entropy of the polymer system on the formation of an n-F atomic constituent. The above inequality must be satisfied if the system is to consist of a mixture of small molecules with a relatively large configuration entropy, rather than one giant molecule.

In deriving this theory, multiply-connected polymer structures, i.e., ones containing closed loops, were ignored because of the mathematical difficulties. Therefore, the theory cannot be applied directly to alloys with an average valence per atom V>2, since multiply-connected structures are then undoubtedly important. However, it has been possible to extend BET to apply to n-F _molecular_ constituents as well as _atomic_ constituents.[5] An n-F molecular constituent is defined as a group of atoms bonded together in a specific configuration, with n unspecified bonds by which it can make external linkages to other constituents in the system. If one starts with an arbitrary multiply-connected molecular structure, such as one shown in Figure 1, it is possible to reduce it to a simply-connected structure if each irreducible multiply-connected part is treated as a molecular constituent. An irreducible multiply-connected unit is defined as the smallest group of atoms among which a bond can be cut without separating the molecule into two pieces.

This extension of BET makes it possible in principle to deal with network-forming alloys, i.e., alloys with V>2, such as As-Se. However, the theoretical expression for a molecular constituent contains additional factors reflecting the statistics and energetics of the internal structure (akin to a molecular partition function) which are difficult to determine _a priori_. Therefore, the extension is practical only if one can establish the validity of certain simplifying principles which make it possible to consider only a few kinds of molecular constituents in calculating a solution. Their validity is at present a conjecture and is the subject of a continuing investigation. If they are correct, they have important implications for the equilibrium structure of network-forming alloys. They suggest that the dominant structure consists of many polymer molecules containing relatively small molecular constituents in an otherwise simply-connected configuration, such as the one illustrated in Fig. 1. This contradicts the continuous random network hypothesis which is often used as a model for covalent alloys.

The purpose of this paper is to present a review of BET which can provide a basis for understanding its implications for the structure of glasses. In Section 2, we start by showing the

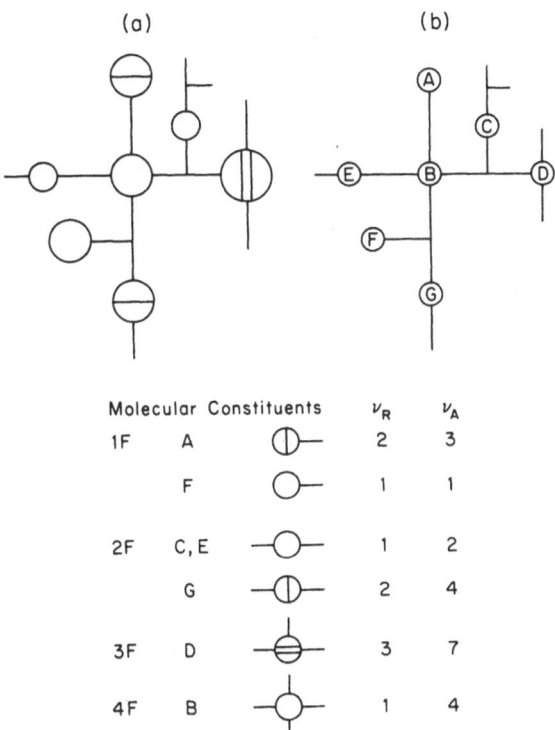

Fig. 1   (a) Diagram of a multiply-connected polymer molecule in
         which lines indicate chains of 2F atoms.   (b) Simply-
         connected equivalent molecule with molecular constituents.
         (c) List of molecular constituents classified according to
         the number of external links.   The number of rings ν_R and
         the number of 3F atoms ν_A are shown for each molecular
         constituent.

relationship of the BET to the LMA for atomic constituents, and
then discuss the implications for the behavior of bond defect atoms
in the context of liquid Se, using a model in which ring structures
are ignored.   This has the advantage of developing the concepts of
the polymer factors and their implications in a simple situation,
as well as yielding insights into the behavior of bond defects which
are believed to be significant constituents in glasses.   Then the
extension of BET to molecular constituents is discussed in Section
3 in the context of dilute alloys of As in Se.   That is the context
in which the theory is currently being developed, and it serves well
to illustrate the significant concepts and the problems in application to network-forming alloys.   Finally, in the last section the
difficulties and limitations of the BET approach in understanding
the structure of glasses are reviewed, and the relationship of the
BET model to some other models is discussed.

BOND EQUILIBRIUM IN SELENIUM

Let us first consider a simple derivation of the Law of Mass
Action (LMA) for the equilibrium concentrations c_γ of constituents
γ in Se. This is done by minimizing the free energy of the mixture
of the constituents in Se, using an expression for the configurational
entropy which assumes that the constituents are in a random mixture
rather than parts of polymer molecules. An atomic constituent γ is
defined as a type of atom which is distinguished by the chemical
species (only Se in the present case), the number of bonds, and its
charge. Some important constituents in Se are 1F neutral dangling
bond atoms ($D*$), 1F negative ions (D^-) and 3F positive ions (D^+).
If the free energy of formation of an atomic constituent Γ from the
reference state (assumed to be the normally bonded 2F atom) is g_γ,
the free energy of the system G is given by

$$G = \sum_\gamma c_\gamma (g_\gamma + q_\gamma E_F) - T S_C \;, \tag{1}$$

where c_γ is the concentration and q_γ is the charge number of Γ, E_F
is the chemical potential of an electron (i.e., the Fermi energy),
and S_C is the configuration entropy. (All energies are per constituent, and the concentrations are normalized to the concentration of
atoms N_a.) S_C is related to the number of ways W of arranging the
various atoms in a unit volume by the Boltzmann formula

$$S_C = (k/N_a) \ln W \;. \tag{2}$$

S_C is given by the formula for ideal mixtures:

$$S_C = - k \sum_\gamma c_\gamma \ln c_\gamma - k c_S \ln c_S \;, \tag{3}$$

where

$$\sum_\gamma c_\gamma + c_S = 1 \;, \tag{4}$$

and c_S is the concentration of normally bonded Se atoms. Now if
Eq. 3 is substituted into Eq. 1, and G is minimized with respect
to c_γ subject to the equation of constraint of Eq. 4, the result is

$$c_\gamma = X_S \exp \left[-\beta (g_\gamma + q_\gamma E_F)\right] \;, \tag{5}$$

where $\beta = 1/kT$. In Eq. 5, c_S has been replaced by X_S, the fugacity
of Se, in order to make it consistent with later formulas. One can
see that Eq. 5 is the usual LMA expression for the equilibrium

$$Se = \Gamma + q_\gamma \text{ electrons} \;, \tag{6}$$

using the thermodynamic expression for the equilibrium constant K:

$$K = \exp \left[-\beta (g_\gamma + q_\gamma E_F)\right] \;. \tag{7}$$

Eq. 5 is valid for a system such as a crystal, in which the constituents are lattice defects of various types. It has a useful form for intrinsic constituents, which can be conveniently thought of as thermal excitations of the matrix atoms.

If we now introduce the consideration that the constituents are components of chainlike molecules, the expression for the configuration entropy S_C becomes much more complicated. S_C is calculated by means of Eq. 2, in which an expression for W is used which takes into account all the possible distributions of sizes and topologies of polymer molecules that can be formed from a given set of concentrations of atomic constituents, as well as the number of ways of arranging these molecules in a condensed phase.[2] For the latter purpose, the molecules are arranged with bonded atoms on adjoining sites of a quasilattice. It is easy to see that the distribution of sizes of polymer molecules and their topologies will depend only on the total concentrations c_n of n-F constituents. The characteristic parameters of the polymer mixture are the concentration of polymer molecules c_p, where

$$c_p = (c_1 - c_3)/2 , \tag{8}$$

and the average number of chain branches per molecule λ, where

$$\lambda = 2c_3/(c_1 - c_3) . \tag{9}$$

It is assumed that the possible values of n are 1, 2, or 3. The mathematical problem, though more complicated, is solved in the same way as before, by minimizing G with respect to each c_γ. The result depends on the numbers of bonds, which is indicated by a superscript n:

$$c_\gamma^{(n)} = (c_2 p_n/S_2) \exp [-\beta(g_\gamma + q_\gamma E_F)] , \tag{10}$$

where

$$S_2 = 2c_2/(c_1 + 2c_2 + 3c_3) . \tag{11}$$

It is seen that the results differs from the LMA expression only in an extra factor $c_2 p_n/S_2$, which reflects the statistical effects due to the polymeric structure. The expressions for the polymer factors p_n are:

$$p_1 = (2 + \lambda)c_2 \left[\frac{(2B/h)(1+\lambda)}{(1+2\lambda)(c_1+2c_2+3c_3)}\right]^{1/2} , \tag{12}$$

$$p_3 = \frac{bc_2}{(1+2\lambda)} \left[\frac{2(1+\lambda)(c_1+2c_2+3c_3)}{Bh(1+2\lambda)}\right]^{1/2} , \tag{13}$$

and $p_2 = 1$. B and h are parameters arising from constraints on
the dihedral angle in Se chains. Assuming a simple cubic quasilattice, B=6 and h=2. b is a slowly varying function of λ.   It is
equal to 1 for $\lambda < 4$, and increases to 2.6 when $\lambda \to \infty$.[2]

In the situation of interest for Se, $c_2 \simeq 1$, and c_1 and c_3
are both <<1, which simplifies the expressions for p_1 and p_3. The
physical significance of the polymer factors can be seen in their
behavior in the extreme limits of λ. If $\lambda << 1$, p_1 and p_3 become
constants, and the result is essentially similar to the LMA equa-
tions. But if $\lambda >> 1$, $p_1 \propto \lambda$ and $p_3 \propto \lambda^{-1}$. In this case, there is an
interaction between the formation of 1F constituents, which termi-
nate polymer chains, and 3F constituents, which generate chain
branches. Since $c_1 \propto p_1$ and $c_3 \propto p_3$, $c_1/c_3 \propto \lambda^2$ in this limit, and
Eq. 9 ensures that a mathematic solution occurs in which $c_1/c_3 > 1$.
This corresponds to a situation in which the polymer concentration
$c_p > 0$ (see Eq. 8). It reflects the fact that the configuration
entropy decreases without limit when $c_1/c_3 \to 1$.

In order to get some insight into the use of the BET equations,
let us consider some problems for liquid Se. The problem is
simplified in the sense that we ignore the formation of simple ring
molecules, and chain molecules which are attached to loops by 3F D^+
centers. These structures occur in Se, but their concentrations
are low enough so that the properties are not strongly affected.[3]

First, consider the concentration of dangling bond atoms (D*).
Using the small letter d* for the concentration and g* for the free
energy of formation, we have from Eq. 10

$$d^* = \pi_1 \exp [-\beta g^*] \; , \tag{14}$$

where we write π_n for $(c_2 p_n/S_2)$. Thus, d* depends on the two
parameters, (generally determined empirically) the enthalpy e*
and entropy s* in g* = e* - Ts*, and on the value of π_1, which
we've noted, is a function of λ if $\lambda \overset{\sim}{>} 1$. The significance of
this result is that if d* is determined experimentally, say from
the magnetic susceptibility, it may be used in some circumstances
to determine changes in λ through the dependence of π_1 on λ.[3]

Let us consider next the concentrations of intrinsic ions D^+
and D^-. From Eq. 10,

$$d^- = \pi_1 \exp [-\beta(g^- - E_F)] \; , \tag{15}$$

$$d^+ = \pi_3 \exp [-\beta(g^+ + E_F)] \; . \tag{16}$$

The condition of electrical neutrality requires that $d^- = d^+$, so that one can eliminate E_F to get

$$d^+ = d^- = (\pi_1 \pi_3)^{1/2} \exp [-\beta(g^+ + g^-)/2] \ . \tag{17}$$

Note that $\pi_1 \pi_3$ is independent of λ when $\lambda \gg 1$ as well as when $\lambda \ll 1$, so that the ion concentration is not sensitive to the value of λ. One can also solve for E_F with the result

$$E_F = (g^- - g^+)/2 + (kT/2) \ln (\pi_3/\pi_1) \ , \tag{18}$$

It is seen that E_F depends on the polymer factors, and it decreases as $-kT \ln \lambda$ when $\lambda \gg 1$. The fact that $\pi_3 \ll \pi_1$ at all times is probably one of the reasons why p-type transport behavior is usually observed in chalcogen liquids.

The overall solution for the concentrations at a given T must be obtained self-consistently from a set of equations for concentrations c_γ^n of the various constituents, each of which depends on the value of a polymer factor π_n in addition to the two free energy parameters (e_γ and s_γ). The polymer factors, in turn, are functions of the concentrations c_γ^n, in a form which depends only on n. Thus, if there are several constituents which have the same number of bonds, and are similar in other properties (charge and spin), one can usually ignore all but the one which is dominant in a given range of T. In the example given for Se, we have $c_1 = d^+ + d^-$ and $c_3 = d^+$. Since $d^+ = d^-$, the important parameter $c_1 - c_3 \equiv d*$, i.e., the average molecular size depends only on the concentration of dangling bonds, even though there may be branched chains.

Finally, it is worthwhile to make one further point in order to clear up a common misconception arising from a difference between liquids and amorphous solids. In the literature on amorphous chalcogens, there has been much interest in the equilibrium reaction associated with the so-called negative correlation energy:

$$2D^* = D^+ + D^- \ . \tag{19}$$

This reaction is possible in the solid because it requires merely electron exchanges and slight movements of the atoms. Because the atoms cannot move farther than a bond length in the solid, there is no separate equilibrium between the 2F (C_2) atoms and the individual $D*$, D^-, and D^+ centers, so that Eqs. 14, 15, and 16 are not applicable (with their polymer factors). But as the result of Eq. 19, one can write the LMA relation

$$d^- d^+ / (d^*)^2 = K(T) \ . \tag{20}$$

In a liquid, on the other hand, there is complete equilibrium. Although an equation like Eq. (20) where K now contains the polymer factors) also applies, the equilibrium is more general in a liquid

because the D^* centers and the D^+D^- pairs can be formed independently
from C_2 atoms, in accordance with Eqs. 14 and 17.

ARSENIC SELENIUM ALLOYS

The extension of BET to include molecular constituents in the
polymer structure[5] makes it possible to use the same basic approach
in determining S_C. The molecular constituents (which includes
atomic constituents as a special case) are labeled by an index m in
addition to γ. m describes all internal topological factors such
as the number of atoms of each chemical species of the alloy (say
As and Se), the number of loops, chain branches and external bonds.
The expression derived for the concentration is

$$c_{m\gamma}^{(n)} = \pi_n O_m P_m X_A^{\nu_{Am}} X_S^{\nu_{Sm}} \exp[-\beta(g_\gamma + q_\gamma E_F)] . \tag{21}$$

As compared to the expression for an atomic constituent (Eq. 9),
there are two additional factors O_m and P_m, related to the internal
statistics of the cluster, as well as a fugacity factor X_A for each
of the ν_{Am} As atoms in the cluster, and a fugacity factor X_S for each
of the ν_{Sm} Se atoms. The parameter P_m ($<<1$) takes into account the
reduction in the number of ways of arranging the atoms in configuration space as compared to a simple chain as a result of the constraints
due to the closed loops. The parameter O_m (>1) allows for the multi-
plicity of arrangements within the cluster due to different distributions of the 2F atoms between the chain branches. The internal
statistical parameters, particularly P_m, introduce a difficulty in
describing cluster behavior which will be discussed later.

The important element in Eq. 21 which we consider now is
$\pi_n = (c_2 p_n / S_2)$. These are the same functions of c_1, c_2, and c_3,
given in Eqs. 12 and 13, which describe the behavior of atomic
constituents. But the concentrations c_n now refer to the total
concentrations of n-F clusters. Atomic constituents are counted
individually only if they are not parts of molecular clusters.
For instance, the three As atoms in the 1F molecular constituent A
in Fig. 1 together make a single contribution to the value of c_1
and no contribution to c_3. The conclusions concerning the relation
of c_3 to c_1 when $\lambda >> 1$, discussed in Section 2, are valid also for
polymers constructed from cluster units. That means that as V
increases above 2 in an alloy such as $As_x Se_{1-x}$ (where V=2 + x) more
As atoms must be forced into n-F clusters, with smaller values of n
preferred. This is the key to understanding the main implication
of BET with respect to the equilibrium configurations of covalently
bonded alloys with V>2.

It should be noted at this point that 0-F clusters, such as
free rings are also possible. Eq. 21 describes also the concentration of 0-F clusters, with $\pi_0 = 1$.

It is also necessary to consider cases where $n > 3$. The mathematical problem of counting configurations of branched chain polymers which contain n-F atomic constituents, where $n > 3$, has not yet been solved. So alloys such as Ge-Se are excluded from consideration. But even in alloys such as As-Se, where the maximum atomic valence is 3, n-F clusters with $n > 3$ are possible, as shown in Fig. 1. An extension to cases where $n > 3$ has been made by applying the method of induction to the observation that $\pi_n \propto \lambda^{2-n}$ for $n=1$, 2, and 3 when $\lambda \gg 1$.[5] First, the definitions of c_p and λ are extended as follows:

$$c_p = c_1 - \sum_{n=3} (n-2)c_n ,$$

(22)

and

$$\lambda = [\sum_{n=3} (n-2)c_n]/c_p .$$

(23)

Then using the inductive process, the estimate is made that

$$\pi_n = \pi_3 (\pi_3/\pi_1)^{(n-3)/2} .$$

(24)

The accuracy of this estimate is not worrysome for application of BET to $As_x Se_{1-x}$, because the solution of the problem is expected to be dominated by the behavior of c_n with $n \leq 3$.

In applying BET with cluster constituents to an alloy such as $As_x Se_{1-x}$, one is led to group the clusters into topological classes, such as those indicated in Fig. 1 The members of a given topological class differ from each other only in the number of 2F Se atoms, which determines the lengths of the chains between the junctions provided by the 3F As atoms. Then the number ν_A of As atoms in a given class is a constant determined by the number of internal loops ν_R and the number of external links n:

$$3\nu_A = n + (\nu_R/2) .$$

(25)

For a given class K the concentration c_K is given by an infinite sum of the form

$$c_K^{(n)} = \pi_n X_A^{\nu_A} \sum_{\nu=\nu_m}^{\infty} X_S^{\nu-\nu_A} P_\nu 0_\nu$$

(26)

In writing this expression, it is assumed that the cluster contains no bond defect atoms, and that there is no appreciable strain in the bonds, so that the argument $g + q E_F$ in the exponential is zero for each term. This restricts the lower limit of summation index to the minimum number ν_m of Se atoms that will form a strain-free cluster. As noted earlier, 0_ν is a combinational factor, and in the limit of large ν it will be of the form ν^a where a is a positive

constant. The value of P_ν, which characterizes the reduction of
configuration space available to the cluster because of the internal
loops, is hard to determine when ν is in the vicinity of ν_m. When ν
is large, one can use random walk theory in a reasonable approximation
to get $P_\nu \propto \nu^{-3n_L/2}$. The significant result is that in the limit
large ν, $0_\nu P_\nu$ varies as ν raised to some constant power. Consequently,
the sum in Eq. 26 always converges if $X_S < 1$.

In view of the large number of possible classes of clusters as
suggested by the example in Fig. 1, is it feasible to carry out a
calculation using BET? Earlier work on the effects of simple ring
structures in Se-Te alloys[3] suggested that a physical principle may
be valid which can make such calculations practical. This principle
is that the most important clusters are ones which have a minimum
number of atoms consistent with steric constraints. (The steric
constraints are due to the limitations on bond lengths and angles.)
This would tend to be true if X_S is considerably less than 1, so that
only the first few terms of the sum in Eq. 26 are important for each
class of clusters. One expects in addition, that one or two classes
of n-F clusters to dominate for a given value of n, and these will
be the smallest ones which can be formed with a composition approximating the average composition of the alloy.

To test the validity of these ideas, a BET calculation was
made for an alloy $As_x Se_{1-x}$, in which the only clusters considered
were 0F, 1F, 2F, and 3F simple rings (i.e., with $\nu_R = 1$). This can
be expected to be a good model if x is not too large. In addition
to the n-F rings (R_n) with concentrations r_n, the only other constituents are the 3F atomic As constituents with concentration a_3, and
2F Se atomic constituents with concentration s_2. In order to simplify
the evaluation of the sums (Eq. 26) for the R_n constituents, the
random-walk expression for P_ν corresponding to an ν-atom ring was
used even though it is expected to be accurate only when ν is very
large. This gives

$$P_\nu = A\nu^{-3/2} , \tag{27}$$

where the value of $A = .023$ was obtained from experimental information determined from 8-atom sulfur ring molecules. The minimum
number of atoms in a ring was assumed to be 8, so that

$$r_n = \pi_n A X_A^n \sum_{\nu=8}^{\infty} \nu^{-3/2} 0_\nu X_S^{\nu-n} . \tag{28}$$

The combinational factor 0_ν is 1 for n=0, and 2 for n=1. For n=2,
$0_\nu = \nu-1$, and for n=2, $0_\nu = 2(\nu-1)(\nu-2)/3$. In addition the atomic
constituents have concentrations

$$a_3 = \pi_3 X_A , \tag{29}$$

$$s_2 = \pi_2 X_S . \tag{30}$$

The polymer factors are functions of c_1, c_2, and c_3, where

$$c_1 = r_1 , \tag{31}$$

$$c_2 = r_2 + s_2 , \tag{32}$$

$$c_3 = r_3 + a_3 . \tag{33}$$

This leaves two degrees of freedom related to the values of X_A and X_S. X_A is constrained by the compositional relation

$$X = a_3 + r_1 + 2r_2 + 3r_3 . \tag{34}$$

A similar compositional relation for Se requires determination of the average number of atoms A_n for an n-F ring. Then

$$1-x = s_2 + r_0 A_0 + r_1 (A_1 - 1) + r_2 (A_2 - 2) + r_3 (A_3 - 3) . \tag{35}$$

It is easy to see that

$$A_n = \sum_{\nu=8} \nu T_\nu^{(n)} / \sum_{\nu=8} T_\nu^{(n)} , \tag{36}$$

where $T_\nu^{(n)}$ is the summand term in Eq. 28.

Eqs. 28-36, together with the expressions for $\pi_n = c_2 p_n / S_2$ in Eqs. 9, 11, 12, and 13, constitute a complete set of transcendental equations which need to be solved at a given composition x. This can be done on a small computer with the use of a search procedure. Since there are no thermally excited bond defect atoms or strained ring structures in the model, the solution is independent of temperature.

Calculations were made in the composition range $0<x<0.15$, and the behavior of the various parameters is shown in Fig. 2. As expected, λ increases with x, causing p_1 to increase and p_3 to decrease. At 10% As, $c_3 \overset{\sim}{<} .025$, so that only one quarter of the As atoms have effectively produced chain branches. The rest of them are forced into R_1 and R_2 constituents for the most part, so that c_1 remains greater than c_3.

The behavior of X_S is most significant in relation to our problem. As expected, it is less than 1, and decreases with x, reaching a value $\sim.8$ at $x = .10$. Corresponding to this, the average number N of atoms in a ring decreases from ~20 at $x = .01$ to ~10 at $x = .10$. That is close to the minimum value 8. This

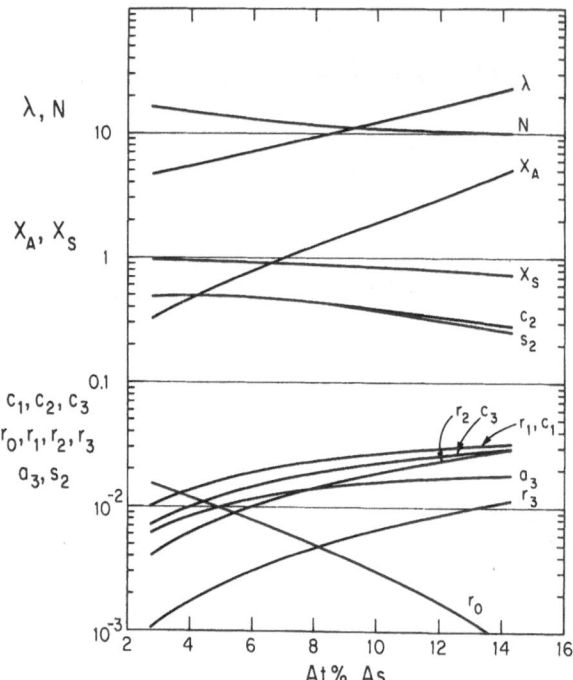

Fig. 2 Result of a model calculation for As_xSe_{1-x} assuming that the
        only molecular constituents are 0F, 1F, 2F, and 3F simple
        rings ($\nu_R = 1$).

encourages the belief that at larger As concentrations, where more
complicated multiple ring structures need to be considered, the sum
for each topological class can be restricted to a few terms. In view
of the crude approximations made in determining P_ν when $\nu \approx \nu_m$, one
would seek to replace the sum by a simple expression of the form

$$c_\kappa^{\,n} = \pi_n B_\kappa X_S^{\,\nu_{S\kappa}} X_A^{\,\nu_{A\kappa}} . \tag{37}$$

where the constant B_κ is empirically determined, and $\nu_\kappa = \nu_{A\kappa} + \nu_{S\kappa}$
corresponds to the minimum number of atoms in the cluster.

DISCUSSION

       The biggest problem in making a useful BET calculation for
As_xSe_{1-x} at larger x is the identification of the important cluster
classes, and evaluation of their statistical factors. As x increases,
the arsenic fugacity X_A increases (see Fig. 2), favoring the formation of clusters with larger ν_A, because of the factor $X_A^{\,\nu_A}$ in Eq.
21.[6] This in turn leads to a larger number of external bonds n, or
a larger number of internal loops. Each internal loop is expected

to decrease P_m by a factor << 1, while larger values of n cause the polymer factor p_n to be smaller. It is not hard to see that the possibility of making relatively small clusters (i.e., with a small ratio ν_S/ν_A) with large loop number ν_R, would have a strong influence on the BET solution, for instance, in determining how large λ gets to be at large x. This would be especially important for 1F and 2F clusters. Also, one should not neglect the possible role of 0F clusters, such as possible cage-like As_4Se_6 molecules, in reducing the value of λ. This is expected to depend strongly on topological possibilities for formation of low-strain loop configurations, which are determined by the equilibrium bond lengths and bond angles and their force constants.

Efforts which have been made to determine P_m for simple rings illustrate the difficulty of the problem. Semlyen has made calculations of statistical weights by determining the fraction of the rotational iosmers of a chain molecule which have their ends close enough to form a ring-closing bond.[7] The rotational isomers arise from the fact that the dihedral angles can have two values at $\pm 90°$. This results in a very irregular dependence of P_m on the number of atoms in a ring when the number of atoms is relatively small strain energies, so that this approach may yield misleading results. It seems possible to make useful calculations on this problem by using a Monte Carlo method, and this is currently under investigation.[8]

Another important question concerning the application of BET is what experimental information can it be related to? Part of the same question is what experimental information can be used to set the values of parameters of the theory which need to be determined empirically? In current efforts at testing theoretical structures, such as random network models, one approach has been to determine diffraction structure factors which are to be compared with experiment.[9] This may be a possibility with BET results, particularly if the Monte Carlo approach to deriving cluster information is successful. But a more effective approach probably is to determine experimentally the behavior of p_1 and p_3 through its effect on the Fermi energy and the equilibrium concentration of paramagnetic Se dangling bond centers. This can be done through the study of the magnetic susceptibility and electronic transport.[3] The calculation described in the preceding section is appropriate for determining p_n even though bond defects were ignored, as long as their concentrations make relatively small contributions to c_1, c_2 and c_3. If necessary, it is of course possible to broaden the BET model to include consideration of bond defect atoms.

Finally, it is useful to discuss the relation of the present work to some of the other models and approaches for the structure of glasses. The most common accepted model seems to be the continuous random network (CRN) model.[10] The conclusions inferred here from BET seem to be in direct conflict with the CRN concept. This is clearly the case in As_xSe_{1-x} at $x \gtrsim .10$, where the assumptions of the calculation of Section 3 seem to be valid. How can one explain physically the failure of the intuitively reasonable concept of chains linking randomly to form an infinite network of loops? An answer may possibly be found if one regards the network structure as being the average result of dynamic equilibrium in which chains are continually breaking and the broken links are reforming randomly. The small cluster conclusion of BET indicates that in this process of dynamic equilibrium, the ends of broken chains are likely to coil and form closed loops on themselves. The coiling of long chain molecules and the relatively short average distance between ends is a well known property of random chains.[11]

We should note, however, that BET cannot be used to prove that equilibrium results in many small molecules ($c_1 > c_3$) rather than a single large molecule ($c_1 = c_3$). The mathematical structure of the solution of the BET equations is similar to what occurs in the cluster expansion theory for imperfect gases. As the clustering tendency increases due to an increase in X_A, it is possible that the number of important cluster classes can increase to the point where the summations in Eqs. 34 and 35 diverge. This would correspond to a situation in which the overall free energy G (Eq. 1) has another lower minimum in addition to the one which is the basis of the BET equations.

Another important theoretical approach attempts to determine relatively dense, i.e., space filling structures (SFS) for random networks.[12] This approach emphasizes the secondary forces which exist in addition to the bonds, and which tend to create compact space-filling structures in glasses. In the BET approach, these secondary forces are ignored. It is tacitly assumed that when the liquid is cooled quickly to form a glass, the chain molecules are squeezed together, and some relatively compact structure develops with inevitable voids and strained bonds. The SFS studies focus on glasses with particular compositions such as Si or As_2Se_3. It has been recognized that SFS models may not be appropriate for amorphous solids with $V \simeq 2$ such as Se, and it is possible that different structural models are appropriate for different amorphous solids depending on the value of V and the method of formation.

The BET concepts have interesting similarities and differences when compared to models for glass structures introduced by J. C. Phillips.[13] The similarities lie in the emphasis on relatively small clusters as building blocks for the structure of the glass, and the emphasis on the topological constraints in relation to the average atomic valence V. BET differs in that these constraints are incorporated into a statistical mechanical treatment of the molecular configurations. Another difference, which results from this, is that one is led to expect a wider variety of clusters to be represented among the building blocks. The choice of the basic

units suggested by Phillips is guided by their existence in crystalline systems of similar composition. The problem of identifying
the important BET clusters has been discussed earlier, and the occurrence of units in crystals is indeed an important way of finding
relatively small strain-free clusters. But it seems likely that
many other small strain-free clusters exist with stoichiometries and
shapes which do not fit well into a lattice. It should also be
mentioned that molecular clusters which have double linkages in a
given direction to form a ribbon shape, such as the unit proposed
for $GeSe_2$ by Phillips, would be a less likely unit in the BET frame
of reference as compared to a freely rotating unit on a single link,
because of the greater constraint in the number of available molecular configurations.

ACKNOWLEDGMENT

This work was supported by the National Science Foundation,
grant No. DMR 80-23682.

REFERENCES

1. A. C. Wright and A. J. Leadbetter, Physics and Chemistry of
    Glasses, 17:122 (1976).
2. M. Cutler, Phys. Rev., B20:2981 (1979).
3. M. Cutler and W. G. Bez, Phys. Rev., B23:6223 (1984).
4. M. Cutler, J. de Physique, 42:C4-201 (1981).
5. M. Cutler, Phil. Mag., B47:11 (1983).
6. It is to be noted that X_A becomes larger than one, which might
    seem to be impossible because pure arsenic would separate from
    the system. The explanation is that the reference state used
    here for As is a hypothetical non-physical one in which the
    atoms are bonded in a single linear chain, but yet each As atom
    has three completed bonds.
7. J. A. Semlyen, Trans. Faraday Soc., 63:2342 (1968).
8. M. Cutler, work in progress.
9. G. A. N. Connell and R. J. Temkin, Phys. Rev., B9:5323 (1974).
10. D. E. Polk, J. Non-Crystalline Solids, 5:365 (1971).
11. P. J. Flory, Statistical Mechanics of Chain Molecules,
    Interscience, New York, (1969).
12. J. F. Sadoc and R. Mosseri, J. de Physique, 42:C4-189 (1981).
13. J. C. Phillips, J. Non-Crystalline Solids, 35-36:1157 (1980).

CURVED SPACE APPROACH TO THE RELATION BETWEEN LONG RANGE DISORDER

AND LOCAL ORDER IN AMORPHOUS TETRACOORDINATED SEMICONDUCTORS

R. Mosseri*, and J. F. Sadoc**

*Laboratoire de physique des solides, CNRS, 92190
Meudon principal Cedex, France; ** Laboratoire de
physique des solides, Universite Paris-Sud Batiment
510 91405 Orsay-Cedex, France

ABSTRACT

We present the curved space description for disordered tetracoordinated semiconductors. For a given, non-crystalline, local
configuration, the ideal structure is first defined in a space of
constant curvature where it can freely propagate. Topological
defects are necessary in order to lower the space curvature and
recover a realistic euclidean model. The excitation spectra of
such models are also described.

I. INTRODUCTION

Solid state physics theoretical methods used in the study of
crystalline materials often call for the periodicity property of
the atomic arrangements. For example, Bloch theorem allows considerable simplifications for the calculation of elementary excitations (electrons, phonons ...). From a strict point of view this
periodicity disappears in the presence of defects, topological
(vacancy, dislocation, walls ...) as well as chemical (disordered
alloy). The physicist will nevertheless try to keep as close as
possible to the ideal crystal case and treat the defect in a
perturbation scheme. This becomes meaningless when the defect
density reaches a high value and thus it cannot be applied to the
amorphous material case for which not only the long range order (LRO)
is completely absent but the local order can be appreciably
different from the stable crystalline phase. However, in both
cases (crystal and glass) the solid may show similar properties,
at least qualitatively. One of the most remarkable ones for the
amorphous semiconductors is the persistence of a gap in the
electronic density of states, separating occupied states (the

valence band) and empty states (the conduction band). This property
can be explained, without referring to the absence or presence of
periodicity, using some simple consideration on the local atomic
structure, for instance in a tight binding Hamiltonian model [1].
A good knowledge of the short range order (SR) appear as crucial
and has been the subject of many experimental work, mostly using
diffraction techniques; X-ray [2,3], electrons [4], and neutrons
[5].

        Starting with the experimental interference function, it is
possible to calculate, by Fourier transform, the radial distribution function (RDF) of the material, which gives the mean number of
atoms located at a distance r from an (arbitrary) reference atom.
At this state it is important to stress the insufficiency of such
an information in order to perfectly specify local configurations.
Their good description requires the knowledge of atomic positions
in 3-dimensional (3-D) space while the RDF only provides an
"unidimensional" information, the distance r replacing the three
spatial coordinates. It is formally possible to reobtain the
atomic arrangements with the help of the successive distribution
functions for n-uplets (triplets, quadruplets...) which cannot
however be experimentally derived. So there is no univoque
transformation between the interference function and the atomic
position in space, and this imposes a constant exchange between
experiment and modelling. We shall now briefly recall the
principle results concerning the amorphous silicon structure as
well as the models which more or less successfully describe it.

        In aSi the first coordination shell is very similar to that
in the crystal. The first neighbor distance is very close to the
crystalline value and the mean coordination number (obtained
by integration under the first RDF peak) equals 4. The bond
angles (derived from the respective values of first and second
neighbor distance) are distributed around the crystalline value
(109°28) with maximum deviation of about 10°. This allows us to
define a first (very local) "ideal" image for the SRO in amorphous
silicon: an atom at the center of a tetrahedron with its four
neighbors at the vertices. One has then to specify in which way
the sites are connected in the network. Experimentally one observes
a rather flat distribution of the third neighbor distances which
manifests dihedral configurations in the amorphous material.
Recall that in the diamond structure all the dihedral configuration
are "staggered" while the wurtzite structure contains a mixing of
"staggered" and "eclipsed" configurations. It is somehow difficult
to establish a strict correspondence between dihedral angles and
ring size. More detailed information about the ring configurations
would be very desirable. Indeed, two distinct families of 4-fold
coordinated structures can merge according to which type of rings
are present. If one tries to limit bond angles variations around
the ideal value, the smallest cycles are (almost flat) pentagons.

It is possible to generate arrangements of tetravalent sites with
a predominance of pentagons which fill the 3-D space. Such structures, of "clathrate" type, have low density and can be analyzed
as a tiling of polyhedral cages. A silicon sodium alloy (with
only a few percent of sodium) presents such a structure in its
crystalline form [6]. In such "caged" tetravalent structures, one
edge (bond) belongs to exactly three polyhedra and a vertex is
common to six polygons. A statistical analysis of the disordered
tiling is then possible using the Euler-Poincare relation (applied
to a polyhedron) and some simple hypothesis [7,8]. One finds that
the mean ring size is approximately 5.1 which, with our assumption
of low bond angles deviation, implies a high concentration of
pentagons. A simple way to generate such a structure is to take
the dual of a tetrahedral compact tiling, the new vertices being
the center of the tetrahedral cells.

      The second family of tetravalent networks consists in an
arrangement of larger polygons in which all the sites are not in
coplanar configuration, the puckering of the ring allowing the
bond angles to stay near the tetrahedral value. The simplest example is the regular diamond structure where the smallest rings are
"chair shaped" hexagons. It is possible to generate many different
regular 4-fold nets [9], as well as topologically disordered ones
often called "Continuous Randon Networks" (CRN). We have inten-
tionally separated the disordered "clathrate" like structures and
the CRN models. One of the most fundamental differences between
these two families is probably the fact that for the first one
there is a natural and non-ambiguous way to operate a cellular
division (with the polyhedral cages) while such partition is often
difficult if not impossible for the second ones.

      In the case of amorphous silicon, the experimental results do
not allow a decisive choice in favor of one family of models.
However, the relatively high value of the material density (difficult to obtain with "caged" structures) and a more or less flat
distribution of third neighbor distances lead to a more probably
CRN structure. A mixed structure with CRN matrix and polyhedral
inclusions cannot be completely excluded. Finally, even the CRN
hypothesis do not lead to a unique structure but to a rather
differenciated family of models characterized for instance by the
presence or not of a given parity for the ring size. Notice for
example, the Polk model [10] which contains even and odd-membered
rings and the Connell-Temkin model [11] where all the rings are
even. This last model has been initially proposed to describe
III-V amorphous alloys (aGaAs, aGaP...) where the chemical order
hypothesis imposes a "bicolor" model. Recently, Dixmier, et al.
[12] have shown that it is possible to distinguish experimentally
the Polk and Connell-Temkin-like models by looking to the ratio
of the successive interference function maxima positions over the
first peak position. They confirm that, if III-V alloys seem well

described by the Connell-Temkin model, amorphous silicon correspond
better to the Polk model.

Let us now discuss the notion of defects in a non-crystalline
system. It is clear that speaking of a defect makes sense only
in reference with an "ideal" structure, which must be well characterized. It is also clear that one takes advantage in defining this
reference structure as close as possible to the real structure.
This is the reason why we have rejected the idea of treating the
amorphous system as a highly defected crystal. Indeed, in many
cases, it is possible to show that the number of defects required
would be then of the order of the number of atoms. A better choice
consists in using the CRN model. We must first make the distinction
between a particular CRN model (Polk, Connell-Temkin, ...) and the
"generic" CRN model (e.g., the more general element in the set of
all CRN models), that is the model about which one cannot say more
than the fact that it satisfies the finite set of rules which define
a CRN. With respect to the generic CRN model, it is then possible
to specify two kinds of defect. The first type of defect is metrical
by nature and consists in the deviation of lengths and angles from
their tetrahedral values (e.g., all first neighbor distances
being equal, and all bond angles being $109^{o}28' = \cos^{-1}(-\frac{1}{3})$). Note
that available CRN models already contain such defects which
manifest for instance in their calculated RDF by the finite width
of the peaks. The second kind of defect is topological in the
sense that it cannot be removed by continuous changes (without
breaking bonds) in order to recover the reference configuration.
Here the difference between the generic and a particular CRN model
can be made clear: if the reference structure is, for instance,
the Connell-Temkin model, an odd membered ring will be called a
defect, while no particular type of ring can be called a defect
when the reference model is the generic one. In both cases, a
dangling bond is a defect since, by definition, sites of ideal CRN
models are supposed to be fully coordinated. This discussion may
appear as trivial but our aim is to stress that the notion of
defect is a relative one, which depends on the prior definition of
a reference state. The generic CRN model has proven to be very
successful in allowing the description of many properties of the
real material. However, it is very difficult to use it in order
to study the medium range order which is this very important region
where the local configurations begin to loose their own coherence.
The difficulty comes from the fact that the propagation of local
configurations is mainly governed by the rings, about which the
generic CRN model says nothing. In order to go one step beyond,
it seems natural to try to define a reference model where the local
configurations can propagate in the long range without restriction.
The "curved space model" [13], presented in the next section,
follows this line. From a geometrical point of view the existence
of a (metastable) amorphous structure is associated with the presence of a rather well defined local order, incompatible with

crystalline periodicity. It is shown that allowing for the free
propagation of a configuration means allowing for space curvature.
In the same way that crystalline structures are characterized
by their symmetry group (one of the 230 extended space groups of the
3-D euclidean space), the new ideal structures are described by a
symmetry group in curved space. In the case which will be considered
here, the underlying 3-D space will be given a constant positive
curvature and becomes the hypersphere (noted $S3$). When the hypersphere S3 is imbedded in E_4 (with equation $X^2+Y^2+Z^2+T^2 = R^2$),
the symmetry groups of regular objects inscribed on S_3 are subgroups
of the orthogonal group in E_4:$O(4)$. The list of such finite groups
is known for long [14], and the regular structures are called
"polytopes", an extension in higher space of the polygons and
polyhedra. A clear description of polytopes can be found for
instance in the book by Coxeter [15]. We shall describe several
polytopes associated to different local configurations. The second
step consists in coming back to the euclidean space R3 by lowering
to zero the curvature of the underlying space. This procedure
requires the introduction of topological defects which are described
in Section III. It will be stressed that a disclination line is
a very natural defect in this context.

We show how to introduce a single disclination in a polytope,
the main effects being to decrease the polytope curvature and to
change the topology in the vicinity of the line. In Section IV
we calculate the excitation spectrum of the polytope, in a simple
first neighbor tight binding model, using an idea proposed in this
context by [21], which allows the definition of a "pseudo" Brillouin
zone where the polytope eigenstates are characterized as "allowed"
values of the "pseudo" momentum. This method consists, in short,
in considering a cyclic subgroup of the total polytope symmetry
group and building Bloch-like sums of the basis functions which
respect this symmetry. In order to study the effect of topological
defects on the excitation spectrum, it is interesting to choose,
for the location of the disclination line, the precise symmetry
axis associated to the cyclic subgroup. The formal expression which
gives energy eigenvalues remains unchanged (with and without the
defect), the main effect of the disclination line being a change
in the allowed k values for the pseudo-momemtum. Note that, since
the polytope consists in a finite set of vertices, its eigenspectrum is discrete, and the degeneracy of the eigenvalues is
simply related to the polytope symmetries. It is then very natural
to propose that, for a disordered structure whose local configurations are well mapped on a given polytope, the real band structure
can be approximized by the polytope spectrum where the discrete
eigenvalue are given a finite width (due to topological disorder).
Note that elastic distortions and finite size effects (the fact
that local order propagation is restricted to coherence regions
in the real material) also contribute to this width. We also give
a short description of the electronic part of the polytope cohesive

energy, e.g., the sum over the occupied states of one-electron
energy eigenvalues. It is shown that, in this approximation,
"non-crystalline" configurations (for instance, associated with
a five-fold symmetry) are not unfavored compared to crystal-like
(diamond lattice) structures.

II. THE CURVED SPACE MODEL

The ambition of this model is twofold: 1) to try a systematic
approach in order to more precisely define the notions of order and
defects in non-crystalline materials, 2) to obtain efficient and
controlled algorithms for the generation of such structures.

At present both goals have been partially reached but more
work remains to be done, especially for the second one. The basic
assumption is that the existence of amorphous materials is generally
related to the presence of a rather well defined local order,
incompatible with crystalline periodicity. For the sake of clarity,
we shall present the curved space model (CSM), when applied to the
description of amorphous metals as it was first proposed by one
of us [13]. The reason to do this is that amorphous metal
structural problems can be mapped, in a first approximation, into
the hard sphere close packing problem which is probably the simplest one to visualize. Also geometrical transformations exist
between close packed and covalent configurations (see for example
Ref. [16] and [17]) which will be used in the following.

A. The space filling problem - The polytope {3,3,5}

If one tries to pack spheres in a dense way by a discrete
aggregation process, one easily finds that the regular tetrahedron
(where a sphere is placed at the tetrahedron vertices) (Fig. 1a)
is the best solution with $N = 4$ spheres. Five tetrahedra can share
a common edge but a void necessarily remains between two triangular
faces (Fig. 1b).   This is due to the fact that the tetrahedron
dihedral angle ($\sim 70°$) is not a submultiple of 2π.   This misfit
angle manifests itself when one tries to propagate the tetrahedral
local configuration and completely surrounds a given vertex.
An imperfect tetrahedron is then obtained (Fig. 1c). Note that
amorphous metal structure is well described by the so-called pseudo
icosahedral ("compact or polytetrahedral") models [18]. It is
desirable to define an ideal model in which the space can be
perfectly tiles by tetrahedra. Let us consider a simple analogous
situation in 2D. It is well known that the euclidean plane
cannot be tiled with regular pentagons, because the pentagon
vertex angle ($108°$) is not a submultiple of 2π (Fig. 2a). But
we can tile with pentagons a surface of constant positive curvature,
the usual sphere (noted S2), and thus obtain a dodecahedron (Fig.
2b). As it is possible to tile a sphere with pentagons, it is
possible to tile, with tetrahedra, a hypersphere (noted S3) which

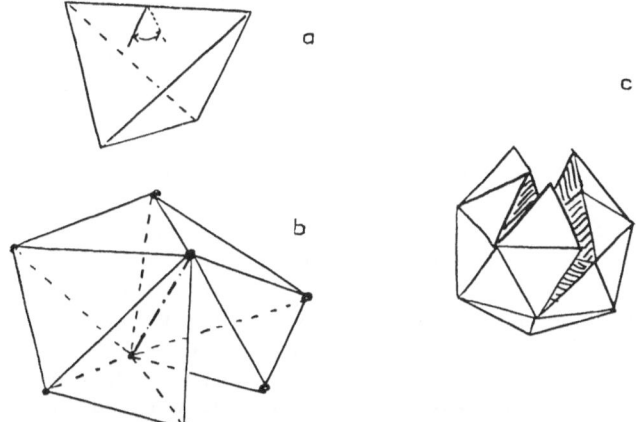

Fig. 1 Packing tetrahedra in 3D space: a) a regular tetrahedron,
the dihedral angle is not a submultiple of 2π , b) 5 regular
tetrahedra sharing a common edge, c) 20 regular tetrahedra
sharing a common vertex: irregular icosahedron.

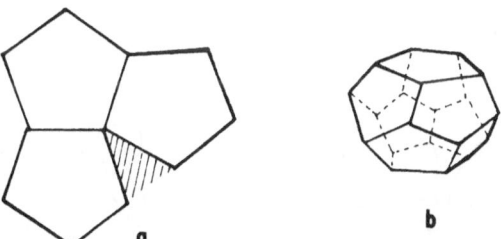

Fig. 2 Pentagons tiling in 2D: a) 3 regular pentagons sharing a
common vertex in the euclidean plane, b) perfect pentagonal
tiling of the sphere: the pentagonal dodecahedron.

can be simply embedded in the 4D euclidean space with equation

$$X^2 + Y^2 + Z^2 + T^2 = R^2 \qquad\qquad (1)$$

Note that only 3 of the 4 coordinates are independent, S3 being a
3D (curved) manifold. The perfect tetrahedral packing on S3 is
called a "polytope" (the analogous of a polyhedron in higher
dimension). This polytope is a finite structure (S3 is finite) and
contains 120 vertices. Exactly 5 tetrahedra share a common edge
and each vertex has 12 neighbors in a perfect icosahedral configuration. This polytope is called {3,3,5} using the standard
Schlaffli notations and is well described in the books by Coxeter
[15] and Du Val [19]. We now process to present its structure as
simply as possible. Here again one can use a 2D analogy. Suppose
one tries to represent, on the euclidean plane, geometrical configuration belonging to the surface of a sphere (it is the main
problem in cartography!). The simplest way to do it is to generate
an orthogonal mapping. If the plane is tangent to the north pole,
the set of parallels (in the geographical sense) is mapped into a
bundle of concentric circles. As long as the mapped region remains
small (in any case restricted to the northern hemisphere), the configuration on the plane is a rather faithful image of the geometry
on the sphere. In the case of a hypersphere S3 orthogonally mapped
on a tangent hyperplane at the "north" pole, one gets a bundle of
concentric spheres centered at the pole. So if the polytope is
orientated in such a way that one vertex coincides with the pole,
the set of successive coordination shells are recovered after the
mapping. For example, the first two neighboring shells are
represented in Fig. 3. The polytope {3,3,5} has proved to be a

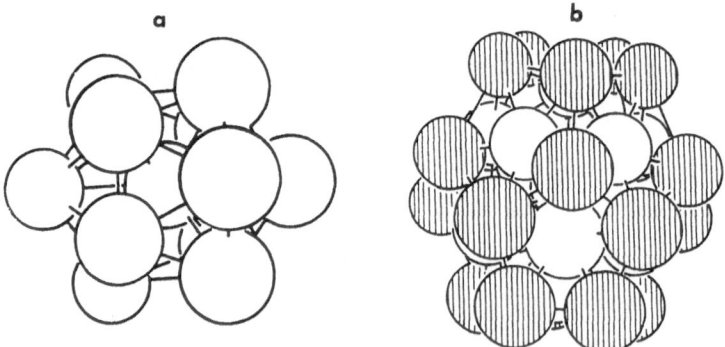

Fig. 3   Orthogonal mappings of polytope {3,3,5} in the euclidean
         3D space: a) one site and its icosahedral first-neighbor
         shell, 2) the next shell (darkened spheres) which form a
         dodecahedron.

good ideal model for amorphous metal structure. It can also be
used for the description of metallic and rare gas agregates [20].
Its symmetry group is very rich and contains 14400 elements (including
both direct and indirect transformations). When the hypersphere is
embedded in E4 (with axes x, y, z, t) the most interesting symmetry
operations are rotations, labelled for example R_{xy} or R_{zt}, which leave
a plane invariant, and also screw operations which correspond to
simultaneous rotations about completely orthogonal planes (which
intersect only at the hypersphere center). The importance of a particular screw depends on the degree to which it persists (after
mapping) in the "realistic" euclidean structure. We have previously
discussed [21] a particular screw axis, noted { 30/11}, whose signa-
ture is often visible in real space model. It is possible to
realize, in the euclidean space, an infinite linear arrangement of
regular tetrahedra (Fig. 4), which is a simple example of incommen-
surate structure. Indeed the vertices belong to a cylinder and their
angular coordinate (in the cylindrical coordinate system) can take
any value between $0°$ and $360°$. The 31th vertex has roughly the same
angular coordinate as the first one, with a small difference of $5°40'$
[22]. This angular misfit disappears on S3 for the polytope {3,3,5},
the straight euclidean line being transformed into a great circle
and the two vertices being then identified. Note that limited
(quasi) linear arrangement of spheres (called "collineations") has
been described in disordered hard sphere packing by Bernal [23].

B.   Tetracoordinated polytopes [24]

        As it has already been mentioned, the problem is complica-
ted because of the directional character of the bonds and the ability
of almost free rotation along a particular bond (almost free dihedral
angle). The information required in order to specify a medium range
configuration cannot be completely extracted from experiments. The
first topological ingredient in such description is the local ring
configuration. This is the reason why we now present three ideal
polytope models which precisely differ by their rings size and
arrangement.

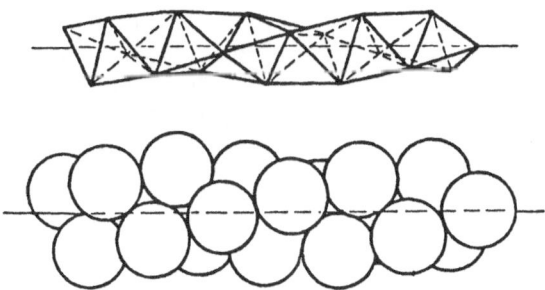

Fig. 4   Linear arrangement of tetrahedra (after Ref. 20).

C.   The polytope {5,3,3}

        If one puts a vertex at the center of each of the 600
tetrahedral cells of the {3,3,5}, and forgets the 120 original
vertices, one obtains the dual polytope called the {5,3,3} [15].
It contains 600 vertices and 120 cells which are regular pentagonal
dodecahedra.   Each vertex is fourfold coordinated, belongs to 4
dodecahedra and to 6 pentagons.   The local configuration corresponds
to the famous "vitron" of Tilton [25] and "amorphon" of Grigorovici
[26].   The first two shells of {5,3,3} vertices are drawn on Fig. 5.
The sites have been orthogonally mapped with the north pole located
at the dodecahedral cell center.   The {5,3,3} Radial Distribution
Function (the hypersphere geodesic distance is used) is shown in
Fig. 6a.

D.   The polytope "3000"

        If one replaces each vertex of the {5,3,3} by a small
centered tetrahedron of defined edge-length, one obtains a tetracoordinated polytope with 3000 vertices.   The local order corresponds
to the hand built model of Dandoloff, Dolher and Bilz [27] and consists of packing of dedecahedra and small clusters of 15 vertices
called "barrelans" (Fig. 7a).   One can also view it as the most
compact tiling (respecting the four fold coordination of the sites)
of a smaller 8 vertex cluster called "small barrelan" (Fig. 7b).
The particular stability of this cluster (from an electronic
point of view) had been previously demonstrated by one of us [28].
Fig. 8 shows 100 vertices of polytope "3000" after a restricted
orthogonal mapping.   The central shell is a regular dodecahedron with

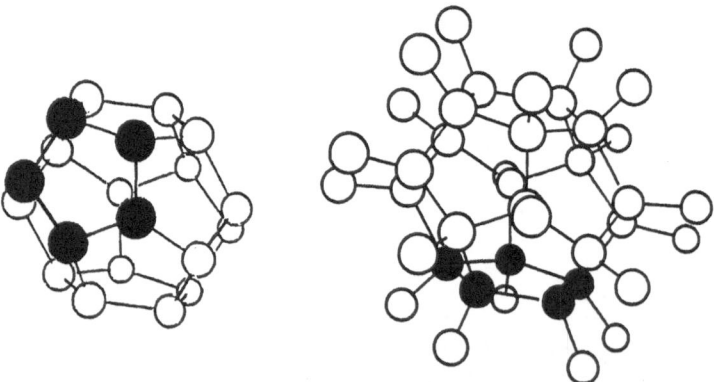

Fig. 5   Orthogonal mappings of polytope {5,3,3}.   The two first
        shells are shown and a pentagon is distinguished.

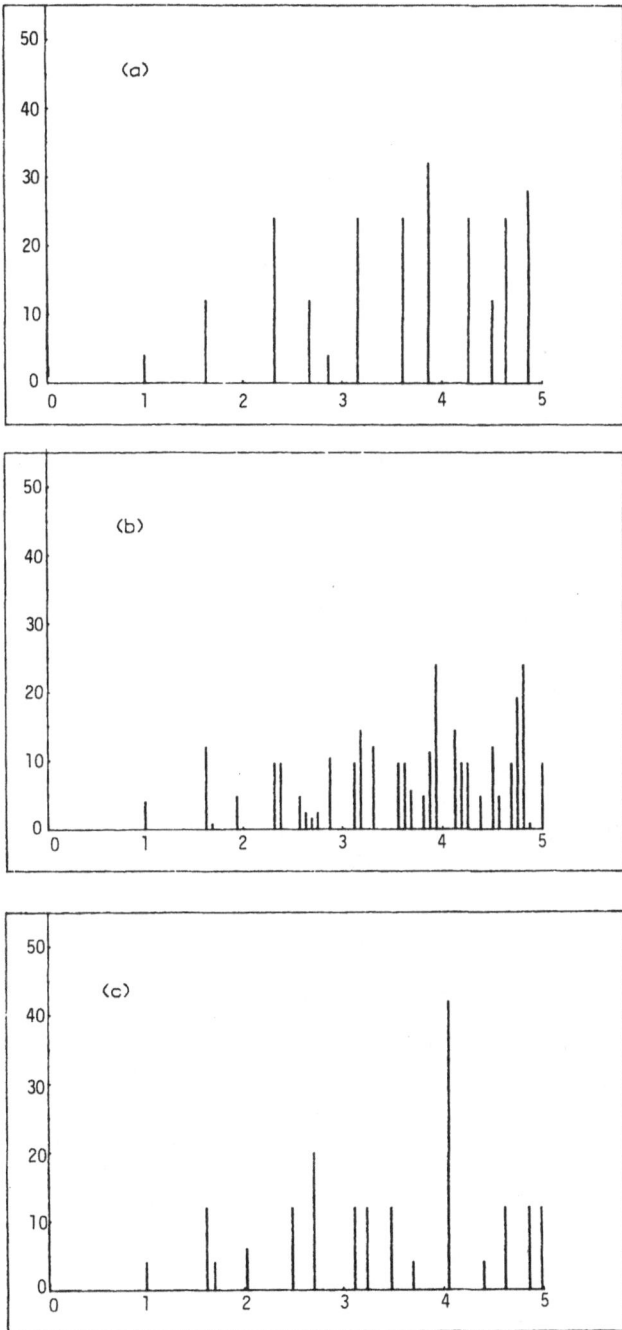

Fig. 6   The Radial Distribution Function of the polytopes, limited
         to 5 times the first neighbor distance: a) polytope {5,3,3},
         b) polytope "3000", c) polytope "240".

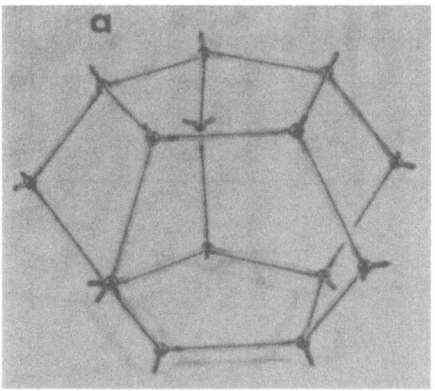

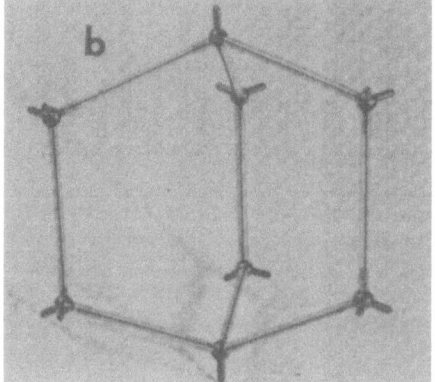

Fig. 7   a) a (pentagonal) barrelan, b) a small barrelan.

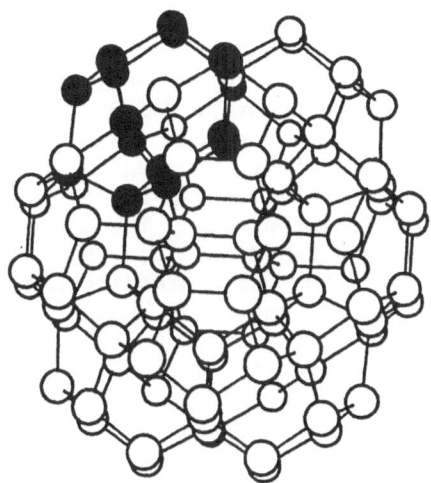

Fig. 8   Orthogonal mapping in E3 of 100 sites from polytope "3000".
         A pentagonal barrelan is darkened.

each of its pentagonal face covered by a barrelan. This polytope contains a mixing of pentagons and of hexagons in "boat" configuration.
Note that there are two distinct types of sites. The less numerous
(600) only belong to hexagons and are the loci where 4 "small barrelans"
are joined. The others (2400) belong to the surface of the dodecahedra and participate to pentagons as well as hexagons. Polytope
"3000" shares the same symmetry group with polytopes {3,3,5} and
{5,3,3}. However, the curvature of the hypersphere paved by polytope "3000" is very low (if one takes the first neighbor distance
as unit length) compared to the other polytopes ({5,3,3} is already
less curved than {3,3,5}). It is the reason why one can build, in
the euclidean space, an important part of polytope "3000" (up to
several hundred of sites) before too high distortions appear. The
polytope "3000" R.D.F. is displayed on Fig. 6b.

E.   The polytope "240"

        We now describe a new polytope which, in contrast with the
preceding ones, does not possess all the {3,3,5} symmetry operations
but only a subgroup. The building rule is similar to the one which
generates the diamond lattice starting from the CFC lattice. The
diamond lattice can be analyzed as being formed by two CFC replica,
displaced one from each other in the (1,1,1) direction. The polytope "240" contains two {3,3,5} replica which are related by a
(spherical) screw symmetry operation. Another way to obtain it is
to place new vertices at the center of some {3,3,5} tetrahedral
cells (in fact one over five cells will be centered).

Let us describe its structure more in detail. Each vertex of
one {3,3,5} replica is surrounded by 4 vertices belonging to the
other replica. Consequently, polytope "240", with its 240 tetracoordinated vertices, is a "bicolor" structure with only even membered
rings. The smallest cycles are hexagons in a twist "boat" configuration which preserves the perfect tetrahedral value for the bond
angles but is associated with a dihedral angle value intermediate
between the staggered and elipsed ones. The local configuration is
perfectly tetrahedral as one can see on Fig. 9, which shows orthogonal mapping of polytope "240" subsets of increasing size. In
close resemblance with the way how three "boats" like hexagons com-
bine into a small barrelan, three polytope "240" hexagons give rise
to twisted small barrelans which are arranged in the most compact
way (while preserving the tetracoordination for the sites). They
are punctured by "channels" very similar to those encountered in
diamond and wurtzite structure. These channels, whose symmetry is
related to the above-mentioned {30/11} screw, are bordered by a
triple helix of sites and follow great circles of S3 (Fig. 10).

The local order and connectivity of polytope "240" are very
close to those observed in the Connel-Temkin model [11]. One way
to become convinced of it is to compare the number of returning
walks of n steps in both structures. Such count is displayed on
Table I for several models. The data for the diamond, Polk and C.T.
models are given by Gaspard [29]. The path count equals that of the

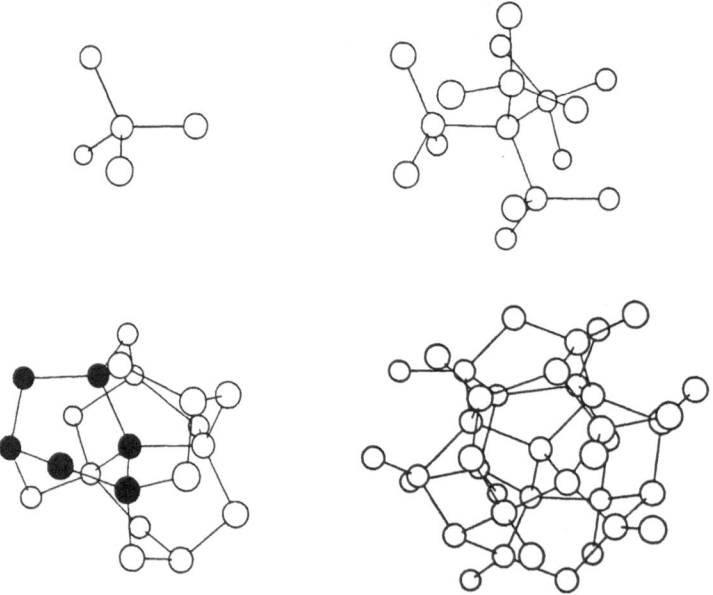

Fig. 9 Orthogonal mappings in E3 of subsets of increasing size from
        polytope "240". A "twisted boat" six fold ring is darkened.

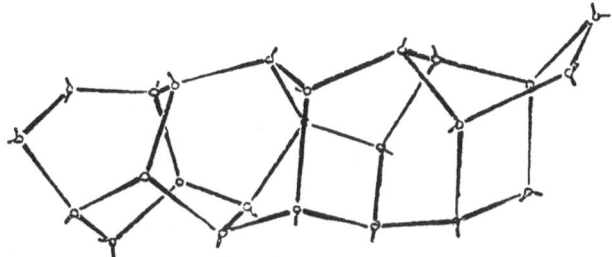

Fig. 10   The local bonding arrangement of sites around the 30/11 channeling axis.

n	BETHE LATTICE	DIAMOND LATTICE	POLK MODEL SITE N, 223	CONNELL TEMKIN MODEL CENTRAL SITE	POLYTOPE "240"	POLYTOPE >5.3.3<	POLYTOPE "3000" SITE 1	POLYTOPE "3000" SITE 2
1	0	0	0	0	0	0	0	0
2	4	4	4	4	4	4	4	4
3	0	0	0	0	0	0	0	0
4	28	28	28	28	28	28	28	28
5	0	0	4	0	0	12	0	6
6	232	256	244	268	268	232	256	244
7	0	0	90	0	0	252	24	120
8	2092	2716	2416	2982	3004	2188	2644	2434
9	0	0	1564	0	0	3960	768	1932
10	19864	31504	26016	35895	36784	23140	29224	26536

Table 1.   Number of returning walks of n steps in several tetra-coordinated networks.

Bethe lattice (which has no cycles) as long as rings are absent. The fact that n_6 = 268 for both the central site of C.T. model and the polytope "240" means that in both cases the sites belong to 18 hexagons.   One can be convinced easily that the configuration which allow for such a high number of hexagons is poorly degenerate and in fact probably unique (up to some trivial symmetry operations) It is realized as a 27 vertices cluster whose cohesive energy (evaluated with a very simple Hamiltonian model) is lower than the best agregates of comparable size taken from the diamond lattice [30]. The polytope "240" RDF is displayed on Fig. 6c.

III. MAPPING AND DEFECTS

     In the preceding paragraph, we have done a "crystallographic"
study of regular arrangements on S3. We now look at the way of
producing realistic 3D structures by an appropriate mapping on the
3D euclidean space E3. There are several mapping procedures which
can take a curved space into a flat space of equal dimension. We
must pick out, in the set of all possible mapped structures, those
which are reasonable from a physical point of view. This requires
that one is able to define simple parameters which faithfully describe
the cohesive energy. It was one motivation of a previous calculation
of the small silicon agregates cohesive energy [28]. In short, one
will try to limit the change in coordination number and in the
first neighbor distance, even at the prize of some angular distortions. For example, an orthogonal mapping from a S3 hemi-(hyper)-
sphere onto E3 generally gives rise to large distortions of bond
lengths and bond angles, and thus cannot be used here. An isometric
mapping (which preserves lengths) would be very attractive but it
can be shown that it is impossible if the two spaces have different
curvatures. Therefore, in a first step, we shall use a "star mapping"
in order to minimize these distortions.

A. The star mapping

     A simple example of star mapping in 2D is the peeling of
an orange on a Table (Fig. 11). One chooses on S2 several geodesic
lines crossing at both poles (meridians). This permits a partition
of S2 into (2D) lobes associated with the lines, each lobe containing the S2 points closer to a given line than to the others. Each
lobe follows its line in a geodesic mapping restricted to the lines,
creating cuts at the lobes borders. In 3D one has an analogous
procedure: the star mapping of S3 gives rise to lobes in E3 all
of which share a common point, the image of one S3 pole. The
opposite pole gives rise to several replica, one at each end of a
lobe.

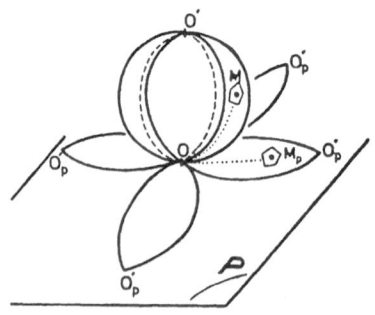

Fig. 11 Star-mapping in 2D.

The polytope {3,3,5}, "star" mapped along 4 different directions
in E3 is shown in Fig. 12. Three of the four lobes are clearly
visible in the drawing. Inside one lobe, the spheres configuration
is close to the original polytope local order and we have distinguished
some local pentagonal symmetries. The prize to pay for that is the
creation of surface cuts which limit the lobes. Structures built
on positively curved space are of finite size; consequently, the
mapped structures are also finite. It is necessary to fill large
regions of E3 in order to obtain realistic models. This can be
achieved by adding new replica of the mapped polytope close to the
first one. It is always possible to keep some local order continuity
between the first and second maps in a limited region. One then gets
a hard sphere connected matrix, where pentagonal local arrangements
dominate, which is pierced by internal cut surfaces where the local
order has discontinuities.

The situation is more complex in the case of tetracoordinated
polytopes. It is always possible to "star" map a given polytope.
For example, one can see on Fig. 13 one lobe of polytope "240" mapped
along four directions. This stick-and-ball structure has been
constructed using the nearest-neighbor relationships of the compu-
tationally mapped polytope. Difficulties arise when one tries to
add new replica of the mapped polytope in order to fill more densely
the space. Because of the directional character of the bonds in
the real materials, one has to connect two lobes, arising from
different replica, along the common direction of dangling bonds
belonging to these two lobes. This could be achieved by translating
and rotating the whole lobe but, as one easily imagines, this can
become rather rapidly a difficult task. It is the reason why we
have, for the moment, only constructed small euclidean structures
with this method. This has to be contrasted with the metallic (compact) case where angles are less important and lobes connection is
easier.

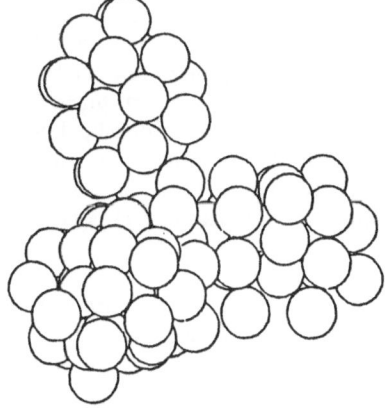

Fig. 12 The polytope {3,3,5} star-mapped along 4 tetrahedral
         direction. Three lobes are clearly visible.

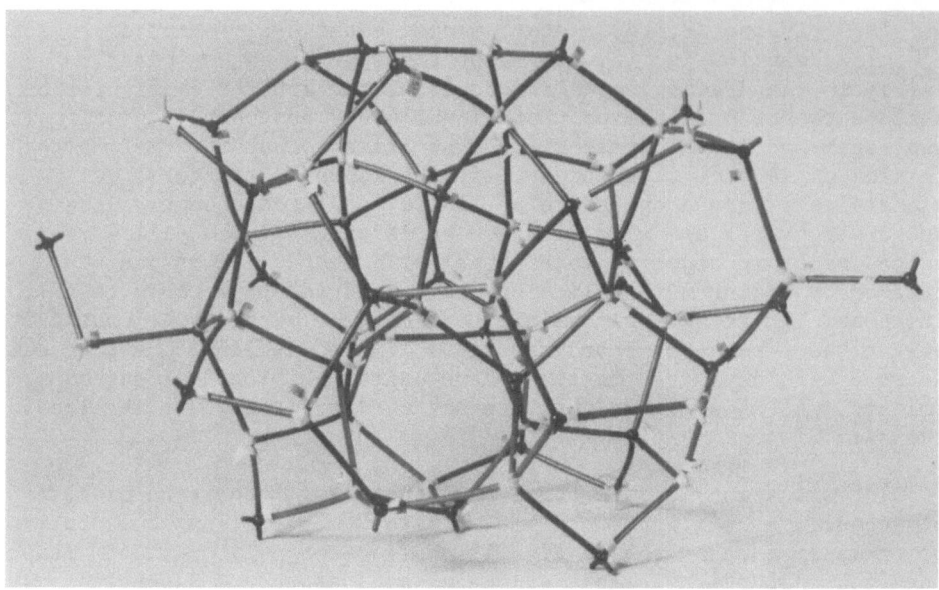

Fig. 13   One lobe of polytope "240" after a star mapping along
          four tetrahedral directions.

Nevertheless, the star mapping procedure can be used for
qualitative discussion of amorphous tetracoordinated structures.
The final structure (after mapping) will consist in a connected
(almost) tetravalent matrix (the lobes network) with internal
surfaces (resulting from the cuts) covered by dangling bonds. Inside the matrix, the local order is similar to that of the polytope
with some elastic distortions added. It is interesting to note
the relation between the number of dangling bonds and the matrix
distortion: a large internal surface area is associated with low
elastic distortions (and vice versa). This is the same as the
current idea that internal surfaces "relax" the structure. Also
this can be used as a qualitative reason why the excitation spectrum band tails (whose width is generally related to the amount
of elastic disorder) are often larger in pure amorphous silicon
than in the hydrogenated material. Indeed, hydrogen atoms in the
best aSi:H materials, which can contain up to 15% of hydrogen,
require a larger internal surface area in order to be bonded with
the silicon matrix which is then more relaxed.

A last application concerns the density variation in $aSi_{1-x}H_x$.
Several authors [31] have observed that the density displays two
different regimes as a function of x, and this despite rather
scattered experimental data which are represented in Fig. 14. The
solid line has been arbitrarily drawn through the experimental points
as a guide for the eye. Roughly speaking, the general tendency can
be described as follows:

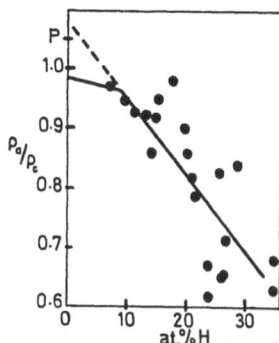

Fig. 14   Density of aSi:H (scaled to the crystalline Si density)
          versus the hydrogen content. P represents the density of
          polytope "240".

–   For x > 0.1, the density exhibits a pronounced decrease as the H
content increases. A remarkable point is that the extrapolation of
this variation to the zero H content leads to a value higher than
the crystalline silicon one.

–   For x < 0.1, we suppose a smooth variation between the two experimental values for x = 0 and x = –1.

     These results have to be explained by topological arguments
since the Si-Si first distance is nearly constant for all compositions.
The Curved Space Model provides a qualitative explanation for this
density variation. We have shown that defects are necessarily present in the euclidean structure, as internal surfaces or as linear
defects (discussed in the next paragraph). Variable amounts of
dangling bonds are associated with these topological defects, the
precise value being governed by energy considerations. Even the
best optimum structure will be able to incorporate a given amount
of hydrogen atoms in order to saturate these dangling bonds.
Moreover, post-hydrogenation experiments on aSi [32] have shown
that it is possible to incorporate up to 10% of hydrogen in pure aSi
in such conditions that large reorganization in the Si matrix is
unlikely. We understand then why density variations remain low
for x < 0.1.

     For x > 0.1 one must (in a star mapping approach) introduce more
cut surfaces in order for all hydrogen atoms to be bonded. All the
dangling bonds on the new cut surfaces will be H-saturated and one
can suppose a more pronounced monotonous decrease of the density.
In such models the Si atoms are gathered in dense regions where the
polytope local order prevails. This explains why the x = 0 extrapolation of the density value is different from the crystalline case
and even the pure aSi case. The extrapolated value refers to that
of the ideal structure, without defects, of the polytope and the

difference with the pure aSi density is related to the minimum
density of defects introduced in order to flatten the polytope.
Among the three tetracoordinated above described polytopes, only
one, the polytope "240", has a density (slightly) higher than the
crystalline one and is then a good candidate for the description of
aSi:H structure. Dihedral configurations intermediate between the
eclipsed and staggered ones are associated with the "twisted boat-
like" polytope "240" hexagons. Note that x-ray diffraction experiments [3] support the presence of such dihedral configurations in
aSi:H.

Using the same kind of considerations, we have described here,
again qualitatively, the behavior of the specific volume of simple
metals in their amorphous and liquid phase, as a function of temperature [33].

B. Mapping by disclinations

        It has sometimes been proposed in the past to describe
noncrystalline structures as highly dislocated crystals. As said
before, the failure of such models may be the result of a bad
initial choice: defining the ideal structure as the crystal
whose local order is not close enough to the amorphous one, necessarily induces the presence of an important defect density.

        This criticism disappears with the Curved Space Model since
one takes by hypothesis an ideal model whose local order is as close
as possible to the real material. The ideal model, the polytope,
has no translation order but contains a high number of rotational
symmetries. Rather naturally, one then asks whether it is possible
to describe the noncrystalline euclidean real material as an ideal
non-euclidean model containing disclinations (dislocation-rotation).
We give a positive answer to this question and this can be simply
modellized in the so-called "corrugated space" approach [34]. In
the case of compact tetrahedral packing it is even possible to
derive good approximations for the defect density. This approach
will not be described here and we shall restrict to the description
of how disclination lines can be used to lower the polytope curvature.

        A simple disclination can be generated "a la Volterra" by
cutting the structure and adding (or removing) a wedge of material
between the two lips of the cut [35]. As dislocations introduce a
strain field in the material, so do the disclinations. But in this
case, it is also possible to describe the induced deformation by a
change in the space curvature. Adding a wedge of material in a
structure defined in a spherical space decreases the curvature,
but topological defects appear along the edges and faces of the wedge.
Nevertheless, if the two faces of a wedge are equivalent under a
rotation belonging to the polytope symmetry group, the defects are
confined near the edge. Thus, a perfect disclination is created.

A 2D example of disclinations in a perfect hexagonal lattice (honey-
comb) is presented in Fig. 15. In this case, it is easy to observe
how the disclination changes the curvature, and also modifies the
connectivity (in terms of ring size) only in the immediate neighboring of the point defect. In 3D the disclination defects become
linear. In Fig. 16, one can see the effect of a disclination line
on an isosahedral shell. In 3D as in 2D the disclinations can modify
the ring size or the coordination numbers, depending on their location
with respect to the atomic positions. Figure 17 shows the local
effect of a disclination line threading a series of pentagons in
polytope {5,3,3}. It is possible to introduce easily two such defects
which appear as great circles belonging to completely orthogonal
planes and contribute to decrease the polytope curvature. The new
structure contains 864 vertices (compared to the 600 vertices of the
original polytope) and a small number of hexagons which surround the
disclinations. In the case of covalent structures it is necessary

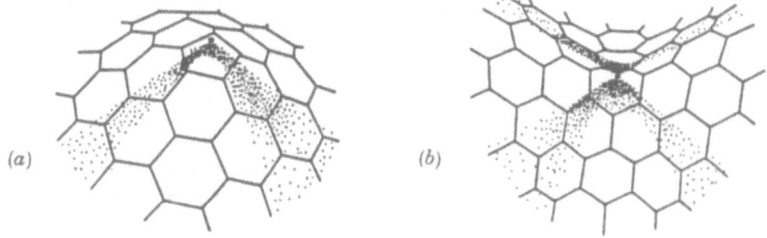

(a)   (b)

Fig. 15   2D illustration of the relation between disclination, curva-
          ture and ring size. The reference structure (free of defect)
          is the hexagonal tiling: a) positive curvature. The point
          is located at the pentagon center. b) negative curvature.
          The point disclination is located at the heptagon center.

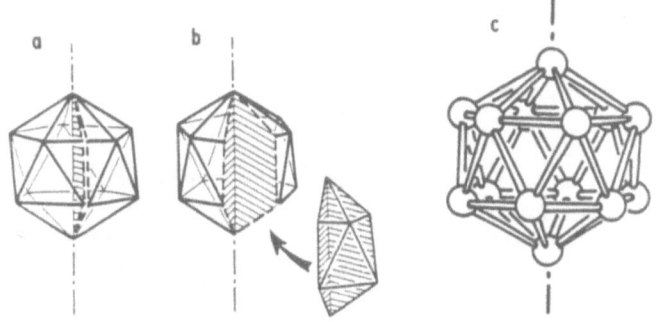

a   b   c

Fig. 16   Local representation of a disclination along a 5-fold axis
          of polytope {3,5,5}. We only draw the first-neighbor shell
          of a site located on the axis. a) perfect icosahedral shell
          in the polytope {3,3,5}, b) generation of the disclination,
          c) the first-neighbor shell after the disclination being
          created.

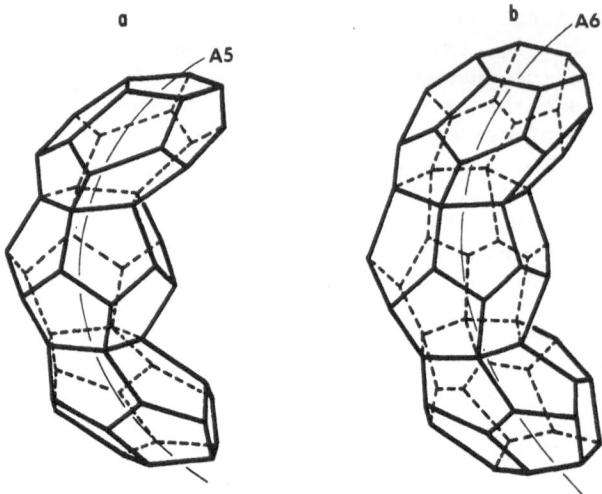

Fig. 17   Local representation of a disclination along the same axis
          as Fig. 16 for the polytope {5,3,3}. a) perfect dodecahedral
          cells of polytope {5,3,3};   b) the disclination has changed
          the 5-fold axis into a 6-fold axis.

that the disclination lines evitates the sites in order to disallow
an increase in the coordination number. On the opposite, changing
the ring size does not cost too much in energy and leads to more
realistic structures.

    The technical procedure for the introduction of such linear
defects in the polytope uses the symmetries of spherical torus and
has been described elsewhere [36].   There are today some unsolved
problems concerning the possibility of completely decreasing the
polytope curvature by the introduction, step by step, of single
disclination lines.   In a recent work [37] we have shown how it is
possible to bypass these difficulties by creating, step by step,
disclination networks.   The obtained structures display a hierarchical
defect organization and provide new ideal models of non-periodic systems.

    Notice that the disclination defects described in this paragraph
are of different type from those introduced in the same context by
N. Rivier [38]. The Rivier lines only thread the odd-membered rings
and do not "carry" curvature. A connection between the Rivier lines
and the disclinations associated with the C.S.M. would be very
desirable in the future.

    A rather remarkable consequence of this work on defects in
non-crystalline materials has been the description of many complex
metallic structures (like the Frank-Kasper phases) as non-euclidean
structures where the curvature is lowered to zero through the presence
of periodic disclination networks [39,40]. Such a description can be

easily extended to tetravalent structures with big primitive cells (clathrate-like structures).

IV. EXCITATION SPECTRUM OF POLYTOPES

A. The Hamiltonian model

In this section, we calculate the polytopes density of states in a simple tight-binding framework with Hamiltonian:

$$H = \Delta \sum_{\substack{i,J,J' \\ J' \neq J}} |iJ> < iJ'| + \beta \sum_{\substack{i,i',J \\ i,i' \text{ neighb}}} |iJ> <i'J| \qquad (1)$$

The one electron wave function reads

$$| \psi > = \sum_{i,J} a_{iJ} | iJ > \qquad (2)$$

$|iJ>$ is an sp^3 orbital which is centered on site i. Δ and β are respectively the intra- and interatomic transfer integrals. The zero of energy is taken as

$$\frac{E_s + 3 E_p}{4} = 0$$

This model had been used by Leman and Friedel [41] in order to study the diamond band structure and by Weaire and Thorpe [42] in their approach of amorphous tetracoordinated systems. Let us recall briefly well known results concerning this tight binding model. The eigen-spectrum of H divides in two sets:

- 2 flat "p" bands with energy:

$$E = - \Delta \pm \beta \qquad (3)$$

- 2 wide "sp" bands with energy:

$$E = \Delta \pm (4 \Delta^2 + \beta^2 + \Delta\beta\varepsilon)^{1/2} \qquad (4)$$

where ε are the eigenvalues of a simplified Hamiltonian H'

$$H' = \sum_{\substack{i,i' \\ \text{neighb}}} |i > < i'| \qquad (5)$$

This is the famous 1 band↔2 bands transformation which simpli-fies the calculation. Note that this model suggests an isolating or semiconductor behavior when $|\beta/\Delta| > 2$.

The fact that we use a sp³ basis for calculation on S3 is not
a serious hindrance. We are not interested in the calculation of
the "true" excitation spectrum of an atomic system on S3 but rather
more in the effect of the connectivity on the spectrum, the spherical
aspect being interpreted here as particular boundary conditions.
The same meaning has to be given to band structure calculation on
exotic structures like the Bethe lattice and the Husimi cactus.
Notice that these last structures can be embedded in a negatively
curved surface [43]. Furthermore, since we assume a β transfer
integral which does not depend in the intersite distance, the
metrical nature of the network disappears. One has to calculate the
H' excitation spectrum which is nothing but the spectrum of the
graph [44] spanned by the sites and bonds of the original structure.
In the following we shall use the metrical regularity of the polytopes
in order to introduce a "pseudo" reciprocal lattice. However, the
excitation spectrum would not change under any distortion which
respects the connectivity.

B.   The spherical torus method

        We are interested in the "s" band excitation spectrum of
polytopes on S3 whose regularity is associated with large symmetry
groups. In principal, one can use this group in order to calculate
the eigenspectrum, the eigenvectors being then labelled by the
irreducible representations of the group. This has recently been
done in an elegant manner by Nelson and Widom [45]. Here we present
another method which can be considered as semi-analytical and presents,
in some aspects, several advantages. In short, it consists in
selecting in the total symmetry group, a (abelian) cyclic subgroup
related to a given rotational or screw symmetry operation. We define
an equivalence relation between sites which link any two sites when-
ever they can be put in correspondence by an element of the subgroup.
One has then partitioned the set of polytope vertices into a family
of vertices globally invariant under the subgroup operations. This
will considerably simplify the excitation spectrum calculation by
allowing to write Bloch-like sums of basis states. We are not going
to enter into the details of the method which is described elsewhere
[21,46]. The name of the method ("spherical torus") refers to the
fact that the family of points belongs to special torus surfaces on
S3, whose relation with polytopes has been mentioned in Chapter III
[36].

        Let us first describe a simple example in lower dimension. We
want to calculate the eigenspectrum of a cube [21]. We a priori
have to diagonalize an 8x8 matrix. A particular 4-fold symmetry
axis is selected (Fig. 18a). The cube can be endlessly "mapped"
on the euclidean plane by rolling and gives rise to an infinite
ladder (Fig. 18b). The 4-fold rotation is transformed into a
translation. Notice that the cube can be topologically recovered
from the infinite ladder by identifying sites separated by 4 times

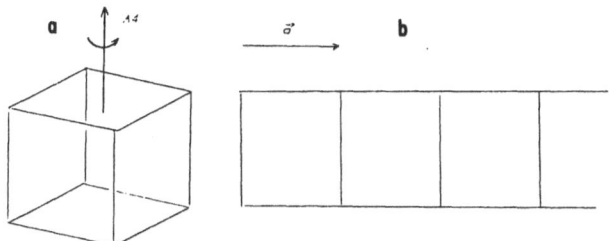

Fig. 18   a) The cube with one 4-fold axis.   b) The ladder obtained
          by a mapping which relates the 4-fold axis and a translation
          in flat space.

the unit translation distance.   The infinite ladder band spectrum
can be easily calculated using the Bloch theorem.   The Hamiltonian
matrix is reduced to a 2x2 matrix

$$\begin{pmatrix} 2 \cos ka + 1 & \\ +1 & 2 \cos ka \end{pmatrix}$$ (6)

which is diagonalized at each point k of the first Brillouin zone
and leads to the eigenvalues

$$E_k = 2 \cos ka \pm 1$$ (7)

The cube eigenvalues are obtained by the above cited identification
of sites, e-g by using Born-Van Karman conditions which select four
allowed k values (Fig. 19)

$$k = \frac{2 \pi n}{4a} \quad n = 0, 1, 2, 3$$ (8)

In order to get the {3,3,5} eigenspectrum we shall use the {30/11}
symmetry axis which was described in paragraph II.A.   It is a 30-fold
symmetry operation which leads to a partition of the 120 {3,3,5}
vertices into four sets of 30 vertices, one of which looks like the
configuration of Fig. 4 with tetrahedra arranged along an S3 great
circle.   The 120x120 initial matrix is then reduced to a 4x4 secular
matrix which has to be calculated at the 30 "allowed" k values:

$$k = \frac{2 n \pi}{30 a} \quad n = 0, \ldots, 29$$ (9)

The band diagram and the density of states (obtained by projection
on the vertical axis) are displayed in Fig. 20.

We can solve the polytope "240" eigenspectrum using the same
{30/11} symmetry operation.   The set of 240 vertices is divided into
eight sets of 30 vertices which leads to an 8x8 secular matrix
calculated at the same 30k values as above.   The "s" band density
of states of polytope "240" is shown in Fig. 21a.   Applying the

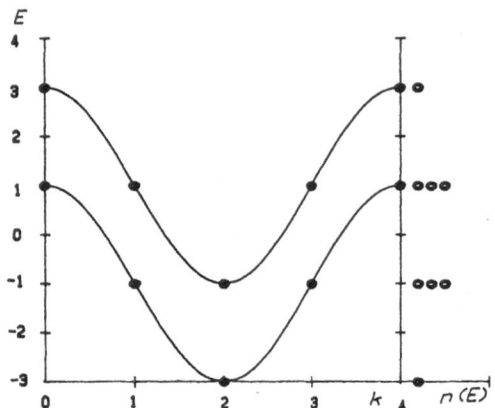

Fig. 19   The "s" band spectrum of the cube.   The continuous spectrum
          corresponds to the infinite ladder.   The cube density of
          states is obtained by a projection on the energy axis (on
          the right of the figure).

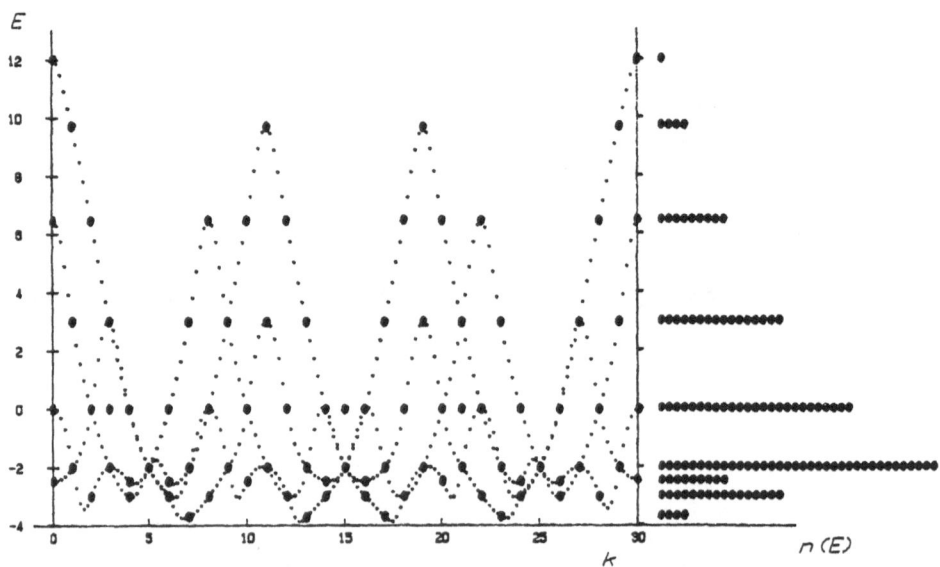

Fig. 20   "s" band of polytope {3,3,5} along the 30/11 axis (whose
          symmetry is that of Figure 4).

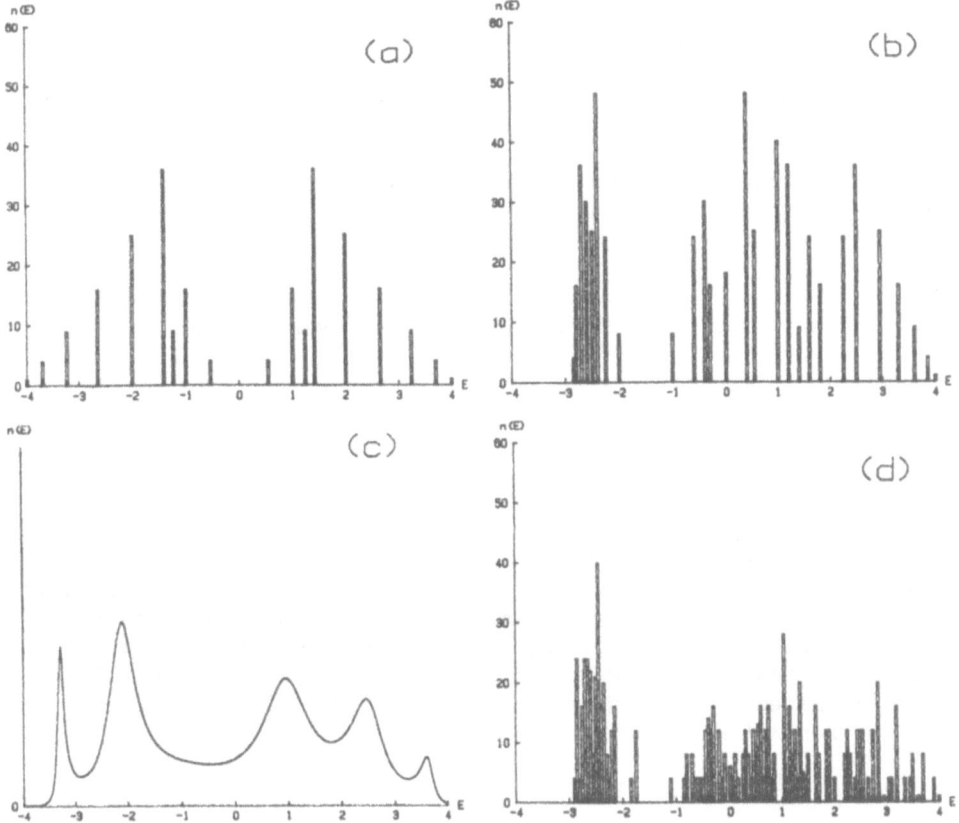

Fig. 21 "s" bands of the tetracoordinated polytopes. a) polytope
"240"; b) polytope {5,3,3}; c) polytope "3000"; d) polytope "864".

1 band ↔ 2 bands transformation, we can easily derive the sp³ band
which is represented in Fig. 22a. Notice that the "s" band displays
a symmetrical shape (around zero) which is characteristic of bicolor
networks (having only even-membered rings).

In order to study the polytope {5,3,3} eigenspectrum we shall
use the combination of two symmetry axis, a 5-fold rotation axis
and a 10-fold screw axis [46]. One obtains a 12x12 matrix which has
to be diagonalized in the following 50 points of a 2D "pseudo"
Brillouin zone:

$$k_x = \frac{n \, \pi}{5 \, a} \qquad n = 0, 1, 2, 3, 4$$

$$k_y = \frac{2 \, m \, \pi}{10 \, a} \qquad m = 0, \ldots\ldots, 9$$

(10)

The "s" and sp³ bands are displayed in Fig. 21b and 22b,

respectively.   In Fig. 21b, we can verify that the antibonding limit-
4 is not reached, which is a well known effect of the presence of
odd-membered rings [1].   A second particularity of this spectrum is
the region between − 2 and − 1 which is free of eigenvalue.   We
expect that this internal gap is also due to the odd-membered rings
but we could not find a simple explanation for it.   In the next
paragraph we shall, however, provide an indirect evidence of this
effect.

In the case of polytope "3000" we have not tried to apply this
method because the complexity increases and even the partition of
the set of vertices should be done numerically.   Instead we have
calculated numerically the density of states with a continuous
fraction technique [29] which gives rise to a continuous band
spectrum (Fig. 21c and 22c).

C.   Effect of a topological defect on the excitation spectrum

Let us begin with the simple example of a cube.   A disclination of angle $\pi/2$ along the four-fold "vertical" axis transforms
the cube into pentagonal prism (Fig. 23).   The polyhedron is topologically related to the infinite ladder by the identification of
sites separated by five unit translations.   Thus, the eigenvalues
are still given by the expression (7), but the allowed k values are
no more those of expression (8) and read now:

$$k = \frac{2 \pi n}{5 a} \qquad n = 0, 1, 2, 3, 4 \tag{11}$$

It is clear that the key idea consists in using the same axis for
the disclination defect and for the labelling of the eigenstates.
This can easily applied to the polytopes.   In section III.B. we
have described a disclinated version of polytope {5,3,3} with the
defect lines located along five-fold symmetry axes.   The excitation
spectrum of this polytope is obtained with the same formal secular
12x12 matrix but, instead of expression (10), the allowed k values
are now given by:

$$k_x = \frac{n \pi}{6 a} \qquad n = 0, 1, 2, 3, 4, 5$$

$$k_y = \frac{m \pi}{12 a} \qquad m = 0, \ldots, 11$$

The correspondence between the number of polytope vertices (864) and
the number of eigenvalues is easily verified.   The "s" band of polytope "864" is shown in Fig. 21d.   The antibonding limit has been
(very slightly) repelled toward the −4 limit.   This is due to the
introduction of some even-membered rings in the polytope.   An
interesting point concerns the region between −2 and −1.   The
presence of hexagon coincides with the occurrence of eigenstates
in this interval.   This is the indirect reason why we attribute this
internal gap in polytope {5,3,3} to the odness of the structure.

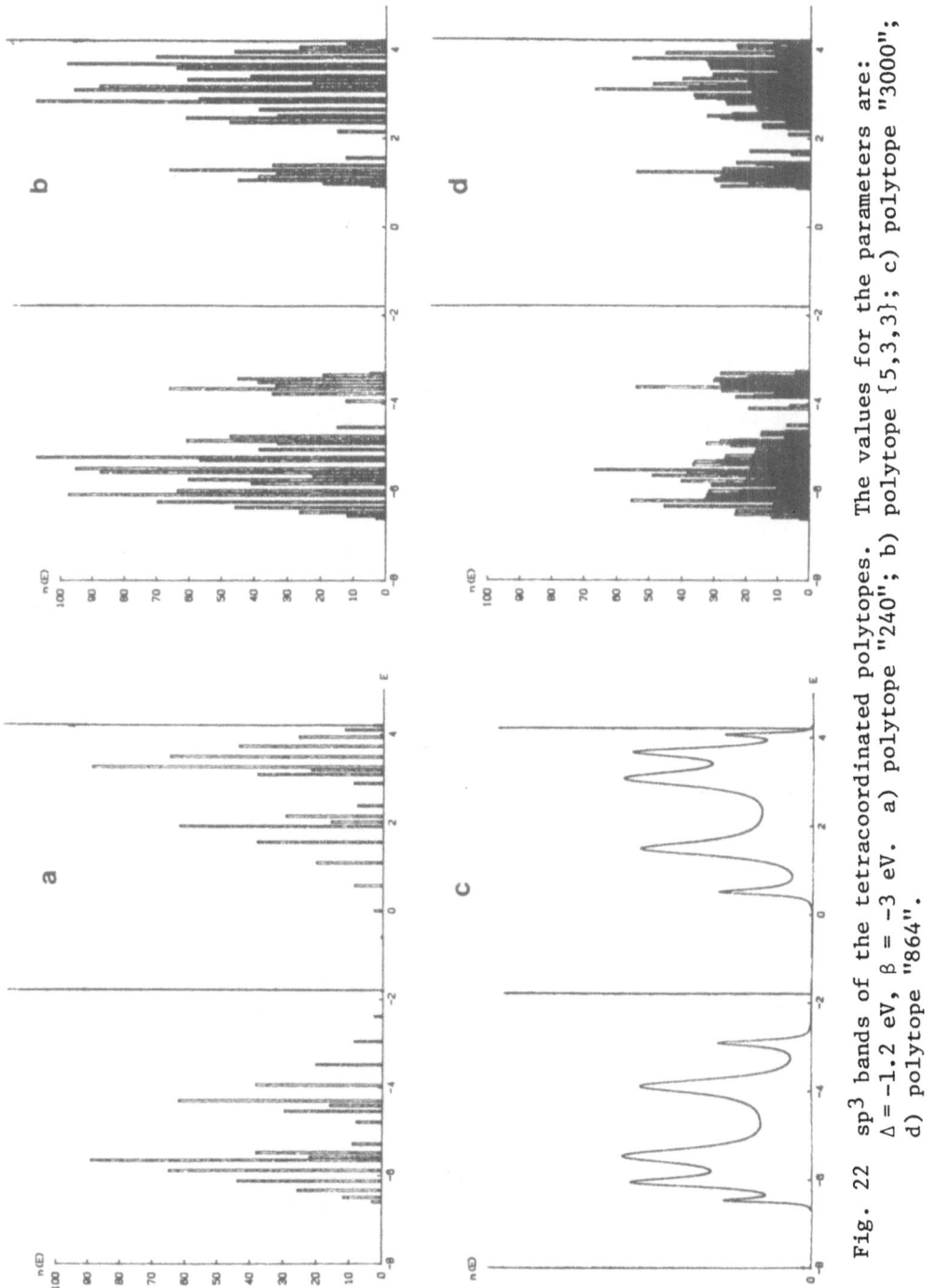

Fig. 22 sp³ bands of the tetracoordinated polytopes. The values for the parameters are:
$\Delta = -1.2$ eV, $\beta = -3$ eV. a) polytope "240"; b) polytope "3000";
c) polytope {5,3,3}; d) polytope "864".

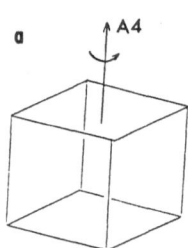 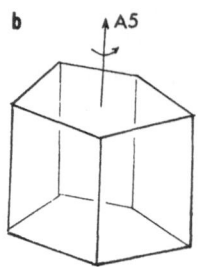

Fig. 23   A disclination along the four-fold axis transforms a cube
           (a) into a pentagonal prism (b).

The "s" band density of states of polytope "864" bears many
resemblances with that of polytope {5,3,3}. At first sight, one
could say that the first one can be derived from the second by some
broadening process. The advantage of the spherical torus method is
that it points out the precise relation of that broadening with the
topological defect.

D.   Cohesive energy of polytopes

We now calculate the electronic part of the polytopes
cohesive energy by summing over the energies of the occupied states
(with two electrons per eigenstate) in the sp^3 density of states.
A discussion of this approximation to the true cohesive properties
of covalent solids can be found in the paper by Friedel [1]. Because
we do the calculation on the ideal polytope structure, the resulting
cohesive energy only depends on the network connectivity. Indeed
all the structures are perfectly four-fold coordinated, with constant
nearest neighbor distance. It has been shown [1,47] that the sp^3
cohesive energy can be expanded in terms of the moments of the "s"
band density of states. These moments are closely related to the
number of returning walks on the networks [29]. An important result
is that paths with an odd number of bonds lower the total energy,
while even rings (or odd-membered rings circumnavigated twice)
raise the energy. Since the smallest rings in our structures are
pentagons, the positive contribution to the total energy, due to
odd-membered rings, occurs at high orders in the expansion, compared to the negative contributions which occur at all odd orders
(starting at order 5). Note that the Bethe Lattice, which is a
cycle free structure, is more stable than even-membered networks
like the diamond structure [47].

We have verified that polytope {5,3,3}, which contains only
odd-membered rings, is more stable than polytope "240", which is
an even structure. Polytope {5,3,3} is also more stable than the
diamond structure, up to the accuracy of our calculation which has
to compare summation over a discrete eigenspectrum (the polytope)
and a continuous one (the diamond structure). Nevertheless, the

differences between all the cohesive energies remain small, which confirm that the stability mainly depend on the four-fold coordination. This may no more be true if one includes a dependence in the distance for the Hamiltonian parameters in the more realistic case of the polytopes mapped in the euclidean space. Indeed, different amounts of distortion can be present according to which polytope and which mapping procedure is used. Therefore, it is rather difficult to draw precise conclusions from the above calculations.

This has to be contrasted with another approach which focuses on the stability of small clusters of sites. One considers small agregates in euclidean space where the atomic positions mimic the ideal polytope configuration (think of a local orthogonal mapping on a tangent euclidean space, for example). There the surface effect (the dangling bonds pointing outward the cluster) is taken into account. The cohesive energy of a number of small clusters has been discussed in Ref. [28]. Its usefulness comes from the fact that the stable clusters are expected to occur frequently during film growth. We have shown that for (not too small) 3D clusters, a rough count of the surface dangling bonds gives a good image of the cluster stability. The dodecahedron (basic unit of the polytope {5,3,3}) and the small barrelan (basic unit of polytope "3000") have proved to be specially stable units. A 27 vertex cluster, with the same local bonding topology as polytope "240" also shares this property [30].

We would like to thank J. P. Gaspard, D. Di Vincenzo, M. H. Brodsky and J. Dixmier for helpful and stimulating discussions.

REFERENCES

1. J. Friedel, Journal de Physique 39:651 (1978).
2. H. Richter, G. Breitling, Z. Naturforsch. 13a:988 (1958).
3. N. Shevchik, W. Paul, J. of Non-Cryst. Sol. 13:1 (1973/1974).
4. S. C. Moss, J. F. Graczyk, Phys. Rev. Lett. 23:1167 (1969).
5. G. Etherington, A. C. Wright, J. T. Wenzel, J. C. Dore, J. H. Clarke and R. N. Sinclair, J. Non-Cryst. Sol. 48:265 (1982).
6. J. S. Kasper, et al., Science 150:1713 (1965).
7. N. Rivier, Journal de Physique C9 43:91 (1982).
8. J. F. Sadoc, R. Mosseri, Journal de Physique 45:1025 (1984).
9. A. F. Wells, Three dimensional nets and polyhedra, Wiley ed. (1977).
10. D. E. Polk, J. of Non-Cryst. Sol. 5:365 (1971).
11. G. A. N. Connell, R. J. Temkin, Phys. Rev. B9:5323 (1974).
12. J. Dixmier, A. Gheorgiu, M. L. Theye, J. of Phys. C17:2271 (1974).
13. M. Kleman, J. F. Sadoc, Journal de Physique Lettres 40:L-569 (1979).
14. Goursat, An. Sci. Ec. Norm. Sup 6:9 (1889).
15. H. S. M. Coxeter, Regular polytopes, Dover ed. (1973).

16. A. C. Wright, G. A. N. Connell, J. W. Allen, J. Non-Cryst. Sol. 42:69 (1980).
17. R. Zallen, The physics of amorphous solids, Wiley ed. (1983).
18. J. F. Sadoc, J. Dixmier, A. Guinier, J. of Non-Cryst. Sol. 12:46 (1973).
19. P. DuVal, Homographies, Quaternaions and Rotations, Oxford University Press (1964).
20. J. Farges, M. F. De Feraudy, B. Raoult, G. Torchet, 4e Ecole d'Ete Mediterraneenne, ed. F. Cyrot-Lackmann (1981).
21. M. H. Brodsky, D. P. DiVincenzo, Physica 117B-118B:971 (1983); D. P. DiVincenzo, R. Mosseri, M. H. Brodsky, J. F. Sadoc, Phys. Rev. B29:5934 (1984).
22. H. S. M. Coxeter, Introduction to geometry, Wiley ed. (1969).
23. J. D. Bernal, Proc. R. Soc. London ser. A280:299 (1964).
24. J. F. Sadoc, R. Mosseri, Journal de Physique 42:C4-189 (1982); J. F. Sadoc, R. Mosseri, Phil. Mag. B45:467 (1982).
25. L. N. Tilton, J. Res. Nat. Bur. Standards 59:139 (1957).
26. R. Grigorovici, R. Manaila, J. of Non-Cryst. Sol. 1:371 (1969).
27. R. Dandoloff, G. Dohler, H. Bilz, J. of Non-Cryst. Sol. 35-36:537 (1980).
28. R. Mosseri, J. P. Gaspard, Journal de Physique 42:C4-245 (1981).
29. J. P. Gaspard, These d'Etat, Orsay (1975).
30. M. H. Brodsky, D. P. DiVincenzo, R. Mosseri, J. F. Sadoc, San Francisco International Conference on the Physics of Semiconductors, to be published (1984).
31. P. John, et al., J. of Physics C14:309 (1981).
32. D. Dieumegarde, D. Dubreuil, N. Proust, Proc. Intern. Vacuum Congress, Le Vide 201:731 (1980).
33. J. F. Sadoc, R. Mosseri, Journal de Physique 43:C9-43 (1982).
34. J. P. Gaspard, R. Mosseri, J. F. Sadoc, The structure of non-crystalline materials 1982, ed. Gaskell, Parker, Davies, p. 550. Same authors, Phil. Mag., to be published.
35. W. F. Harris, Scientific American, p. 130 (Dec. 1977).
36. J. F. Sadoc, R. Mosseri, Topological disorder in condensed matter, ed. F. Yonezawa, T. Ninomiya) p. 30, Solid State Sciences, 46, Springer Verlag.
37. R. Mosseri, J. F. Sadoc, Journal de Physique Lettres, to be published.
38. N. Rivier, Phil. Mag. B40:859 (1979).
39. J. F. Sadoc, Journal de Physique Lettres 44:L-707 (1983).
40. D. Nelson, Phys. Rev. Lett. 13:983 (1983).
41. G. Leman, J. Friedel, J. of Appl. Phys. 33:281 (1962).
42. D. Weaire, M. F. Thorpe, Phys. Rev. B4:2508 (1971).
43. R. Mosseri, J. F. Sadoc, Journal de Physique Lettres 43:L249 (1982).
44. N. Biggs, Algebraic graph theory, Cambridge University Press (1974).
45. D. Nelson, M. Widom, Nucl. Phys. B., to be published.
46. R. Mosseri, These d'Etat, Orsay (1983).
47. M. F. Thorpe, D. Weaire, R. Alben, Phys. Rev. B7:3777 (1973).

ELECTRONIC STRUCTURES OF CRYSTALLINE AND AMORPHOUS SILICON DIOXIDE AND RELATED MATERIALS

W. Y. Ching

Department of Physics, University of Missouri-Kansas City

Kansas City, Missouri 64110

INTRODUCTION

In spite of considerable amount of research work done on the structures and properties of silicon dioxide, both experimentally and theoretically, in the last fifteen years [1-3], our understanding about this remarkable material in both crystalline and amorphous forms is still far from complete. Thus, there is still no consensus with regards to the structures of amorphous SiO_2 (a-SiO_2). Continuous random network (CRN) model [4], mixture of microcrystalline model [5] and model of aggregation of large clusters [6] with internal surfaces, all have their respective supporters. Unlike semiconductors, the measured value of the band gap in SiO_2 has not yet been fully settled [1,7] and the interpretation of various spectroscopic data remains controversial [1,7]. One of the key elements in advancing further understanding about the structure and bonding of SiO_2 is the detailed knowledge of the electronic structures of various forms of SiO_2 and their dependence on the atomic scale structures. Early theoretical work on SiO_2 concentrated on studying molecular fragments [9-13]. Empirical methods have been used to study the electronic structure of bulk materials [14-17]. Only quite recently, conventional band theory using local density functional theory has been used to study the electronic structures of α-quartz [18-19] and idealized structures of β-cristobalite [20,21]. The attempt to study electronic structures of a-SiO_2 is much less numerous and mostly limited only to calculations with model Hamiltonians on artificial structures such as Bethe lattice [17].

In this chapter, we review the results of theoretical calculations of electronic structures on various forms of crystalline (c-SiO_2) and noncrystalline SiO_2 using the first-principles

orthogonalized linear combination of atomic orbitals method
(OLCAO) [22]. These include polycrystalline forms of α-quartz,
β-quartz, β-tridymite, α-crystalobalite, β-crystobalite, keatite,
coesite, idealized forms of β-crystobalite [23], random network
models of a-SiO_2 [24] and SiO_x [25], polycrystalline forms of
alkali silicates such as sodium and lithium monosilicates ($Na_2Si_2O_3$,
Li_2SiO_3) and disilicates (α-$Na_2Si_2O_5$, β-$Na_2Si_2O_5$, $Li_2Si_2O_5$) [26,27].
Because in this method, unlike the empirical tight-binding method,
no approximations are made in the calculation of interaction matrix,
the effects of hybridization of various orbitals are explicitly
taken into account and the calculated band structures and density
of states (DOS) are sensitive to small variations in the bond
lengths and bond angles among various crystalline and amorphous
structures. By studying these different forms of SiO_2 using the
same method and correlating the calculated DOS, a great deal of insight can be obtained about the structures and properties of SiO_2
and related materials. In the study of noncrystalline SiO_2 and
SiO_x, we assume the CRN model for its structure and built realistic
models with periodic boundaries so that similar method of electronic
structure calculation can be applied. Because a-SiO_2 and c-SiO_2
have similar short range order (SRO), knowledge of electronic
structures on various crystalline polymorphs definitely enhances
our understanding about the electronic structures of noncrystalline
SiO_2. The study on aklaki silicates provides additional information on the role of nonbridging oxygens (NBO) and the nature of
electron states of alkali ions in silicate glasses.

     In the next section, I outline the method and approach used
in these studies. The results of calculations are presented and
discussed in section III with particular emphasis on the electron
DOS. In the last section, I discuss these results in the general
context of structure and bonding in silicon dioxide and related
systems.

METHOD

     In order to obtain electronic structures of sufficient accuracy
for amorphous and polycrystalline forms of SiO_2 with very low
symmetry, the method of calculation must meet several criterion:
(1) It must be very efficient so it can be applied to systems
with complicated structures. (2) It should be sufficiently accurate
so as to reflect the rather subtle structural differences among
various phases and models. (3) It must be completely general so
that it can be consistently applied to all systems of interest.
The OLCAO method used in a semi-ab initio approach has proved to
meet these criterion. We start with the construction of the total
charge density of the system ρ(r) by the superposition of atomic
charge densities using free-atom wave functions obtained from
atomic Hartree-Fock self-consistent field calculations [28]. In

accordance with the local density functional theory [29-31], the exchange-correlation part of the one-electron potential $V_{EC}(r)$ can be constructed:

$$V_{EC}(\vec{r}) = \frac{3}{2}\alpha[3\rho(\vec{r})/\pi]^{1/3}$$

$$\rho(\vec{r}) = \Sigma_\ell \rho^A(\vec{r} - \vec{r}_\ell) \tag{1}$$

where α is the exchange parameter and $\rho^A(\vec{r})$ is the charge density of atom A at site $\vec{r}_\ell$. α ranges from 2/3 (the Kohn-Sham value) [30] to 1 (the Slater value) [29]. In the spirit of semi-ab initio approach, we use $\alpha = 0.81$ in the calculation for all polycrystalline forms of SiO_2; since for this value the calculated band gap for α-quartz in 8.8 eV, close to experimental measurement [32]. We used $\alpha = 2/3$ in the calculations for a-SiO_2, a-SiO_x and alkali silicates. $V_{EC}(\vec{r})$ can be decomposed into localized functions centered at each atomic site by a numerical fitting procedure. The Coulomb part of the potential, $V_C(\vec{r})$ is a simple superposition of Coulomb potentials of each atom:

$$V_C(\vec{r}) = \Sigma_\ell V_C^A(\vec{r} - \vec{r}_\ell)$$

$$V_C^A = -\frac{Z_A}{r} + \int \frac{\rho^A(\vec{r}')}{|\vec{r} - \vec{r}'|}\, d\vec{r}' \tag{2}$$

where Z_A is the atomic mass number of atom A. Finally, the total one-electron potential $V(\vec{r})$ can be written as a superposition of atomic-like potentials $V^A(\vec{r})$ centered at each site:

$$V(\vec{r}) = V_C(\vec{r}) + V_{EC}(\vec{r}) = \Sigma_\ell V^A(\vec{r} - \vec{r}_\ell) \tag{3}$$

The decomposed atomic-like potential $V^A(r)$ is similar to the true atomic potential except at large r, where the influence due to the presence of other nearby atoms in the solid becomes appreciable.

In order to facilitate the general evaluation of multicenter integrals occurring in the Hamiltonian matrix elements, $V^A(r)$ is numerically fitted into the following form:

$$V^A(r) = -\frac{Z_A}{r}\bar{e}^{\gamma\, r^2} + \sum_i C_i\, e^{-\beta_i r^2} \tag{4}$$

We next use the method of contraction [33] to obtain a set of
atomic-like orbitals $\{\mu_i(\vec{r})\}$ which are linear combinations of
Gaussian type of orbitals (GTO). A set of even-tempered Gaussian
exponentials $\{\alpha_i\}$ is chosen and is used to form s-type $(\exp(-\alpha_i r^2))$,
p-type $(x\exp(-\alpha_i r^2))$ and d-type $(xy\exp(-\alpha_i r^2))$, Gaussian orbitals.
These GTO are then used as basis functions in solving the single
site Schrodinger equation with a potential $V^A(r)$. The resulting
eigenvalues can be readily identified as the atomic-like states of
1s, 2s, 2p, 3p, 3d, etc. and the corresponding normalized eigenvector components would be the coefficients of each GTO in forming
the contracted atomic-like orbitals $\{u_i^A(r)\}$. These contracted
atomic-like orbitals are qualitatively similar to the free atom
orbitals but are generally shorter ranged and include, to some
degree, the distortion caused by the presence of other atoms in the
solid. Generally, the same set of $\{\alpha_i\}$ is used for contraction
with all types of atoms in the solid as well as for all orbitals
of different angular momenta. This is a very important feature in
the method because it greatly reduces the number of integrals need
to be calculated thus brings a seemingly impossible task of doing
first-principles type of calculation with amorphous solids or
complex crystals to within a manageable level. In the present
calculation, a minimal basis set is employed for both Si and O
atoms and the Si 3d orbital is not included. Earlier work on
Si_3N_4 [34] indicates Si 3d orbitals are not important. Similar
conclusion is also reached in the band structure calculation of
α-quartz using pseudopotential method [19].

The next step is to construct the Bloch sum $b_{i\alpha}(\vec{k},\vec{r})$ from
$\{u_i^A(r)\}$:

$$b_{i\alpha}(\vec{k},\vec{r}) = \sum_\nu e^{i\,\vec{k}\cdot\vec{R}_\nu} \mu_i^A(\vec{r} - \vec{\rho}_\alpha - \vec{R}_\nu) \tag{5}$$

where $\mu_i^A(\vec{r} - \vec{\rho}_\alpha - \vec{R}_\nu)$ is the ith localized atomic-like function of
the αth atom in the R_ν^{th} lattice cell. The Hamiltonian and the
overlap matrix elements are then formed between the Bloch sums:

$$H_{i\alpha,j\beta}(\vec{k}) = \langle b_{i\alpha}(\vec{k},\vec{r})|H|b_{j\beta}(\vec{k},\vec{r})\rangle$$

$$\tag{6}$$

$$= \sum_{R_\rho} e^{-i\vec{k}\cdot\vec{R}_\rho} \langle \mu_i^A(\vec{r}-\vec{\rho}_\alpha)|-\nabla^2+\sum_\sigma V^A(\vec{r}-\vec{R}_\sigma)|u_j^A(\vec{r}-\vec{\rho}_\beta-\vec{R}_\rho)\rangle$$

$$S_{i\alpha,j\beta}(\vec{k}) = <b_{i\alpha}(\vec{k},\vec{r}) | b_{j\beta}(\vec{k},\vec{r})>$$

(7)

$$= \sum_{R_\rho} e^{-i\vec{k},\vec{R}_\rho} <u_i^A (\vec{r}-\vec{\rho}_\alpha) | u_j^A (\vec{r}-\vec{\rho}_\beta - \vec{R}_\rho)>$$

Since both $u_i^A (\vec{r})$ and the atomiclike potentials $v^A(\vec{r})$ are expressed in terms of GTO, the two-center and the three-center integrals occurring in (6) and (7) can be easily evaluated using the technique of Gaussian transformation [34,35]. The lattice sums in (6) and (7) have to be carried to full convergence. Typically, 6th or 7th nearest neighbor interactions are included in a system such as SiO_2.

The index i in the basis Bloch sum $b_i\alpha(\vec{k})$ covers both the core and the non-core states. In order to eliminate the core states by the orthogonalization procedure, we write the orthogonalized Bloch sum as:

$$b_{i\alpha}^{'n}(\vec{k},\vec{r}) = b_{i\alpha}^n (\vec{k},\vec{r}) + \sum_{\ell,\gamma} a_{i\alpha,\ell\gamma}^k b_{\ell\gamma}^c(\vec{k},\vec{r})$$

(8)

where c and n denote the core and non-core portion of the basis and ' indicates the basis is in the orthogonalized space. It is straight-forward to show that the expansion coefficients $a_{i\alpha,\ell\gamma}^k$ are related to the overlap matrix by:

$$a_{i\alpha,\ell\gamma}^k = -<b_{i\alpha}^n (\vec{k},\vec{r}) | b_{\ell\gamma}^c(\vec{k},\vec{r})>$$

(9)

Thus the matrix elements $H'_{i\alpha,j\beta}$, $S'_{i\alpha,j\beta}$ in the orthogonalized representation can be obtained from the non-core portions of $H_{i\alpha,j\beta}$ and $S_{i\alpha,j\beta}$ plus appropriate correction terms involving matrix elements between core and non-core orbitals. Solution of the standard eigenvalue problem:

$$\left| H'_{i\alpha,j\beta}(\vec{k}) - S'_{i\alpha,j\beta}(\vec{k}) E(\vec{k}) \right| = 0$$

(10)

in which the core states are removed gives the energy eigenvalues $E_m(\vec{k})$ and the corresponding eigenvectors $C_{i\alpha}^m$. The one-electron state wave function is then given by:

$$\psi_m(\vec{k},\vec{r}) = \sum_{i,\alpha} C_{i\alpha}^m(\vec{k}) b_{i\alpha}'(\vec{k},\vec{r})$$

(11)

In the case of a-SiO_2 and SiO_x, $\vec{R}_\nu$ in (5) represents the supercell lattice of the quasi-periodic model structure which is

usually quite large, lattice sum convergence is therefore very rapid
and the corresponding Brillouin zine (BZ) is very small.   In this
case, we only need to solve equation (10) for the case of $\vec{k}=0$ where
the matrix equation is real.   Thus effectively speaking, the calculations using large periodic models are $\vec{k}$-independent as it should
be with amorphous solids where $\vec{k}$ is no longer a good quantum number.
We can then drop the label $\vec{k}$ in describing the electron states for
amorphous solids.

Let $\psi_n(\vec{r})$ to be a normalized electron wavefunction of an
amorphous solid from (10).   We can divide this one electron among
different orbitals of different atoms using Mulliken's population
analysis scheme: [36]

$$1 = \int |\psi_m(r)|^2 \vec{dr} = \sum_{\alpha,i} \rho_{i\alpha}^m \tag{12}$$

$$\rho_{i\alpha}^m = \sum_{j,\beta} C_{i\alpha}^{m*} C_{j\beta}^m S_{i\alpha,j\beta}' \tag{13}$$

where $C_{i\alpha}^m$ is the eigenvector of the state m. $\rho_{i\alpha}^m$ is the fractional
charge for state m associated with each orbital i in a given atom.
$\rho_{i\alpha}^m$ and can be used to define an approximate effective charge Q_α^*
on each atom:

$$Q_\alpha^* = \sum_{\substack{m \\ occup}} \sum_i \rho_{i\alpha}^m \tag{14}$$

It can also be used to project out the partial components of the
total DOS curves according to different i or different α.   The
partial DOS (PDOS) is very useful in understanding the bonding
character of the solid.

If the localization or delocalization of a one-electron state
in a noncrystalline solid is measured by the spread of its probability
density over different sites, we can define a localization index
(LI) for each state m as:

$$L_m = \sum_{\alpha,i} (\rho_{i\alpha}^m)^2 \tag{15}$$

It is apparent that L_m lies between the limits $\frac{1}{N}$ where N is the
dimension the matrix equation (10) for a completely delocalized
state with equal fractional charges for each orbital, to 1 for a
completely localized state with charges confined to a single
orbital in a single atom.   With a sufficiently large number of
atoms in the model structure, the procedure outlined above enables

us to make a realistic estimate of the localization of one-electron
states across the entire energy spectrum in a real disordered solid
based on microscopic calculation of electronic structures.

SOME CALCULATED RESULTS

     In this section, we discuss some of the calculated results on
the electronic structures of c-SiO$_2$ and a-SiO$_2$ and related systems
using the method described in the last section. Particular
emphasis is laid on the intercomparison of DOS curves and its implication on the structure and bonding of the materials under discussion. More detailed information can be found in the published
literature [23-27].

(A) α-quartz

          α-quartz is the most common and stable phase of all known
polycrystalline forms of SiO$_2$. The hexagonal lattice contains
three SiO$_2$ molecules in the unit cell forming rigid SiO$_4$ tetrahedral
units [37]. These tetrahedral units are connected at corners to
form 6 member rings with a Si-O-Si bridging angle of 144° and Si-O
band length of 1.61 Å. The band structure of α-quartz has been
studied recently using a first-principles approach [19,20], most
notably, using the self-consistent pseudopotential method [19].
Prior to that, only idealized β-crystobalite structure for SiO$_2$
has been studied [20,21]. The band structure and the DOS for
α-quartz calculated by using the OLCAO method are shown in Fig. 1
and 2 respectively. An indirect band gap of 8.8 eV from K→Γ
is found, which is close to experimental value [32]. There are
several k points with energies at the top of valence band (VB)
higher than that of Γ. Calbreese and Fowler had shown [20] that
the symmetry of the wave function in α-quartz is such that they
are direct forbidden in optical transitions. In the present
calculation, an exchange parameter of α=0.81 was used. Recently,
new measurements indicate that the optical gap in α-quartz may
be as large as 11.50 eV [7]. If this new value is to be trusted,
exchange parameter close to 1 may have to be used in order to have
the calculated gap value match the experiment. From the PDOS in
Fig. 2 (b)-(e), the VB of α-quartz can be well separated into O_{2s}
and O_{2p} bands indicating no hybridization between O_{2s} and O_{2p}
orbitals. The O_{2s} band shows a sharp peak at lower binding
energy and a smaller peak at a slightly higher binding energy.
The O_{2p} bands can be further divided into an upper nonbonding band
and a lower bonding band separated by a gap. Both Si 3s and Si
3p orbitals contribute to the VB DOS with comparatively smaller
magnitudes. This is consistent with the occurrence of sp^3
hybridization of Si orbitals and a significant charge transfer
from Si to O atoms. Effective charge calculation show 0.947
electron for Si atom and 7.526 electron for O atom which translates

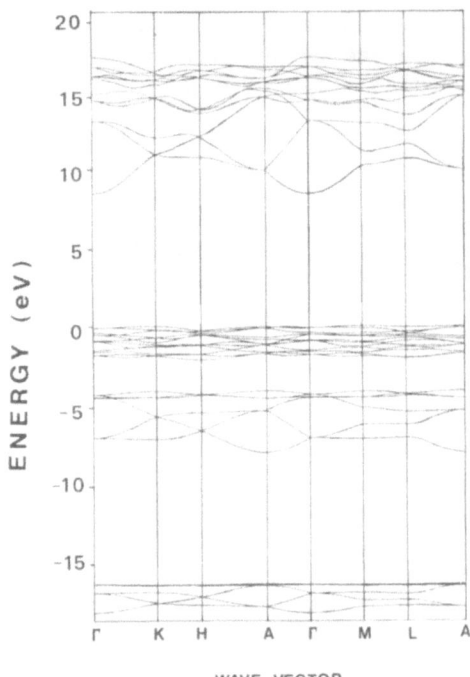

Fig. 1 Band structure of α-quartz along the symmetry directions
        of the Brillouin zone of hexagonal lattice.

into a charge transfer of about 0.76 electron per Si-O bond. Thus
the OLCAO calculation gives more ionic result than previously
estimated [12-13]. Both Si_{3p} and Si_{3s} orbitals and to a lesser
extend, O_{2p} orbitals contributed to the lower conduction bands (CB)
in α-quartz. The electron effective mass at Γ is estimated to be
about 0.74 of free electron mass.

(B) Other 4:2 Coordinated Polymorphs of SiO_2

        The band structures and DOS of other polycrystalline forms
of SiO_2 with well determined crystal parameters are calculated using
the same atomiclike potentials and basis functions as used in α-quartz.
These include the hexagonal β-quartz, hexagonal β-tridymite, tetragonal α-crystobalite, cubic β-crystobalite, tetragonal keatite, monoclinic coesite and two idealized forms of β-crystobalite (Ideal A
and Ideal B) with f.c.c. structure. Ideal A structure was studied
by Ciraci and Batra [21] which has the same mass density as the
real cubic β-crystobalite. Ideal B structure was studied by Schneider
and Fowler [18] in which the Si-O bond is scaled to 1.61 Å, same as
in α-quartz. Thus Ideal B has very low mass density and Ideal A
has very short Si-O bond length because the Si-O-Si angles are
restricted to 180° in both forms. The DOS calculated for all these
crystals are shown in Fig. 3 together with that of α-quartz. The

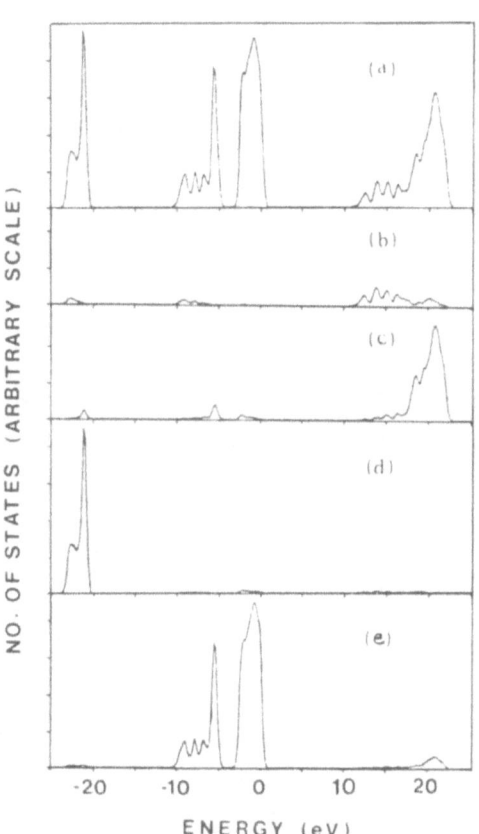

ENERGY (eV)

Fig. 2 DOS, PDOS of α-quartz
(a) Total; (b) Si_{3s};
(c) Si_{3p}; (d) O_{2s};
(e) O_{2p}

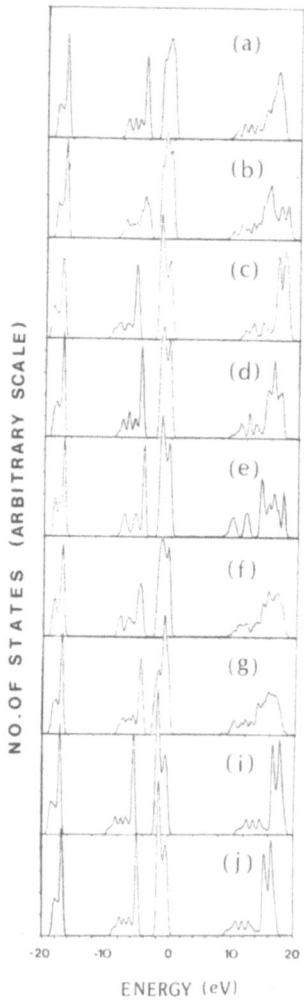

Fig. 3 DOS of 4:2 coordinated polymorphs: (a) α-quartz; (b) β-quartz; (c) β-tridymite; (d) α-crystobalite; (e) β-crystobalite; (f) keatite; (g) coesite; (i) ideal A; (j) ideal B.

high pressure phase of tetragonal stishovite will be discussed in
the next section. At the first glance, all the DOS for the 4:2
coordinated polymorphs appear to be similar, detailed examination
shows considerable differences in substructures for both the lower
O_{2p} bands (σ bands) and the upper nonbonding O_{2p} bands (lone pairs).
These differences are due to different bond lengths and bond angles
in these crystals [23]. Let us compare the DOS of O_{2p} band of
α-quartz with that of β-crystobalite as an example. For α-quartz,
the upper band has a major peak at -0.90 eV and the lower band
consists of a sharp peak at -4.24 eV and three smaller subpeaks at
-5.20, -6.02 and -6.98 eV, respectively. For β-crystobalite, the
upper band consists of a major peak at -1.44 eV and a minor peak
at -0.32 eV; while the lower band has only three structures: a
sharp one at -4.25 eV and smaller but well-resolved ones at -5.51
and -7.34 eV, respectively. These differences in DOS must result
from the differences in the short range order in these two crystal
forms. As discussed in Ref. 23, there are four different Si-O bond
lengths in β-crystobalite with an average of 1.635 Å and two different Si-O-Si bridging angles (180° and 137.2°) with an average of
147.9°. Both crystals have 6-member ring structures. The DOS of
both of the idealized forms of β-crystobalite are quite different
from the true β-crystobalite structure. Most noticeable is that
the Ideal A and Ideal B structures both have four peaks in the
lower O_{2p} bonding band in contrast to only three peaks in the
real β-crystobalite structure.

     In Fig. 4(a) and (b), we plot the minimum gap Eg of all the
polycrystals studied against the average Si-O bond lengths and the
minimum Si-O-Si angles. The horizontal error bars in Fig. 4 indicate
the maximum and the minimum Si-O bond and the maximum and the minimum Si-O-Si angle, respectively. There is an excellent linear
correlation between Eg and the average bond length. The same is
true for Eg vs minimum Si-O-Si angle if the data points for Ideal A
and Ideal B are deleted. This indicates that bond length has a more
dominating effect on the electronic structure than the Si-O-Si bond
angles. Most of the above discussed polymorphs have 6 member ring
structures. Keatite has a 5 member ring structure and coesite has
a 4 member ring structure. However, both the DOS spectra of Fig. 3
and the band gap correlation analysis of Fig. 4 and further effective charge calculations [23] fail to reveal any correlation with
the ring structures. It appears that the ring structure per se
is irrelevant to the electronic structure in SiO_2. More detailed
discussion about the electronic structures of these polycrystals
can be found in Ref. 23.

(C) Stishovite

     Stishovite is a high density (4.287 gm/c/c) phase of
SiO_2. The tetragonal lattice has two SiO_2 molecules per unit cell
in an octahedral coordination [38]. The Si and O atoms are 6:2

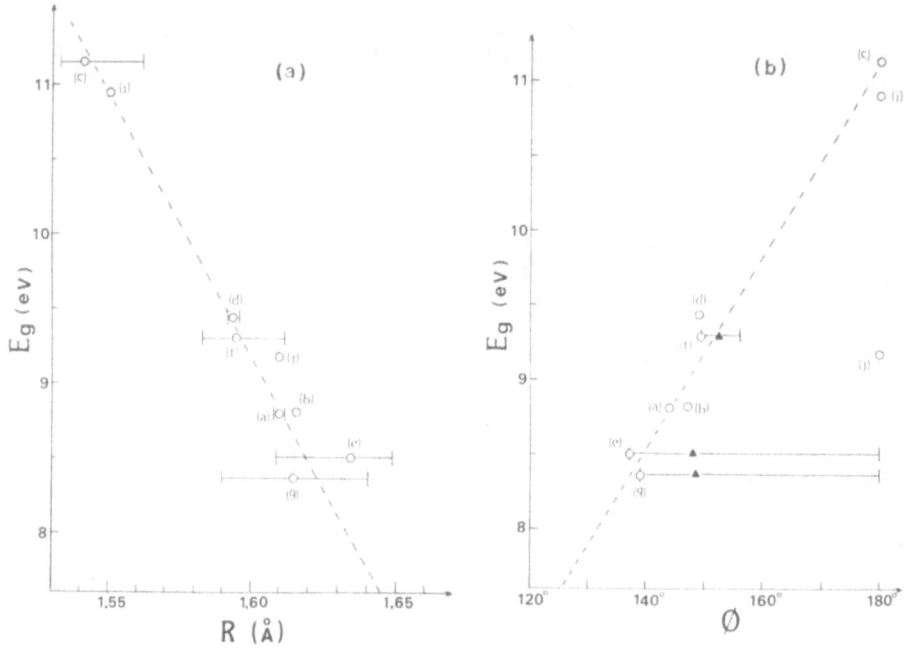

Fig. 4 Correlation plot of band gap Eg vs: (a) Si-O bond length;
(b) Si-O-Si bond angle. The notations are the same as
Figure 3.

coordinated with two types of Si-O bond lengths (1.757 Å and 1.810 Å).
The structure is unstable except at very high pressure close to 100
k bar. This is in big contrast to GeO_2 where octahedrally coordinated stable phase exist. The structure and bonding in stishovite
is significantly different from the 4:2 coordinated crystals discussed in (A) and (B). The first valence band structure of stishovite was calculated by Rudra and Fowler [39] using a simple tight-
binding method. The complete band structure of stishovite including
the CB calculated by the OLCAO method is shown in Fig. 5 and the
corresponding DOS in Fig. 6. The bands are suffciently complicated
to result in multiple structures in the DOS. A direct band gap of
7.51 eV at Γ is obtained which is smaller than the gap values of any
of the 4:2 coordinated polymorphs discussed above. The width of
O_{2p} band (8.95 eV) and O_{2s} band (3.44 eV) are much larger than those
in α-quartz and other polymorphs. These can be explained by the
increased interaction due to increased coordination number of each
atom in stishovite.

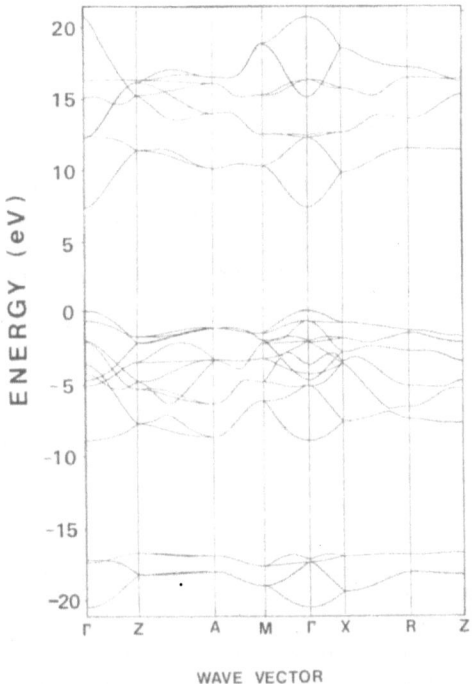

Fig. 5 Band structure of stishovite along the symmetry directions
of the BP of the tetragonal lattice.

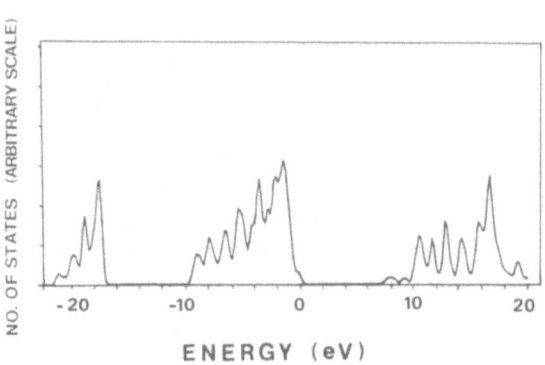

Fig. 6 DOS of stishovite.

Similar to earlier calculation of Ref. 39, we found no gap between the bonding and nonbonding O_{2p} bands in contrast to the 4:2 coordinated polymorphs where well defined gap within the O_{2p} bands always exists. The calculated effective charges per atom are also quite different from α-quartz and other polymorphs. We found $Q*_{Si}$ = 1.383 electron and $Q*_O$ = 7.293 electron, while the largest $Q*_{Si}$ and the smallest $Q*_O$ of all the 4:2 coordinated polymorphs are 0.979 electron (coesite) and 7.423 electrons (β-crystobalite), respectively. This clearly demonstrates an increased covalent and less ionic bonding character in stishovite. The calculated electron effective mass at Γ is similar to that of α-quartz.

(D)   Random Network of a-SiO$_2$

Construction of appropriate mechanical models for a-SiO$_2$ based on the CRN theory began some twenty years ago [40]. Almost all of these models are cluster type with free surfaces. We had constructed several CRN quasi-periodic models (QPM) for a-SiO$_2$ [41] which are generically derived from the corresponding models for amorphous Si [42,43] so that the difficulty of matching the periodic boundries is circumvented. The radial distribution function of these models are in good agreement with x-ray and neutron scattering experiments [44-48]. The characteristics of these models were discussed in detail [41]. There are several advantages of using sufficiently large QPM to represent the structures of amorphous solid in order to study its electronic structures. First, it represents an infinite array of network of atoms free of surface effects. Second, the mass density of the model can always be constrained to the measured value. Third, the structures and characteristics of each model can be quantitatively analyzed and linked to the electronic structures. Fourth, and most importantly, methods developed for crystalline solid state calculations can be readily applied to QPM.

In this section, we review the result of such calculations using a QPM containing 54 nearly tetrahedrally coordinated SiO$_2$ molecules [41]. The initial topological configuration of the model was derived from a periodic model for a-Si due to Guttman [43-49] and the model was systematically computer-relaxed under a Keating-type of potential [50]. The final relaxed structure has a density of 2.2 gm/c.c. The average and the root mean square values of Si-O bond length are 1.62 Å and 0.011 Å, respectively. The corresponding values for O-Si-O tetrahedral angle and Si-O-Si bridging angle are 109.3° ± 4.8° and 147.2° ± 13.8°, respectively. Thus the distortion of SiO$_4$ tetrahedral unit is quite small which must be attributed to the flexibility of the Si-O-Si angles. Fig. 7 shows the DOS of a-SiO$_2$ obtained from such a calculation. The band gap is about 0.7 eV smaller than that calculated for α-quartz. The gross feature of the DOS is similar to α-quartz and other 4:2 coordinated polymorphs described before and is also in

good agreement with photoemission measurement on vitreous silica [51].
But many of the substructures present in the DOS of c-SiO$_2$ are now
smeared out, although the identification of O$_{2s}$ band, O$_{2p}$ bonding
band and O$_{2p}$ nonbonding band remains intact. A deep dip, instead
of a gap, is developed between the bonding and the nonbonding bands
of O$_{2p}$ orbitals. These features are due to a continuous distribution of bond lengths and bond angles that exist in a CRN structure.
The average charge transfer from Si to O atom in a-SiO$_2$ is found
to be similar to c-SiO$_2$.

     In Fig. 7(a), we show the LI for each state in a-SiO$_2$ calcu-
lated according to (15). This is one of the most important quantity
that distinguishes the electron states in an amorphous solid from
its counterparts in the crystalline phase. In the VB region, states
are localized at the band edges and delocalized at the center of
the band in accordance with the formal theory [52]. A mobility
edge of about 0.2 eV can be estimated for the top of VB. It appears
that as far as localization is concerned, the O$_{2p}$ bonding and non-
bonding bands manifest as separate bands. The localization is also
stronger for states in s bands than those in p bands. The most
conspicuous fact is the complete delocalization of states near the
CB edge which may be attributed to over 95% of s character in the
wave functions of these states. The complete delocalization of the
CB edge states explains the experimental fact that the drift
mobility of an injected electron is very high in a-SiO$_2$ [53,54].

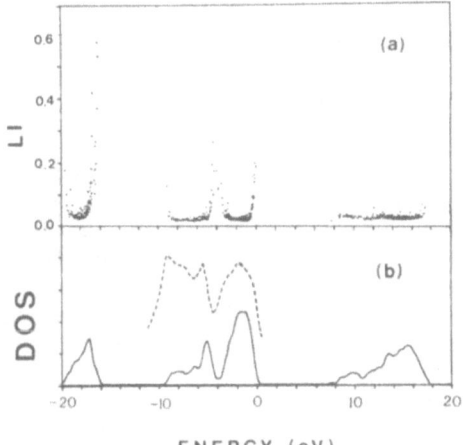

ENERGY (eV)

Fig. 7   (a) The localization index LI for each of the states a:SiO$_2$
         (b) DOS of a-SiO$_2$ using CRN model with periodic boundaries.
         The dashed lines is from experimental photoemission measure-
         ment of Ref. 51.

More detailed analysis of these results indicate that the electronic
structures of a-SiO$_2$ has a much more sensitive dependence on the
Si-O bond lengths than on the Si-O-Si bond angles [24]. This is
contrary to the conventional belief which tends to emphasize the
role of the bridging angles.

(E) Random Bond Models of SiO$_x$

        SiO$_x$ structure with x varies between 2 (a-SiO$_2$) and 0
(a-Si) is an important system because the atomic scale structure
at the Si-SiO$_2$ interface may be described, to a certain extent, as
a graded region of SiO$_x$. Up to a few years ago, there were two
competing models for the structure of SiO$_x$: (1) the microscopic
random-bond (RB) model [55] in which the Si-Si and Si-O bonds are
statistically randomly distributed throughout the SiO$_x$ structure
and (2) the random mixture (RM) model [56] in which the tetrahedrally bonded units of a-Si and a-SiO$_2$ are randomly dispersed, each has
a domain size of a few tetrahedral units. Based on model construction, model analysis and subsequent electronic structure calculations
on these models and comparing the results with experiments [57,58], it
was concluded the RB model is more appropriate to describe the
structures of SiO$_x$ than the RM model [25,41]. Early study actually
favored the RM model which was based on rather imprecise analysis
[56]. Accordingly, in this section, we discuss results of electronic
structures of SiO$_x$ based on RB models. The calculations were similar to a-SiO$_2$ except that the structural models for SiO$_x$ have much
bigger bond length and bond angle distortions than those of a-SiO$_2$
because of the random mixing of the bonds. In Fig. 8, the DOS
of SiO$_x$ with x=1.5, 1.0 and 0.5 are shown together with a-SiO$_2$ and
a-Si results. The number of Si atoms in each model of different x
are the same, i.e. [54]. The shaded portion corresponds to the
PDOS of Si atoms. From this figure, it can be seen that the evolution of the DOS of SiO$_x$ from x=2.0 to x=0.0 can be regarded as a
progressive filling in of Si states into the gap of SiO$_2$. Near
x=0.5, both the CB and the VB already resemble that of a-Si. From
x=2.0 to x=0.5, the O$_{2s}$ peak becomes broader and shifts to lower
binding energy, reflecting an increased degree of distortion in
the structural models. The O$_{2p}$ peaks which can be readily resolved
into bonding and nonbonding peaks gradually coalesce to form a
single O$_{2p}$ peak that widens as x is decreased. For x equals to
neither 2.0 nor 0.0, the DOS near Fermi level exhibit a deep dip
instead of a gap. The depth of this dip decreases as x decreases.
The absence of actual intrinsic gaps in SiO$_x$ is an artifact which
is related to the considerable amount of bond-angle and bond-length
distortions present in these models. However, the size of any
intrinsic gap, if present, should not be larger than that of a-Si.
In a real a-SiO$_x$ film or at the Si-SiO$_2$ interface, severely distorted bonds and various forms of defect structures definitely
present, so the possibility of a finite value of DOS at the Fermi
level should not be ruled out. The DOS calculated for SiO$_x$ is in

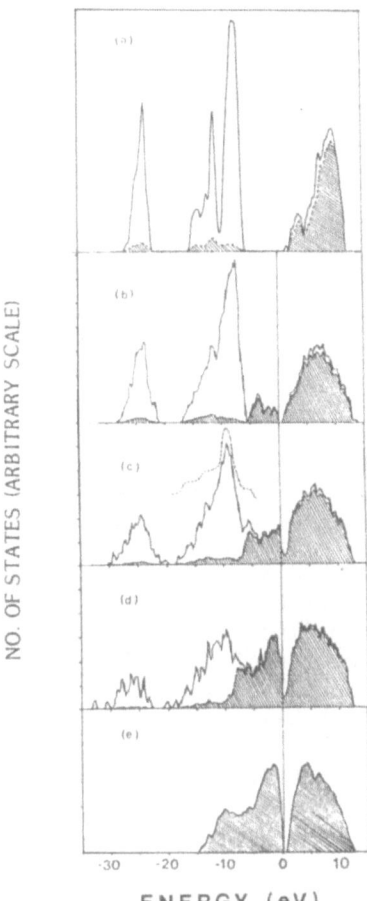

NO. OF STATES (ARBITRARY SCALE)

-30   -20   -10   0   10

ENERGY (eV)

Fig. 8   The DOS of RB models of
SiO$_x$:   (a) x=2.0, (a-SiO$_2$)
(b) x=1.5, (c) x=1.0, (d)
x=0.5, (e) x-0 (a-Si). The
shaded area corresponds to
the Si portion of the local
DOS.

reasonable agreement with photoemission experiment [59,61]. Calculation of LI shows that the states in SiO$_x$ are much more localized
than those of a-SiO$_2$ [25]. For states immediately below the Fermi
level, the degree of localization is proportional to x and is
related to the abundance of S-O bonds. So if we view the Si-SiO$_2$
interface as consisting of a graded region of SiO$_x$, we expect the
occupied VB states to be more localized on the SiO$_2$ side of the
interface and less localized on the a-Si side. The calculated joint
DOS near threshold for SiO$_x$ is also in good agreement with optical
absorption experiments [55,61], especially when the effects of the
localization is taken into account [25].

The electronic structures of SiO$_x$ discussed in this section
are entirely based on the concept of CRN model similar to that of
a-SiO$_2$ and assuming no roles played by defects or impurity ions.
Any disorder that enters into the calculation is purely stoichiometric in nature with the associated statistical distribution of

distorted bonds and angles.   In real SiO_x films or $Si-SiO_2$ interfaces, impurity ions and defects play a major role in determining
the electronic structures.   It will be of high interest to study
the electron states of impurities and defects at the $Si-SiO_2$
interface.

(F) Crystalline Alkali Silicates

          Alkali silicate glasses is a very important class of
material of many practical applications and the science of silicate
glasses has been studied since ancient times.   However, the study
of electronic structures of silicate glasses in terms of modern
band theory is almost nonexistent.   The major reason behind is
that the atomic scale structures of silicate glasses formed by dis-
solivng a controlled amount of alkali oxide into silica is extremely
complicated.   Fortunately, there occurs in nature a large class of
mineral crystalline forms of alkali silicates.   The structures of
many of these crystals are well determined.   Since we may expect
that the SRO of the glassy and crystalline alkali silicates to be
similar; it is logical to study the electronic structures of these
complicated alkali silicate crystals in order to gain some insight
about the structure and bonding in silicate glasses.   In this section,
we discuss the calculated electronic structures of some alkali
silicate crystals.   These include the orthorhombic sodium metasilicate and lithium metasilicate (Na_2SiO_3, Li_2SiO_3), orthorhombic
alpha-sodium disilicates ($\alpha-Na_2Si_2O_5$), monoclinic beta-sodium disilicate
($\beta-Na_2Si_2O_5$), and monoclinic lithium disilicate ($Li_2Si_2O_5$).   These
are crystals with complicated chain-like and sheet like structures
of tetrahedral units with bridging oxygens (BO) and nonbridging
oxygens (NBO) [62-66].   The structural arrangements of the atoms
which range from 24 to 36 per unit cell, have been discussed in
detail [26,27].   The Si-BO bond lengths are generally longer than
Si-NBO bond lengths.   Contrary to simple-minded view based on
charge compensation model, the NBO is not tied up by a single
alkali ion, but by four alkali ions at comparable distances of
separation.   Furthermore, alkali ions close to BO can also be
identified [26,27].   Thus we expect the electronic structures of
these crystals to be fairly complicated.   In Figs. 9 and 10, we
show the DOS of sodium silicates and lithium silicates calculated
using the OLCAO method with the exchange parameter α set equal to
2/3.   The most discernible feature is the well resolved O_{2s} double
peak separated by about 2-3 eV in Na silicates which clearly
reflects the different chemical environments of the BO and the NBO.
The DOS of O_{2p} VB is also quite different from those of SiO_2
crystals because of the presence of the NBO oxygens and the alkali
ions.   The calculated DOS of sodium silicates are in good agreement with photoemission experiments [26,27] especially when the
theoretical PDOS curves are weighted by the intensity factors
calculated by Scofield [68].   However, the agreement is less
satisfactory for lithium silicates [27].   Furthermore, the splitting

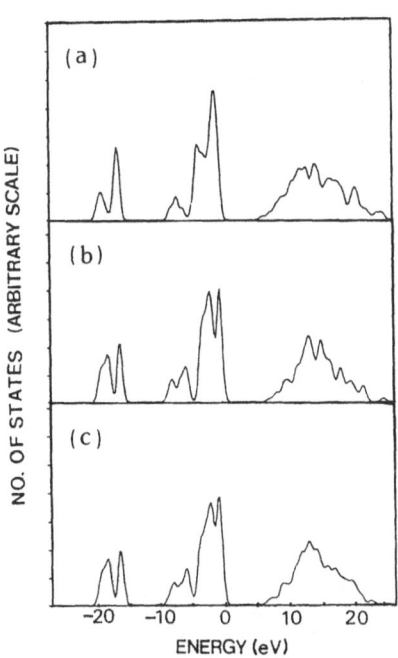 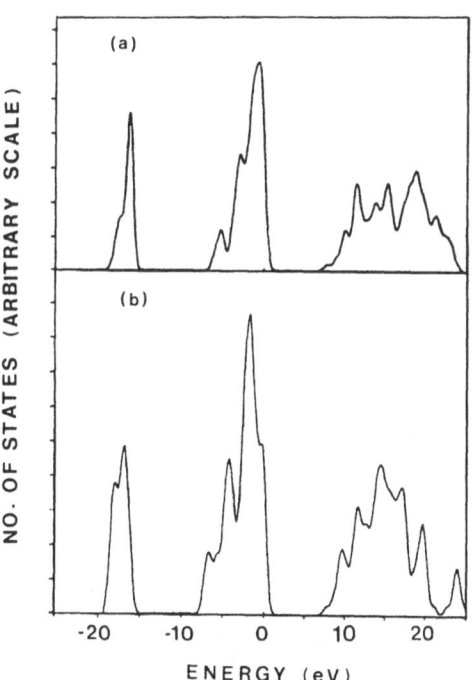

Fig. 9 DOS of sodium silicates: Fig. 10 DOS of lithium silicates:
(a) α-Na$_2$SiO$_3$; (b) (a) Li$_2$SiO$_3$; (b) Li$_2$Si$_2$O$_5$.
α-Na$_2$Si$_2$O$_5$; (c) β-Na$_2$Si$_2$O$_5$.

of O$_{2s}$ bands due to BO and NBO is less prominent in Li silicates.
Experimentally, the photoemission spectra of Na and Li silicates
are very similar. It is not clear at this moment whether this is
due to the impure samples in the experiment or some subtle different
bonding mechanism in the case of Li ions. The later situation cannot
be ruled out since PDOS calculation in both Na and Li silicates does
reveal some dissimilarities. Incidentally, current resolution of
photoemission experiment is not high enough to reveal any difference
in the spectra between glassy and crystalline samples of silicates
with comparable alkali compositions. This may also indicate that
the SRO in the glassy and crystalline alkali silicates are very
similar.

The band gaps calculated for these alkali silicate cyrstals are
generally smaller than that of c-SiO$_2$ or a-SiO$_2$. They are 6.5 eV,
7.2 eV, 6.9 eV for Na$_2$SiO$_3$, α-Na$_2$Si$_2$O$_5$ and βNa$_2$SiO$_5$, respectively;
and 7.3 eV and 7.5 eV for Li$_2$SiO$_3$ and Li$_2$Si$_2$O$_5$. Calculation of
effective charges shows that the alkali ions have completely lost
its single electron, while the charges on the NBO is only about
0.1 electron more than the BO. It appears that the charges supplied
by the alkali ions is more or less redistributed throughout the
lattice and each positive alkali ion cannot always associate with

a particular NBO site. More detailed discussion about the electronic
structures of these alkali silicate crystals can be found in the
Ref. 26 and 27.

DISCUSSION

We have studied the electronic structure and bonding of both
crystalline and amorphous SiO_2 as well as some other related systems
using the same method. The calculations, which are first-principles
in nature and capable of accurately accounting for the positional
dependence of the atoms in the structure, allow some meaningful
comparisons to be made. Thus in the 4:2 coordinated SiO_2 polycrystals, linear correlations are established between the band gaps
and the average Si-O bond lengths as well as Si-O-Si bond angles.
The drastic difference in the band structure and DOS of the 6:2
coordinated stishovite and the 4:2 coordinated polymorphs is clearly
demonstrated. The similarities and contrasts in the electronic
structures of $c-SiO_2$ and $a-SiO_2$ are also fully exploited. Of
particular significance is our ability to analyze each electron
state in $a-SiO_2$ in terms of a LI. Although in the present calculation, the CRN model is adopted for the structure of $a-SiO_2$ and
$a-SiO_x$, it must be pointed out that our numerical calculation can
be equally well applied to other alternate models. In fact, the
calculated DOS in conjunction with high resolution photoemission
experiments, may be used to determine the appropriateness of each
suggested model. It is the opinion of the author that in $a-SiO_2$ and
related systems, the electronic structures and bonding is mostly
affected by the Si-O bond lengths, and to a lesser extent, by the
Si-O-Si bond angles. These two quantities, however, are highly
correlated. For example, the shorter bond lengths may be associated
with larger Si-O-Si angles. There is no evidence, whatsoever, that
the ring structures play any crucial role in determining the electronic
structures, other than those already reflected in the bond lengths
and bond angles. This viewpoint is in sharp contrast to those
earlier theoretical investigations [15-17] which accentuate the
importance of ring statistics.   Part of the reason may be due to
the fact that in these calculations, there were no accurate ways
of taking positional dependence of the atoms into the calculations,
thus the ring statistic becomes the only parameter to distinguish
the structures of crystalline and amorphous forms. On the other
hand, there is some experimental evidence that the ring structure
may play veritable role in the vibrational spectra of $a-SiO_2$ [69].

The study of structures such as SiO_x is of particular importance since this may be related to $Si-SiO_2$ interface structures
relevant to the design and fabrication of MOS devices. Current
results favor RB model over RM model for SiO_x. Calculations can
be further extended to some other more imaginative model structures
for the interfaces or with specific built-in defect structures which
may provide very useful information about the electronic properties

of the interfacial states and their potential influence on the device performance. The study of the alkali silicates is only at its very beginning. The present result demonstrated the unique role played by the NBO in alkali silicates and also revealed some unexpected findings. There exists a rich variety of many mineral crystals with well-determined crystal parameters. These structures all involve SiO_4 tetrahedral units together with other metal and non-metal ions in different bonding configurations; therefore can provide a good test ground to disclose more penetrating details about the electronic structures of silicate glasses.

The present calculations, which are essentially based on the band theory of solids give results quite different from the quantum chemical studies of molecular fragments. For example, in the present calculations, the effect of Si 3d orbitals is found to be unimportant; the effective charge calculation shows more ionic character for SiO_2 and the conclusions regarding the relative importance of Si-O bonds and Si-O-Si angles on the electronic states are also different. These clearly indicate that some serious conceptual differences exist in the ways of studying the structure and bonding in noncrystalline solids. While it is still polemical as to which approach is more reliable, the method and approach used in the present study clearly have several advantages: (1) the solid is represented by infinite array of atoms; (2) meaningful CB states and well defined band gaps can be obtained; (3) the one-electron wave functions obtained can be used to illustrate important concepts such as localization and delocalization of the electron states in a disordered solid. We expect to apply the current method to study the electronic structures of other Si-based glasses and complex crystals, and to test the validity of structural models put forward for a variety of noncrystalline solids.

ACKNOWLEDGMENTS

This work was supported by the Office of Basic Energy Research of the U.S. Department of Energy (contracts DE-AC02-79ER10462 and DE-FG02-84ER4570).

REFERENCES

1. D. L. Griscom, J. of Non-Cryst. Solids 24:155 (1977).
2. S. T. Pantelides, Ed. The Physics of SiO_2 and Its Interfaces, Pergamon, New York (1978).
3. G. Lucovsky, S. T. Pantelides and F. L. Galeener, Eds., The Physics of MOS Insulators, Pergamon, New York (1980).
4. R. J. Bell and P. Dean, Philos. Mag. 25:1381 (1972).
5. Y. Bando and K. Ishizuka, J. Non-Cryst. Solids 33:375 (1979).
6. J. C. Phillips, Phys. Status Solidi B 101:473 (1980); Phys. Rev. B 24:1744 (1981).
7. R. Evrard, A. N. Trukhin, Phys. Rev. B 25:4102 (1982).

8. M. Rossinelli and M. A. Bosch, Phys. Rev. B 25:6482 (1982).
9. G.A.D. Collins, D.W.J. Cruickshank and A. Breeze, J. Chem. Soc. Faraday Trans. II 68:1189 (1972).
10. J. A. Tossell, D. J. Vaughan and K. H. Johnson, Chem. Phys. Letters, 20:329 (1973).
11. J. A. Tossell, J. Phys. Chem. Solids 34:307 (1973).
12. T. L. Gilbert, W. J. Stevens, H. Schrenk, M. Yoshimine and P. S. Bagus, Phys. Rev. B 8:5977 (1973).
13. K. L. Yip and W. B. Fowler, Phys. Rev. B 10:1391;1400 (1974).
14. A. J. Bennett and L. M. Roth, J. Phys. Chem. Solids 32:1251 (1971).
15. S. T. Pantelides and W. A. Harrison, Phys. Rev. B 13:2667 (1976).
16. R. N. Nucho and A. Madhukar, Phys. Rev. B 21:1576 (1980).
17. R. B. Laughlin, J. D. Joannopoulos, and D. J. Chadi, Phys. Rev. B 20:5220 (1979).
18. P. M. Schnieder and W. B. Fowler, Phys. Rev. Lett. 36:427 (1976).
19. J. R. Chelikowsky and M. Schluter, Phys. Rev. B 15:4020 (1977).
20. E. Calabrese and W. B. Fowler, Phys. Rev. B 15, 7122 (1978).
21. S. Ciraci and I. P. Batra, Phys. Rev. B 15:4923 (1977).
22. W. Y. Ching and C. C. Lin, Phys. Rev. B 12:5536 (1975); 16:2989 (1977).
23. Y. P. Li and W. Y. Ching, Phys. Rev. B 31:2172 (1985).
24. W. Y. Ching, Phys. Rev. Lett 46:607 (1981; Phys. Rev. B 26:6622 (1982).
25. W. Y. Ching, Phys. Rev. B 26:6633 (1982).
26. W. Y. Ching, A. D. Murray, D. J. Lam and B. W. Veal, Phys. Rev. B 28:4724 (1983).
27. W. Y. Ching, Y. P. Li, B. W. Veal and D. J. Lam, Phys. Rev. B 32; 1203 (1985).
28. E. Clementi and C. Roetti, Atomic Data and Nuclear Data Tables, [14]:3-4 (1974).
29. J. C. Slater, Phys. Rev. 81:385 (1951).
30. W. Kohn and L. J. Sham, Phys. Rev. 140:A1133 (1965).
31. R. Gaspar, Acta Phys. Hung. 3:263 (1954).
32. T. H. DiStefano and D. E. Eastman, Solid Commun. 9:2259 (1971).
33. J. E. Simmons, C. C. Lin, D. F. Fonquet, E. E. Lafon and R. C. Chaney, J. Phys. C 8:1549 (1975).
34. S. Y. Ren and W. Y. Ching, Phys. Rev. B 23:5454 (1981).
35. R. C. Chaney, T. K. Tung, C. C. Lin and E. E. Lafon, J. Chem. Phys. 52:361 (1970); W. Y. Ching, C. C. Lin and D. L. Huber, Phys. Rev. B 14:620 (1976).
36. R. S. Mulliken, J. Chem. Phys. 23:1833 (1955).
37. R.W.G. Wyckoff, Crystal Structures, Interscience, New York (1965).
38. W. H. Baur and A. A. Khan, Acta Cryst B 27:2133 (1971).
39. J. K. Rudra and W. B. Fowler, Phys. Rev. B 28:1061 (1983).
40. F. Ordway, Science 143:800 (1964); R. L. Bell and P. Dean, ibid. 212:1354 (1966).
41. W. Y. Ching, Phys. Rev. B 26:6610 (1982).

42. D. Henderson, J. Non-Crystal. Solids 16:317 (1974); D. Henderson and F. Herman, ibid 8-10:369 (1972).
43. W. Y. Ching, C. C. Lin, and L. Guttman, Phys. Rev. B 16:5488 (1977).
44. E. H. Henninger, R. C. Buschart, and L. Heaton, J. Phys. Chem. Solids 28:423 (1967).
45. R. L. Mozzi and B. E. Warren, J. Appl. Crystallogr. 2:164 (1969).
46. J. H. Konnert and J. Karle, Acta Crystallogr. Sec A 29:702 (1973).
47. J.R.G. DaSilva, D. G. Pinnati, C. E. Anderson and M. L. Radee, Philos. Mag. 31:713 (1975).
48. C. F. George and P. D'Antonio, J. Non-Crystal. Solids 34:323 (1979).
49. L. Guttman, Bull. Am. Phys. Soc. 22:64 (1977).
50. P. N. Keating, Phys. Rev. 145:637 (1966).
51. T. H. DiStefano and D. E. Eastman, Solid State Commun. 9:2259 (1970).
52. P. W. Anderson, Phys. Rev. 109:1492 (1952).
53. R. C. Hughs, Phys. Rev. Lett. 30:1333 (1973); Appl. Phys. Lett. 26:436 (1975).
54. F. B. McLean, H. E. Boesch, and J. M. McGamity, IEEE Trans. Nucl. Sci. 6:1506 (1976).
55. H. R. Philipp, J. Noncryst. Solids, 8-10:627 (1972); J. Phys. Chem. Solids 32:1935 (1971).
56. R. J. Temkin, J. Non-Crystal. Solids, 17:215 (1975).
57. J. A. Yasaitis and R. Kaplow, J. Appl. Phys. 43:995 (1972).
58. C. F. George and P. D'Antonio, J. Non-Crystal. Solids 34:323 (1979).
59. T. H. DeStefano, J. Vac. Sci. Technol. 13:856 (1976).
60. G. Hollinger, Y. Jugnet, and Tran Mihn Duc, Solid State Commun. 22:277 (1977); G. Hollinger and Tran Minh Duc, Ref. 2, p. 87,
61. E. Holzenkampfer, F. W. Richter, J. Stuke, and V. Voget-Grote, J. Non-Crystal. Solids 32:327 (1979).
62. A. Grund and M. Pizy, Acta Crystallogr. 5:837 (1952); W. S. McDonald and D.W.J. Cruickshank, ibid 22:37 (1967).
63. A. K. Pant and D.W.J. Cruickshank, Acta Crystallogr. Sec. B 24:13 (1968).
64. A. Grund, Bull. Soc. Frane, Miner. Cryst. 77:775 (1954); A. K. Pant. Acta Cryst. Sec. B 24:1077 (1968).
65. K. -F. Hesse, Acta Cryst. B 33:901 (1977).
66. V. F. Leibau, Acta Cryst. 14:389 (1961).
67. D. J. Lam, A. P. Paulikas, and B. M. Veal, J. Non-Cryst. Solids 42:41 (1980).
68. J. H. Scofield, J. Electron Spectrosc. Relat. Phenon. 8:129 (1976).
69. R. A. Bavrio, F. L. Galeener and E. Martinez, Phys. Rev. Lett. 52:1786 (1984).

NON-EXPONENTIAL RELAXATIONS IN DISORDERED SYSTEMS

A. K. Jonscher

Chelsea Dielectrics Group
King's College, University of London
London SW6 5PR, U.K.

ABSTRACT

Experimental evidence is presented which shows conclusively that the time dependence of relaxation of dielectric polarization and of mechanical stress both follow well defined power laws with fractional exponents, rather than the widely expected exponential laws that would apply to the relaxation of independent non-interacting systems. In addition, a recent review of the time dependence of delayed luminescence also shows the same type of power law decay there, once again despite the widely anticipated exponential relations.

It is suggested that the common form of power law found in these very different types of relaxation and in very different materials, is the result of the dominance of many-body interactions in disordered solids. Our growing understanding of the physical processes involved should help to elucidate this remarkable commonality of responses of very different relaxations.

INTRODUCTION

The object of the present paper is to draw attention to a specific aspect of the hehavior of disordered systems which appears to have excaped attention in the past and which may throw significant light on the properties of condensed matter. We refer to the transient response of disordered systems -- at least the non-metallic ones -- after external perturbation of their thermal equilibrium, with particular reference to dielectric, mechanical, photoconductive and luminescent responses. All these transient phenomena are normally expected to follow exponential time dependences resulting from

the simplest first-order perturbation behavior in which the rate
of return of the system to equilibrium is proportional to the
remaining perturbation. These relations would be expected to be
obeyed by systems consisting of non-interacting components, e.g.,
dipoles or charges or luminescent centers, etc.

It is well known, however, that the experimentally observed
transient responses seldom, if ever, follow the classically
expected exponential laws, although this does not prevent the
theoretical interpretations from revolving round the classical
concepts with some evident modifications like "distributions of
relaxation times" (DRTs), which may be plausible at first sight
but which do not stand up to a more critical examination.

We link the replacement of exponential time dependence by
power-law relaxations with the disordered nature of all systems in
which these relaxations can be seen -- be they dielectric, mechanical, photoconductive or luminescent. The close relationship between
many-body processes and disorder arises from the ease of excitation
of disordered systems, as distinct from the energetically much
less favorable excitation of perfectly ordered media where the
breaking of perfect order requires a large input of energy. These
concepts have been developed in recent theoretical papers [1-4]
with particular reference to dielectric and mechanical relaxations.
The intimate connection between disorder and dielectric polarization
has been developed in reference [5], where it was pointed out that
the disorder arises either from inherent structural defects in the
material, or from a random distribution of dipoles or localized
electronic states in the material and interactions are possible
between such dipoles or charges and their immediate surroundings.

In addition to the dielectric and mechanical relaxations
[2,6,7], we will be concerned in the present discussion with the
relaxation of certain excited deep levels in semiconducting and
semi-insulating materials, and these levels are essentially a part
of the prevailing disorder.

The central theme of our argument will be the presence of
interactions between the "active sites" -- dipoles in dielectric
relaxation, mechanical stress and localized levels in semiconductors -- so that they are no longer independent entities
relaxing in complete isolation from one another. Instead, the
amply available experimental evidence will be taken to demonstrate
that there exist strong interactions which are capable of dominating the time- or frequency-dependence of the relaxation response.

EXPONENTIAL VERSUS NON-EXPONENTIAL RELAXATION

A deeply rooted conviction appears to have developed over
many years that all relaxation behavior is due to the perturbation

and subsequent return to equilibrium of assemblies of non-interacting isolated systems, with the inevitable result that the relaxation process is exponential in time. This follows from the nature of the differential equation governing the return to equilibrium of such non-interacting relaxors:

$$d\alpha/dt = -\gamma\alpha \tag{1}$$

where α is the strain in question and $1/\gamma$ is the relevant time constant. The implication of this equation is that the rate of change of strain in proportional to the remaining strain, leading to an exponential time dependence.

In the case of interacting systems, on the other hand, a detailed solution of the relaxation equation demands the knowledge of certain parameters dependent on the properties of the system under investigation, but it is very simple to show that this solution cannot obey the exponential law. To this end we note that if $\alpha_\ell(t)$ is the strain or displacement of the ℓ-th particle or component in an interactive system, then the differential equation for the return to equilibrium may be written in the form:

$$d\alpha_\ell(t)/dt = (i\omega_\ell - \gamma_\ell)\alpha_\ell(t) - \sum_{\ell'} \alpha_\ell(t)\, \alpha_{\ell'}(t)\, V_{\ell\ell'}, \tag{2}$$

where ω_ℓ is a natural frequency of oscillation at site ℓ, γ_ℓ is the corresponding damping coefficient and $V_{\ell\ell'}$ are the coupling potentials between displacements or excitations at site ℓ and ℓ' [1,2]. In the absence of inter-site coupling, i.e., in a noninteracting system, the summation term vanishes and the solution is exponential in time, as in Eq. (1), with the possibility of a damped oscillatory relaxation. With the summation term present, on the other hand, an exponential solution is physically impossible [1,2].

In the following we shall give a brief description of the principal types of time-dependent relaxations observed in a wide range of systems.

DIELECTRIC RELAXATION

The prevalence of power-law time- and frequency-dependence in dielectric relaxation has been very well documented and is known to apply over many decades of time and frequency taking the form of the "universal" relation [4] for the characteristic response function which is proportional to the time-dependent depolarization current after a sudden removal of a steady polarizing field:

$$f(t) \propto t^{-s} \text{ with } 0 < s < 2 \tag{3}$$

which has been known since the turn of the century as the Curie-von
Schweidler law [8]. The time dependence splits up into two ranges
with different values of s. At "short" times, $t \ll 1/\omega_p$, where ω_p
is a characteristic frequency, the exponent falls in the range
$0 < s < 1$, and the corresponding frequency dependence of the dielectric susceptibility is given by the Fourier transform of Eq. (3)

$$X(\omega) \propto [1-i \cot(s\pi/2)]\omega^{s-1} \quad (0 < s < 1, \omega \gg \omega_p) \tag{4}$$

implying that both the real and imaginary components of $X(\omega)$ are
the same functions of frequency except for a constant. At "long"
times, $t \gg 1/\omega_p$, the corresponding Fourier transformations become
the low-frequency susceptibility spectra:

$$X'(\omega) = X(0) - \text{const} \times X''(\omega)$$
$$\quad (1 < s < 2, \omega \ll \omega_p) \tag{5}$$
$$X''(\omega) \propto \omega^{s-1}$$

the time domain relations (3) are shown schematically in Fig. 1,
while the totality of the observed types of frequency responses is
shown in Fig. 2. The principal conclusions from these is that
dielectrics in which the polarization is dominated by dipolar
responses due either to permanent molecular dipoles whose orientations
are randomly distributed, or to "ionic dipoles" consisting of a
pair of ions held together by their Coulombic interactions, show
loss peaks with power law slopes and with peak frequency ω_p. The
slopes $m = s-1$ and $n - 1 = s-1$, at low and high frequencies,
respectively, are both in the range 0 to 1. The ideal Debye limit
is never seen in condensed matter.

A different type of low-frequency long-time behavior is seen
in dielectrics in which hopping electronic or ionic charge carriers
are capable of moving over much larger distances than the interatomic spacings characterizing typical dipolar orientations. The
dielectric response of these materials is characterized by the same
power-law relation as Eq. (3) with the exponent s close to zero,
giving a "quasi-dc" behavior, shown in Fig. 2, and the corresponding
frequency dependence is given by Eq. (4) with both the real and
the imaginary components of $X(\omega)$ going as approximately $1/\omega$. This
is known as strong Low Frequency Dispersion (LFD) and is seen in
very many dielectric systems at sufficiently high temperatures and
low frequencies [2,9-11].

The phenomenon of LFD is not generally acknowledged as a
specific type of dielectric - or mechanical, see below - behavior,
despite its many manifestations and is usually dismissed as an
ill-defined interfacial process known under the name of Maxwell-Wagner
effect. It is clearly present in many volume situations [9-11] where
it is associated with low-dimensional flow, i.e., flow of hopping
charges which is restricted to certain preferred paths arising from

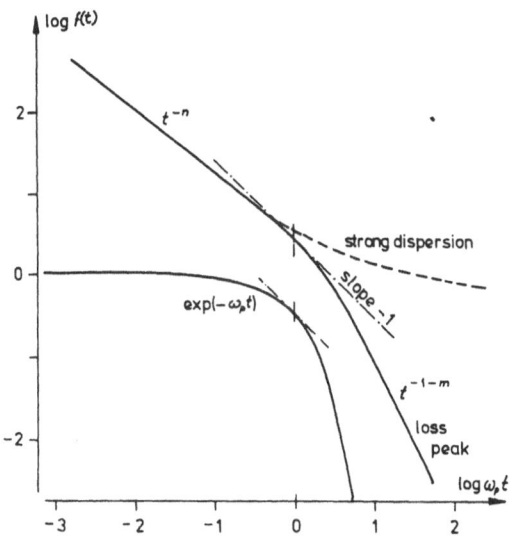

Fig. 1 A schematic representation of the two basic forms of the
characteristic dielectric response function in the time
domain, f(t), which is proportional to the dielectric depolarization current following a sudden removal of a steady
depolarizing field. The general form of this function is
given by Eq. (3), with the exponent s having different values
in the range of "short" and "long" times. At short times
s is set equal to n, at long times m = s-1. Both m and n
are in the range 0 to 1 in dipolar systems which show a
loss peak in the frequency domain. In charge carrier
dominated systems m is effectively negative and close to
unity in absolute value, s tends to zero. The Fourier
transformation of this response function into the frequency
domain gives the frequency dependence of the dielectric
susceptibility $X(\omega)$ which is shown schematically in Fig.
2 for all known types of dielectric responses. From [4].

the existence in the system of inhomogeneities. Equally, however,
there are many situations in which LFD is associated with interfacial processes, for example the generation and recombination of
charges at two-dimensional interfaces. This may apply to ionic
dissociation and recombination in solid or liquid electrolytes in
contact with metallic electrodes and its electronic manifestation
may be seen in Schottky barriers. LFD is also prominent in primary
and secondary batteries. In the sense of the present discussion
these interfacial processes leading to LFD may be traced back to
two-dimensional disorder which produces preferred sites for
association or dissociation of charge carriers of opposite signs.
It is highly significant, however, that these interfacial, i.e.,
essentially two-dimensional processes, should produce the same
form of power law as the other types of many-body interactions and

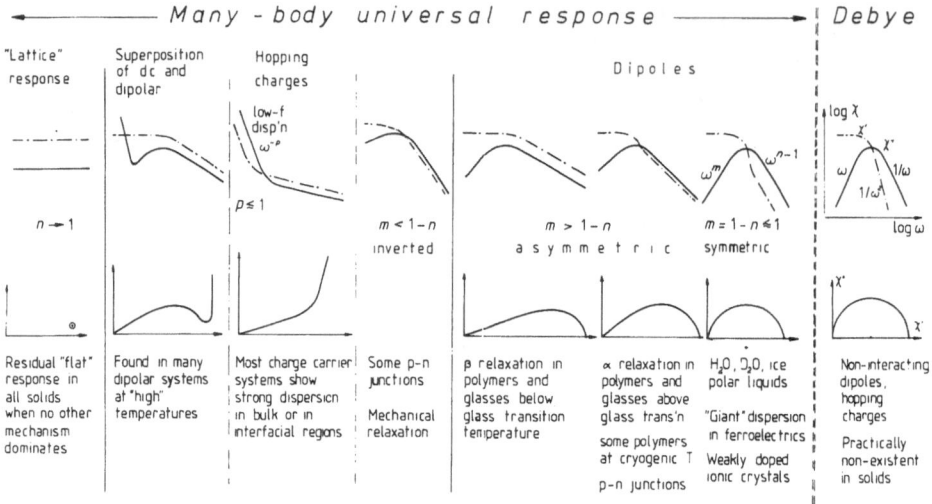

Fig. 2 A general classification of all types of dielectric response
in the entire range of solids and liquids. The upper part
of the figure represents the plots of log X'(ω) as a function
of log ω as the chain dotted lines and log X"(ω) as a function of log ω as solid lines. They range from the ideal
Debye response on the far right, which is very seldom, if
ever, seen in solids, through the α and β peaks in dipolar
materials, the "inverted" peaks observed in p-n junction
generation/trapping processes and in mechanical loss processes, and on to the universal dependence for charge carrier systems. The limiting forms of behavior are represented
by the strong low frequency dispersion for which n→0 and
by the "flat" frequency response of low-loss solids for which
n→1.

The lower diagram presents the corresponding complex susceptibility (Cole-Cole) plots. The principal types of
materials giving the respective responses are indicated,
together with the mechanisms. From Ref. [4].

a proper understanding of this situation should contribute to a
better insight into these many-body processes. Furthermore, it has
been pointed out that a clear connection exists between some aspects
of LFD and electrochemical reactions at interfaces [10].

The traditional interpretations of non-exponential dielectric
relaxation processes are based to a large extent on the concept of
distributions of relaxation frequencies in non-interacting single-
particle systems, or alternatively on the concept of correlation
functions which attempt to take into account particle interactions
in a very limited way. Neither of these approaches seems capable
to account satisfactorily, i.e., from first principles, for the

observed universality of power law relations, with its well defined range of exponents s. These difficulties are particularly pronounced in the case of severe departures from the Debye response which are well documented experimentally [4].

By contrast, the new many-body interpretation [1-4] leads to the derivation of the power law relations in a wholly general manner without the need for any arbitrary assumptions, whose validity extends throughout the observed range of exponents s and which justifies the limitation of this range to values between 0 and 2. The concept of disorder is central to all these considerations.

MECHANICAL RELAXATION

The existence of close similarities between electrical and mechanical relaxations has been recognized for some time, especially in terms of the relation between dielectric and mechanical loss peak frequencies [12] and similar considerations were also extended to Nuclear Magnetic Resonance peaks. What has not been recognized is the close relationship between the detailed forms of time- and frequency-responses of dielectric and mechanical relaxations in terms of the exponents of the universal power law (3) which were found to obey the relation [2,6]

$$s_{diel} + s_{mech} = 3 \text{ or } 1 \qquad (6)$$

respectively, in the low- and high-frequency regions. This is one of the direct consequences of the Dissado-Hill many-body theory [1] and can be understood in terms of the complementarity of dielectric and mechanical energy storage in dipolar systems.

The dielectric LFD processes have their direct counterpart in the well known phenomenon of creep in mechanical relaxations [7].

We do not propose to review here the detailed experimental evidence relating to the power law dependence of the mechanical moduli on time and frequency - the data available are not as extensive as those for the dielectric relaxations but the general trends are unmistakeable [6,7].

It is interesting to note that the accepted interpretations of the known non-exponential mechanical relaxation processes are based on concepts similar to the distributions of relaxation times adopted in dielectric theories and the same fundamental counter-arguments apply in both situations -- the essential arbitrariness of an approach which has to postulate a distribution of some physical parameter, for which there is no particular justification, in order to explain an experimentally observed behavior. One of the difficulties with these types of approaches is the absence of a more general explanation of the reasons for the observed universality of behavior in terms of power laws.

With reference to the main theme of the present paper, we stress again the intimate connection between the phenomena of mechanical relaxation and the concepts of disorder on which the many-body interactions are based.

RELAXATION OF DEEP LEVELS IN SEMI-INSULATORS

To the well-documented power law relations found experimentally in dielectric and mechanical relaxations we now wish to add the not at all recognized but equally valid relaxation in deep levels in semi-insulators and semiconductors. In this instance, we are dealing with the essentially electronic processes of recombination, trapping and release of electrons from deep levels falling in the forbidden gap of these materials and arising essentially from the presence of disorder in them. The principle experimental manifestations of these relaxation processes are the decay of photoconductivity in photoconductors and the emission of light in phosphors -- the latter involving some intermediate trapping processes between the excitation of the system and the emission of light. The processes of photoconduction and of luminescence decay will be discussed separately.

Photoconductive relaxation

Although a large volume of work has been reported on photoconductivity during the past half-century, relatively little of it related to the time dependence of photoconduction after a short burst of excitation and even then there are very few reliable measurements extending over more than three decades of time. This range is the minimum required for unambiguous recognition of the prevailing law of decay and one of the reasons for the limited experimental range lay in the need for a large dynamic range of response of the detectors and also in the inevitable dark currents which limit the signal-to-noise ratio at longer times.

While the experimental evidence is not entirely satisfactory, the existence of departures from simple exponential decay is easily seen and the main thrust of interpretation has gone into the attribution, once again, of suitable distributions of recombination times. The only well recognized form of departure from the exponential law is the bimolecular process which would give rise to a decay of the form $(t + t_0)^{-1}$, where t_0 is a constant. For the rest, although the use of a correction has been proposed which would make the initial recombination process faster as the density is high, there have been no serious attempts to move away from the concept of isolated non-interacting recombination and trapping centers which would make exponential decay the only conceivable law to be expected.

A recently published alternative approach to the usual time-domain measurement of photocurrent decay uses the frequency-domain method which has certain definite advantages and which can produce five or more powers of ten of time range. Measurements made on semi-insulating GaAs are reproduced in Fig. 3 and they show the very marked departure from the simple exponential time dependence [13].

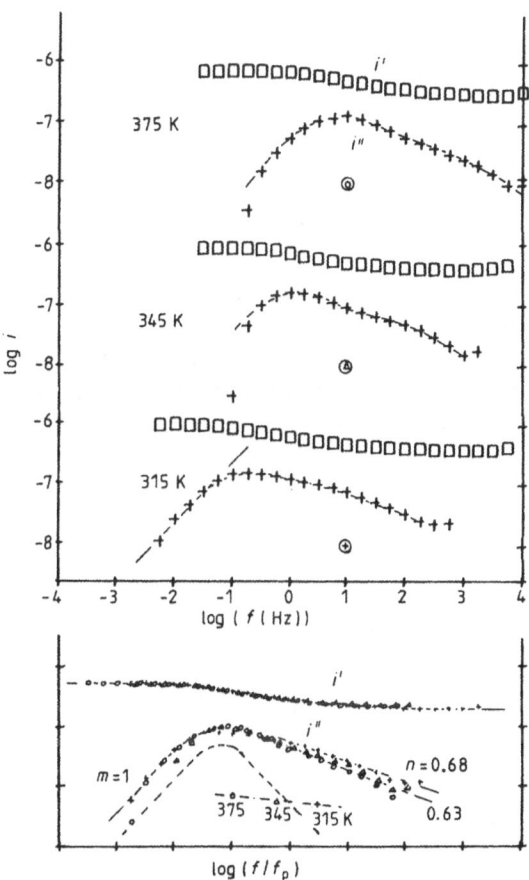

Fig. 3 The three upper diagrams show the print-out from the plotter of the FRA corresponding to the in-phase (real) component of the photocurrent, i'(ω), and the quadrature (imaginary) component i"(ω), as functions of the frequency f=ω/2π, for three temperatures. The lower diagram shows the normalization of these data into a master curve, with the locus of the reference points +, Δ and 0 corresponding to the positions shown in the upper diagrams. The dotted curve shows the shape of a Debye-like loss peak which would correspond to an exponential time decay of the photoconductivity. f_p is the loss peak frequency. From Ref. [13].

Similar conclusions were reached on the basis of admittance spectroscopy of p-n junctions in silicon and GaAs [4], a silicon spectrum
being shown in Fig. 4.

The available experimental data on photoconduction and on trap
release from deep traps in semiconductors is not sufficiently
extensive to draw definitive conclusions regarding the precise form
of time dependence, except to show that they deviate from the exponential relation in most cases and that they approach a power law
in some. With the advent of the new frequency domain technique, it
may be possible to obtain sufficiently detailed information to settle this question in the future. For the present, however, our
most extensive experimental information comes from the studies of
transient luminescence which will be presented in the next Section.

Delayed luminescence relaxation

Just as photoconductivity, delayed luminescence provides direct
information about the rate of radiative transitions following an
initial short burst of excitation. There is ample literature on
this subject and a recent survey by Jonscher and de Polignac [14]
has shown that exponential decay is observed only in those cases

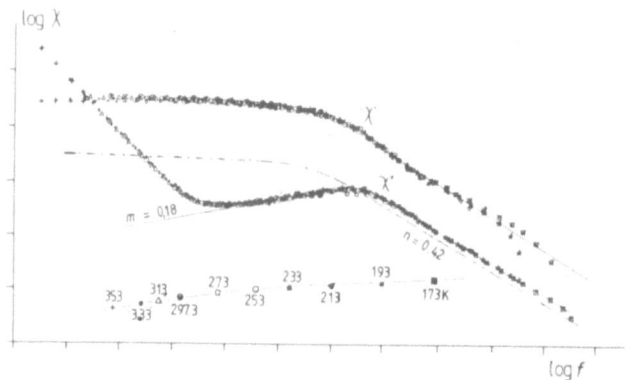

Fig. 4 The dielectric spectrum of a silicon p-n junction, normalized
        for a range of temperatures. The real part X'(ω) is raised
        by one decade with respect to X"(ω) for clarity, its proper
        position is shown by the chain-dotted lite. The low-frequency
        part of loss corresponds to dc conduction. The two slopes
        m and n are indicated. The entire range of some nine decades
        of frequency shows only one single process. This response
        corresponds to the release and trapping of electrons at a
        deep level in the forbidden gap of the p-n junction. From
        Ref. [4].

where "intrinsic" luminescence may be presumed to apply -- where
the excited species remain at the luminescent centers themselves
and radiative recombination occurs in relatively short times of
between nano- and microseconds.   In all other cases where intermediate trapping takes place and the excited electrons are re-emitted
from these traps to undergo delayed luminescent transitions, the
widely observed time dependence follows power laws of the same form
as Eq. (3) but with the exponent in the range $1 < s < 2$.

Some examples of luminescence decay, plotted in log-log
presentation, are shown in the following diagrams.   Figure 5 gives
a remarkable result for an organic glass covering nine decades of
time with electron beam and gamma-ray excitation at short and long
times, respectively.   The data were re-plotted from the original
presentation which gave three separate time ranges, and they show
a uniform slope of -1.06.   Figure 6 shows the response of another
organic compound, stilbene, with gamma-ray, neutron and alpha
particle excitation, which give nearly four decades of uniform
slope of -1.1.   The original presentation of the same data on a
semi-log basis with three times ranges is shown in the inset.

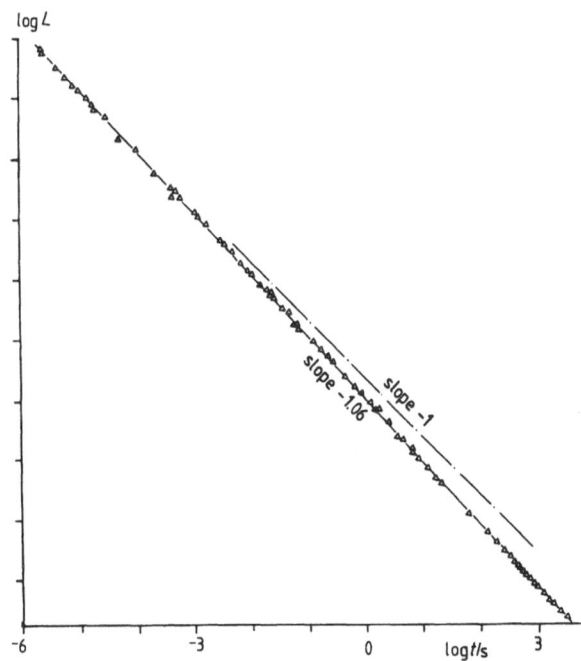

Fig. 5   Luminescence decay in methylcyclohexane biphenyl glass with
         excitation by electron beam at short times and by gamma
         rays at long times.   The combined diagram shows a continu-
         ous slope of - 1.06 throughout the nine decades starting
         at a few microseconds.   From reference [15], adapted to a
         single time scale.   Temperature 77 K.   From Ref. [14].

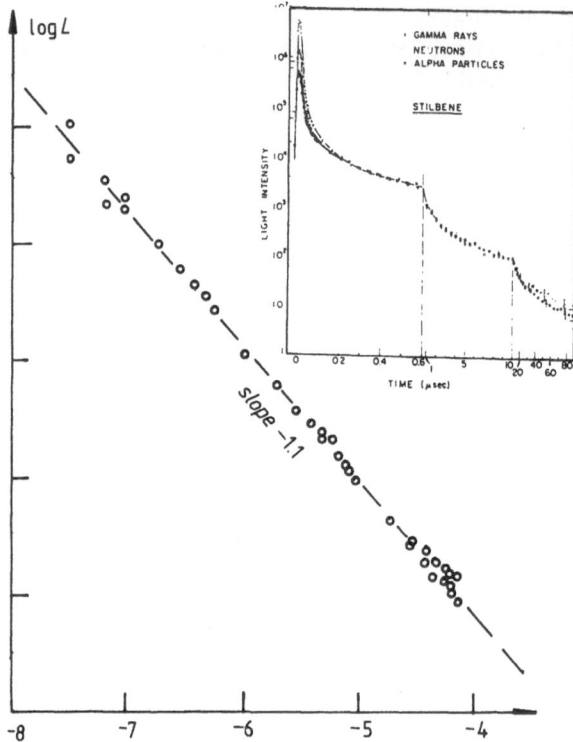

Fig. 6 Luminescence decay in stilbene replotted on a log-log basis
         from the original data from Ref. [16] shown in the inset
         in semi-logarithmic presentation with three consecutive
         linear scales, giving a clear impression of the absence of
         any discernable law. The logarithmic slope is -1.1.
         Room temperature. From Ref. [14].

A set of luminescence decay data for single crystal gallium
phosphide at 1.6 K and normalized for two intensities of the
exciting electron beam are shown in Fig. 7. Here the ultimate slope
is -1.6 but there is an initial region in which the response shows
a continuous curvature. Our last example relates to single crystal
p-type gallium arsenide at 4.2 K with optical excitation, Fig. 8.
Again, the original data given in semi-logarithmic representation
are shown in the inset, while the same results in the log-log
presentation give a clear power law with an exponent of -1.34 over
four decades of time.

The model proposed for this power-law behavior by Jonscher
and de Polignac [14] is shown schematically in Fig. 9 and involves
generation of electrons in the conduction band at a rate g(t)
determined by the external excitation rate. These carriers are

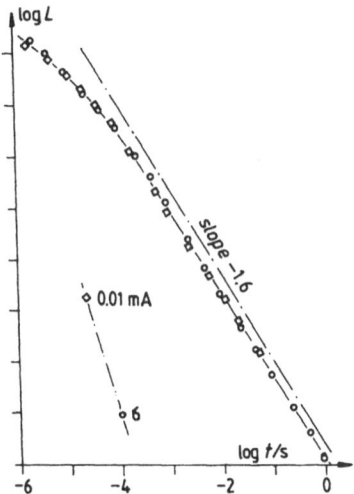

Fig. 7   The decay of cathodoluminescence in single crystal gallium
phosphide at 1.6 K and normalized for two excitation intensities, with an almost vertical translation shown by the
chain -- dotted line.   After an initial curvature a slope
of -1.6 is followed over four decades of frequency.   Original data from reference "]7], adapted for the present
purpose.   From Ref. [14].

rapidly trapped by a process denoted as γ into a set of trapping
levels T from which they are slowly released at a rate $\nu(t)$ either
into the conduction band to be then captured in the luminescence
centers L, giving free-to-bound (F-B) radiation, or into the excited states of the luminescence centers themselves.

   Assuming a rapid trapping process, a delta-function excitation
produces a total number n_{to} of trapped electrons and the trapping
rate is also approximately a delta function, $\gamma(t) = n_{to}\delta(t)$.   The
detrapping rate is then given by the following convolution integral,
which simplifies in the present case to

$$\nu(t) = - \int_0^\infty \dot{h}(t)\, \gamma(t - \tau)\, d\tau = - n_{to}\dot{h}(t) = - dn_t(t)/dt \qquad (7)$$

Here, $h(t)$ denotes the characteristic response function of the
traps which is defined as the probability that an electron trapped
at time $t = 0$ remains trapped at time t.   The dot denotes differen-
tiation.   The implication is that the detrapping process involves
an element of memory of the past history, a concept not known in
the classical approach leading to an exponential detrapping rate.

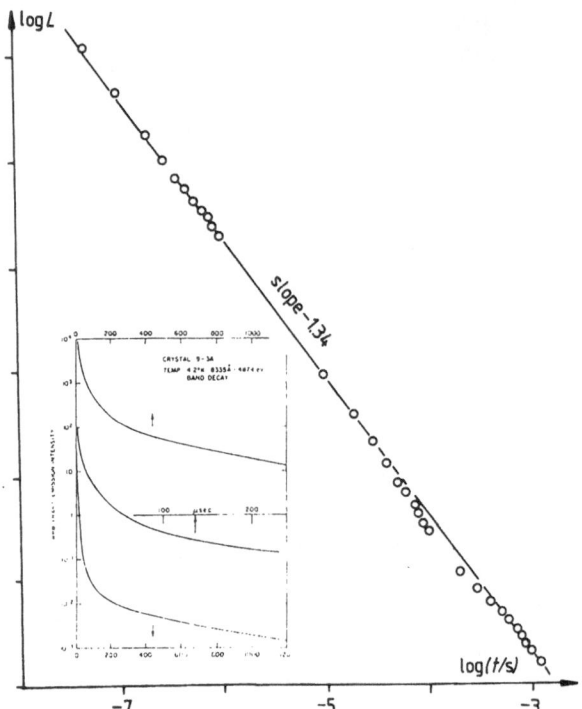

Fig. 8 Luminescence decay in p-type GaAs at 4.2 K, with 7 ns optical
        excitation. The original data in a three-stage semi-logarith-
        mic representation from Ref. [18] are shown in the inset.
        The logarithmic slope is -1.34 over more than four decades
        of time. From Ref. [14].

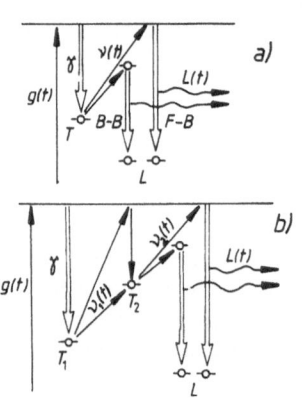

Fig. 9 The proposed model to explain the
        power law decay of luminescence
        found in many materials. In (a)
        a single set of traps is postu-
        lated, marked by T, which rapidly
        trap the excited electrons and
        subsequently slowly releases them
        into the conduction band from
        where they drop into the lumines-
        cent centers, or directly into
        the excited state of these cen-
        ters. In (b) traps T_1 capture
        the excited carriers and slowly
        release them into traps T_2 which
        then act as above. B-B refers to
        bound-to-bound, F-B refers to
        free-to-bound transitions.

The intensity of luminescence is then related directly to the derivative of the characteristic function $\dot{h}/t$ and is a power law in time if the function itself is a power law.

With step-function excitation of finite duration, $0 < t < T$, the trapping rate is given by a step-function

$$\gamma(t) = \gamma_0, \, t < T \qquad \gamma(t) = 0, \, t > T$$

while the detrapping rate is obtained in the following form:

$$\nu(u) \simeq \gamma_0 \, h(u) \qquad \text{(short time after step)}$$

$$\nu(u) \simeq \gamma_0 \, T \, \dot{h}(u) \quad \text{(long time after step)}$$

where u is the time measured from the instant of cessation of the exciting pulse. The conclusion here is that the power-law characteristic may be composed of two parts, if the duration of the exciting pulse is not infinitely short. Alternatively, even with a delta-function excitation rate, it is possible to obtain an effectively extended rate of supply to the luminescent centers, if a second set of trapping states is assumed which rapidly trap carriers and subsequently slowly dispense them into the traps envisaged in Fig. 9. The initial stages of the decay of luminescence may then be very complicated, but the ultimate relaxation always follows the power law corresponding to $\dot{h}(t)$. The complete response is shown schematically in Fig. 10.

More complicated time dependences, especially those where there is a gradual change of slope at short times towards a final slope as above, can be explained by a more complex trapping model described in [14].

Whether this particular model is the ultimate answer to the question of power law decays of luminesence remains to be seen as more detailed theoretical work proceeds. The important point which is beyond reasonable doubt is that there is a wide range of power law decays of delayed luminescence and that no general theory exists at present which would explain their shape in terms of any known physical parameters. The bimolecular law can be seen to lead to a dependence of the form $(t + t_0)^{-2}$ which is not normally seen at short times where one would expect it, while the densities of excited electrons are high.

It is suggested that the observed power law decay of delayed luminescence points strongly to the presence of "memory" in the trapping states, which alone can explain the observed nonexponential dependence on time. The precise mechanism of this memory is not evident at present, but the close analogy with other aspects

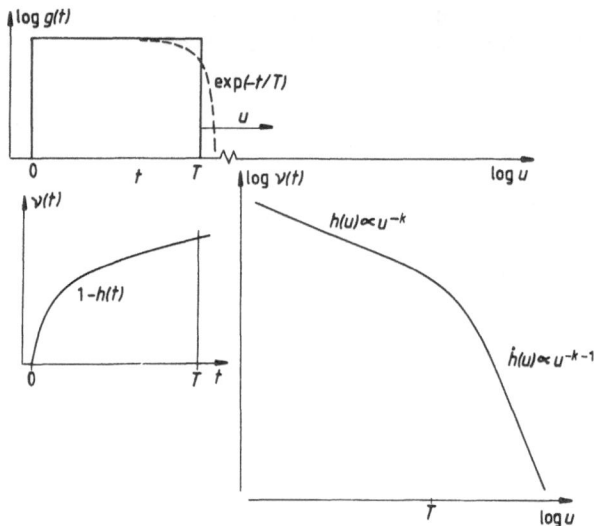

Fig. 10   A schematic representation of the excitation and decay of
          luminescence in a material characterized by a set of
          power-law traps as shown in Fig. 9, under a step function
          generation of duration T. The time of the decaying branch
          is counted from the point T. A power law response
          $h(t) \propto t^{-k}$ is assumed. The diagram of the excitation
          function g(t) also shows a contour of an exponential
          function to emphasize its similarity to square-wave
          generation. From Ref. [14].

of relaxation in disordered systems makes it highly plausible that
such memory is associated in this instance with many-body interactions arising between individual deep traps and the surrounding
crystalline lattice. The emission of an electron from a trap must
be seen as a very drastic perturbation of the system, with a very
rapid transfer of the electron over a considerable distance in
space -- the effect being the same as a rapid reorientations of a
dipole moment equivalent to several hundred Debye units. The
resulting rapidly varying Coulombic fields create a disturbance
which is likely to affect neighboring sites. At the same time,
the local interaction accompanying the capture of an electron in
a deep level produces the "memory" effect which determines the
characteristic function h(t).

CONCLUSIONS

     We have shown on the basis of extensive experimental evidence
that the relaxation of dielectric and mechanical stress and of
trapped charge populations in semiconductors and phosphors all
follow very similar time and frequency dependence involving power
laws, instead of the generally expected exponential relations. We

propose that the common feature of all these relaxations is the dominance of many-body interactions in conjunction with the inherent disorder which is present in all these systems. The theoretical treatment of the dielectric relaxations is now firmly based on a rigorous theory which is capable of explaining the observed range of types of response and which is being used now to extend our understanding of the other relaxation processes. This highly developed state of theory is due in no mean measure to the fact that there exists, in the case of dielectric relaxation, an excellent body of experimental evidence regarding the "universal" power law relations in time and frequency. The situation is less well developed in the case of photoconduction, while in the case of delayed luminescence it is now apparent that a considerable body of experimental evidence exists supporting the power law relaxation processes so that it may be possible to build a more comprehensive theoretical treatment.

The present paper draws together all these types of relaxation, including also the interfacial ionic and electronic generation processes, with the suggestion that they may all have some essential elements in common. We suggest that the common feature is the dominance of many-body effects associated with disorder.

The consequences of the present proposition are significant for our understanding of the trapping and dissociation processes in solids and at interfaces between them. Some of these consequences, in particular the existence of "memory" in trapping reactions, are rather novel and it may take some time to establish a concensus on these matters. We wish, however, to draw a clear distinction between the experimental facts as presented in this paper, about which there should not be much doubt, and the interpretation of the significance of these facts which may be contested. The purpose of the present paper is to stimulate discussion around these rather striking features of different relaxation processes.

REFERENCES

1. L. A. Dissado, and R. M. Hill, Proc. Roy. Soc. London, A390:131 (1983).
2. R. M. Hill and A. K. Jonscher, Contemporary Physics, 24:75 (1983).
3. L. A. Dissado, Phys. Scripta, T1:110 (1982).
4. A. K. Jonscher, Dielectric Relaxation in Solids, Chelsea Dielectrics Press, London, (1983).
5. A. K. Jonscher, R. M. Hill and L. A. Dissado, Phys. Stat. Sol. (b)102:351-356 (1980).
6. R. M. Hill, J. Materials Sci., 17:3630 (1982).
7. R. M. Hill, and L. A. Dissado, J. Materials Sci., 19:1576 (1983).
8. E. von Schweidler, Ann. d. Physik, 24:711 (1907).
9. A. K. Jonscher, Phil. Mag., B38:587-601 (1978).
10. A. K. Jonscher, Maxwell-Wagner Effect and Strong Low-Frequency

Dispersion, Conf. Electrical Insulation and Dielectric Phenomena (CEIDP) Buck Hill Falls, PA, USA. (1983).

11. L. A. Dissado and R. M. Hill, J. Chem. Sci. Faraday Trans. 2, 80: 291 (1984).

12. D. W. McCall, Inst. Phys. Conf. Series No. 58:46, Physics of Dielectric Solids, C. H. L. Goodman, Ed. (1981).

13. A. K. Jonscher, and J-R Li, J. Phys. C. Solid State Phys. 16: L3359-3364 (1983).

14. A. K. Jonscher, and A. de Polignac, J. Phys. C. Solid State Phys. 17:6493 (1984).

15. P. Cordier, J. F. Delouis, F. Keiffer, C. Lapersonne, and J. Rigaut, Cr. Acad. Sci. Paris, 279C:589 (1974).

16. L. M. Bollinger, and G. E. Thomas, Rev. Sci. Instr. 32:1044 (1961).

17. D. G. Thomas, J. J. Hopfield, and W. M. Augustyniak, Phys. Rev. 140:A202 (1965).

18. R. Dingle, Phys. Rev. 184:788 (1969).

CHARACTERISTIC TEMPERATURE AND BOND STRENGTH OF OXIDES IN INORGANIC

GLASSES

Naohiro Soga and Kazuyuki Hirao

Department of Industrial Chemistry
Faculty of Engineering
Kyoto University, Kyoto, Japan

INTRODUCTION

There are several ways of assessing the chemical bond strength
of atoms in solids. One way is based on the concept of electronegativity, but this approach has been sharply criticized [1]. Another
way is based on the effective charge of ions or atoms. Although the
absolute values of the effective charge determined experimentally
are somewhat different depending upon the methods used, the relative
values in the ionicity series are quite useful to interpret various
physical properties of solids.

Heat capacity is one of those properties which reflect the
atomic vibrations associated with the interatomic forces between
atoms or ions in solids, and thus it may be used to assess the
chemical bond strength of atoms by applying a suitable theory of
solid state physics. One of the difficulties to do so for oxide
compounds is that only a limited number of simple solids are available for experiments when one tries to obtain the strength of metal
oxygen bonds in a systematic manner. Another one is lack of a proper theory to deduce interatomic forces in oxide compounds, many of
which have complicated structures and cannot be applied a simple
theory, such as the Debye theory, to interpret the temperature
dependence of heat capacity. The former difficulty may be overcome
by using glasses because different kinds of ions can be incorporated
in large amounts in one type of base glasses. As for the latter,
the three-band theory advanced by the present authors from the
Tarasov's theory has been found quite useful to express the compositional dependence of heat capacity for various oxide glasses, and
to estimate the interatomic forces of ions and atoms constituting
glass structure [2]. This approach was made possible by accurate

119

measurements of heat capacity from low to medium temperature ranges
and data analysis by a large computer [3].

The main purpose of the present paper is to show the validity
of the three-band theory and its usefulness to discuss the following
questions: (1) How strong is the interatomic bond between a so-
called network modifying cation and nearby oxygens? (2) How strong
is the interatomic bond between a network forming cation and a nearby
oxygen? (3) Is there any difference in these interatomic and other
intera·ctions between the crystalline and glassy states?

In order to verify the results obtained by the three-band
theory, the data of far infrared absorption, which is another one of
the properties associated with atomic vibrations of constituent atoms
or ions in solids, were obtained on several oxide glasses and compared
with the characteristic temperatures of the three-band theory.

THREE-BAND THEORY AND BOND STRENGTH

In order to obtain the bond strength from heat capacity of
solids, a usual way is to assume that each atom is a point mass
connected to each other by a spring having an appropriate attractive
force and oscillates its eigenfrequencies. This simple theory of
heat capacity was introduced first by Einstein [4] and then by
Debye [5]. According to the Debye theory, the number of vibrational
modes per frequency interval is proportional to the square of frequency of normal modes up to a certain maximum frequency. This cut-
off frequency is related to the interatomic force, mass and lattice
spacing of constituent atoms, and thus it is possible to estimate
the interatomic forces in solid. Although this Debye theory describes the temperature dependence of heat capacity for a number of
solids, it is not applicable for high polymers with chain or layer
structures. The macromolecular structure of these polymers cannot
be represented by a three-dimensional continuum as postulated by
the Debye theory. So, it must be modified to include the vibrations
of a one-dimensional continuum suitable for these polymers. Several
such modifications have been attempted in the past. For example,
Tarasov [6] developed a theoretical model for substances with chain
or layer structure by considering both of the one- and three-
dimensional continuum distributions.

As for inorganic glasses, the glass forming compounds usually
take some kind of network structure, and thus the temperature dependence of heat capacity of inorganic glasses tends to deviate from
that of the Debye theory. In fact, Westrum [7] and White and Birch [8]
showed that the heat capacity curve of vitreous silica has a
characteristic form different from a normal heat capacity curve of
a simple ionic compound and closely resembles that of a substance
with polymer-like chain structure. To express such data, the three-
band theory has been developed. A brief description of the three-
band theory is given below.

Let us assume that inorganic oxide glasses take a network structure shown schematically in two dimensions in Fig. 1(a). In the Debye model, a solid is usually assumed to be represented by a monoatomic lattice. Thus, in this model the above network structure is modified by one like Fig. 1(b), where all atoms are distributed homogeneously with even spacing. It is clearly seen that the inter-atomic distance in the system (b) becomes larger than the original one (a), and the systems (a) and (b) have different heat capacities. One of the more realistic approximations for the system (a) is to consider the frequency distribution function consisting of two different types of vibrational modes: one due to the strong interaction force between atoms or ions and the other due to the weak interaction force between the repeated units of atomic assemblies, as shown in Fig. 1(c). By following Tarasov's approach, the former is assumed to take the one-dimensional continuum distribution having a total of $3N_1$ high frequency modes from ν_{max} to ν_1 and the latter the three-dimensional continuum distribution having a total of $3N_2$ low frequency modes from ν_1 to 0. Then, the total vibrational energy $U_{1,3}$ of the system is representable by the following equation.

$$U_{1,3} = 3N_1(\nu_{max}-\nu_1)^{-1} \int_{\nu_1}^{\nu_{max}} h\nu[\exp(h\nu/kT)-1]^{-1}d\nu$$

$$+ 9N_2\nu_1^{-3}\int_0^{\nu_1} h\nu^3 [\exp(h\nu/kT)-1]^{-1}d\nu$$

(1)

where h and k are the Planck and Boltzmann constants, respectively. The heat capacity $C_{1,3}$ of the system is obtained by setting $x=h\nu/kT$, $\theta_1=h\nu_{max}/k$ and $\theta_3=h\nu_1/k$, as follows.

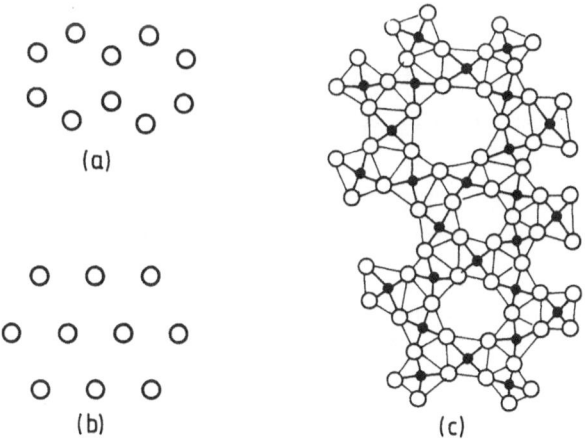

(a)

(b)   (c)

Fig. 1 Schematic model of network structure.

$$C_{1,3} = 3R(T/\theta_1)\int_{\theta_3/T}^{\theta_1/T} [x^2 e^x/(e^x-1)^2]dx$$

(2)

$$+ 9R(T/\theta_3)^3 \int_0^{\theta_3/T} [x^4 e^x/(e^x-1)^2]dx$$

The validity of the above model was examined for various crystalline compounds with different crystallographic structures. When each repeated unit contains only one atom like in the case of MgO, the distance between the repeated units is the same as the interatomic distance. Thus, θ_3 should be equal to θ_1 and these two characteristic temperatures should converge into one value, or the Debye temperature θ_D, even if the heat capacity data were analyzed by Eq. (2). On the other hand, the heat capacity data of a solid with a linear chain structure should be described by θ_1 without θ_3. If a crystal takes a layer structure, Eq. (2) should be modified to include the vibrational modes of the two-dimensional continuum distribution with the parameter θ_2 in place of those of the three-dimensional continuum distribution with θ_3.

The analyses of heat capacity data were made for various compounds, and the results are shown in Table I. As expected, the values of θ_1 and θ_3 become equal for NaCl and MgO, while only θ_1 appeared for Se, As_2O_3 and Bi_2O_3. The data of graphite, As and Sb metals fit for a model of the two-dimensional continuum and the values of θ_1 and θ_2 became equal.

To extend the preceding treatment to various kinds of glasses, additional vibrational energies of other ions (network modifiers) than the network formers must be considered. Since the bond strength of these ions is known to be much weaker than that of network formers and these ions are distributed throughout glass network, it may be assumed that the vibrational modes of these modifying cations are local modes and contribute independently to the heat capacity in the form of an Einstein function with the characteristic

Table I.   Characteristic temperatures for some crystals with simple structures.

Material	Structure	Characteristic Temp.			Data from
		θ_1	θ_3	θ_2	
NaCl	NaCl type	281K	281K	–	/10/
MgO	NaCl type	758	758	–	/11/
Se	Herical chain	375	0	–	/12/
As_2O_3	Chain of Valentinite	1003	0	–	/13/
Bi_2O_3	Chain of Valentinite	661	0	–	/14/
Graphite	Layer	1370	–	1370	/15/
As metal	Layer	331	–	331	/13/
Sb metal	Layer	223	–	223	/13/

temperature θ_E.   Consequently, θ_E depends on the kind of modifying
cations but not on their concentration, nor on the kind of network
formers.   Thus, the representative expression of heat capacity by
the present three-band theory with $y = \theta_E/T$ is

$$Cv = C_{1,3} + 3Ry^2 e^y/(e^y - 1)^2 \qquad\qquad (3)$$

Once the heat capacity of a solid is measured at various temperatures,
θ_1, θ_3, and θ_E can be determined numerically by a computer to fit
the experimental data.   Although the real structure of oxide glass
is unknown, the present model is in accordance with our knowledge
of glass structure, which consists of one, two or three-dimensional
chains of network forming metal polyhedra, similar to those of crystalline states, and network modifying cations.   Thus, the values of
characteristic temperatures θ_1, θ_3 and θ_E should represent the
interatomic bond strength and interactions of ions and atoms constituting glass structure.

CHARACTERISTIC TEMPERATURES FOR VARIOUS GLASSES

     The heat capacity data of various glasses at low temperatures
were determined in the same manner as described in the previous
paper [3,9].   The temperature range was from 77K to 300K for most
cases, but in some cases the measurements were made at lower temperatures than 77K.   All the data were analyzed by the three-band theory
using Eq. (3).   Figure 2 shows an example of the fitness of the
three-band theory with the experimental data for $Na_2O \cdot 2SiO_2$ glass.
The values of the characteristic temperatures for various glasses
thus obtained were summarized in Table II.   For the sake of
comparison, the heat capacity data for various silicates, aluminosilicate, germanate and borate compounds were compiled from literature and were analyzed also by means of Eq. (3).   The results are
given in Table III.

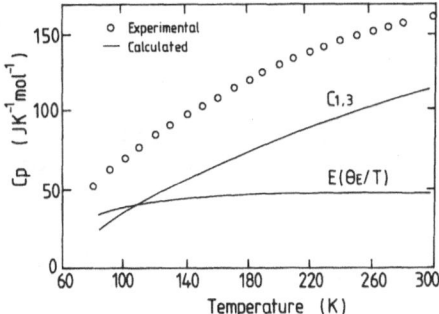

Fig. 2   Temperature dependence of heat capacity for $Na_2O \cdot SiO_2$
         glass.

Table II. Characteristic temperatures for various glasses.

Glass Composition		θ_1	θ_3	θ_E	Glass Composition			θ_1	θ_3	θ_E
Na_2O	$-SiO_2$				Li_2O	$-Al_2O_3$	$-SiO_2$			
0	100	1550	150	-	25	25	50	1210	350	570
14	86	1370	385	265	16.7	16.7	66.6	1220	350	580
20	80	1320	395	250	12.5	12.5	75	1260	340	590
25	75	1320	400	245	Na_2O	$-Al_2O_3$	$-SiO_2$			
33	67	1300	400	230	25	25	50	1195	315	220
50	50	1300	400	200	16.7	16.7	66.6	1200	310	230
K_2O	$-SiO_2$				12.5	12.5	75	1250	310	240
10	90	1260	330	150	MgO	$-CaO$	$-SiO_2$			
15	85	1212	340	150	0	50	50	1180	360	530
33	67	1160	350	145	12.5	37.5	50	1185	360	595
Cs_2O	$-SiO_2$				25	25	50	1190	355	630
10	90	1170	260	60	33.5	16.5	50	1190	360	650
15	85	1050	275	60	37.5	12.5	50	1190	350	665
33	67	970	280	50	Na_2O	$-K_2O$	$-SiO_2$			
Na_2O	$-GeO_2$				11.1	22.2	66.7	1255	380	170
0	100	1250	55	-	16.7	16.7	66.6	1255	385	195
11	89	1315	105	245	22.2	11.1	66.7	1265	395	185
19	81	1340	135	235	K_2O	$\cdot Cs_2O$	$-SiO_2$			
30	70	1260	140	235	10	23.3	66.7	1050	320	75
50	50	1150	150	215	16.7	16.7	66.6	1090	330	95
Na_2O	$-B_2O_3$				23.3	10	66.7	1120	360	120
0	100	1920	50	-	Na_2O	$-Cs_2O$	$-SiO_2$			
8	92	1945	50	240	6.3	27	66.7	1100	330	80
14	86	2000	60	235	16.7	16.7	66.6	1200	360	140
18	82	1990	70	235	20	13.3	66.7	1225	365	145
20	80	1945	75	230	CaO	$-P_2O_5$				
26	74	1910	85	235	30	70		1480	140	430
33	67	1895	100	245	45	55		1420	200	430
Na_2O	$-P_2O_5$				50	50		1380	260	430
30	70	1500	80	210	MgO	$-P_2O_5$				
35	65	1500	90	210	50	50		1400	250	700
45	55	1500	110	210	SrO	$-P_2O_5$				
50	50	1440	110	210	50	50		1420	280	380
55	45	1280	110	210	BaO	$-P_2O_5$				
Li_2O	$-SiO_2$				50	50		1350	180	330
33	67	1410	440	490	K_2O	$-P_2O_5$				
					50	50		1260	100	160
					$Cs_2O-P_2O_5$					
					50	50		1050	90	80

LOW FREQUENCY INFRARED ABSORPTION

According to the lattice dynamic theory, the cut-off frequency of the vibrational modes of an atomic assembly is related to the far infrared absorption band characteristic to the fundamental mode of atomic vibrations. Thus, in order to verify the values of characteristic temperatures θ_1 and θ_E, the far infrared absorption bands were determined for alkali silicate and alkali phosphate glasses by using an infrared spectrometer (Hitachi FIS-3 type). The glass samples were obtained in the form of 50μ thin films by the melt-and-quenching method using a twin-roller, and they were dispersed by the Nujol technique in low density polyethylene at 10∿20 wt%. The far infrared absorption spectra of various alkali disilicate glasses having the composition of $0.33M_2O-0.67SiO_2$ (M=Na, K and Cs) in the region of the cation motion bands are shown in Fig. 3 The peak

Table III. Characteristic temperatures for various crystals.

Crystal	system	lattice dimension a	b	c	θ_1	θ_3	θ_E	Data from
SiO$_2$ (quartz)	hex.	5.00	-	5.46	1450	241	-	/18/
(tridimite)	hex.	5.05	-	8.26	1440	230	-	/18/
(crystobalite)	cub.	7.18	-	-	1480	225	-	/18/
Na$_2$Si$_2$O$_5$	orth.	6.44	15.04	4.96	1360	340	253	/19/
Na$_2$SiO$_3$	hex.	6.08	-	4.83	1320	396	256	/19/
K$_2$Si$_2$O$_5$					1310	315	153	/20/
KAlSi$_2$O$_6$	tet.	13.12	-	13.79	1270	283	150	/16/
CsAlSi$_2$O$_6$	cub.	13.70	-	-	1105	231	50	/17/
MgSiO$_3$	orth.	8.83	18.20	5.19	1250	352	702	/21/
CaSiO$_3$	tri.	6.90	11.78	19.65	1240	303	570	/22/
GeO$_2$	tet.	4.99	-	7.07	1250	500	-	/23/
Na$_2$GeO$_3$	hex.	6.26	-	4.92	1030	425	232	/24/
K$_2$GeO$_3$					890	427	135	/24/
MgGeO$_3$	hex.	4.94	-	13.76	1130	531	761	/24/
CaGeO$_3$	tri.	8.23	7.58	7.34	1125	421	525	/24/
B$_2$O$_3$	cub.	10.06	-	-	1860	240	-	/24/
NaBO$_2$	hex.	11.93	-	6.46	1600	140	210	/25/

positions shift considerably depending on the kind of cation, from
484 cm^{-1}, 210 cm^{-1}, 152 cm^{-1} to 112 cm^{-1} by changing the
kind of alkali metal from Li, Na, K to Cs. The uncertainties in
estimation of band maxima are about $\pm$ 10 cm^{-1}. The similar tendency
to far infrared cation-dependent bands was also observed in other
alkali silicate glasses, containing a higher amount of SiO$_2$ (85
mol%) as shown in Fig. 4 Fig. 4 also includes the results of far
infrared spectra of mixed alkali silicate glasses. The band maximum
positions were at about 130 cm^{-1} for (Cs$_{0.5}$K$_{0.5}$)$_2$O$\cdot$2SiO$_2$ glass and
about 179 cm^{-1} for (K$_{0.5}$Na$_{0.5}$)$_2$O$\cdot$2SiO$_2$ glass, indicating a little

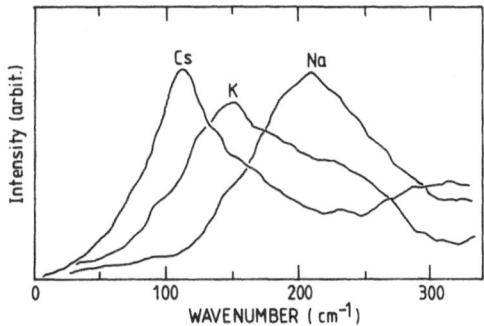

Fig. 3   Far infrared absorption spectra of alkaline metal-oxygen
bonds (M-O) in the composition of 0.33M$_2$O-0.67SiO$_2$ glass
(M = Na, K and Cs).

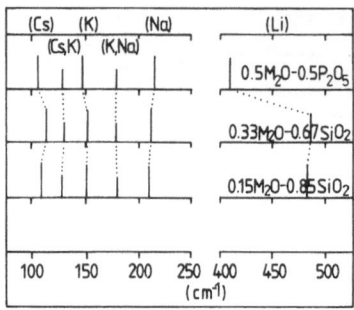

Fig. 4   Far infrared cation-dependent bands in various aklaki silicate
         and phosphate glasses.

negative deviation from the linear additivity of two end members.
Also shown in Fig. 4 are the band maximum positions for alkali
metaphosphate glasses. The values of 215 cm^{-1}, 147 cm^{-1} and 102 cm^{-1}
for $NaPO_3$, KPO_3 and $CsPO_3$ glasses were similar to the results of Nelson
and Exarhos [26]. The constancy of the maximum position of alkali
ions irrespective of base glass compositions, shown in Fig. 4, may
be taken as a proof that the vibrational modes arising from these
alkali ions can be regarded as those of independent vibrators, as
postulated by the Einstein function.

The effect of the kind of alkali ions in silicate glasses on
the maximum frequency of Si-O stretching vibrations was studied on
the same alkali disilicate glasses. It was found that the maximum
frequency existed at 1063 cm^{-1}, 1009 cm^{-1} and 992 cm^{-1} for
$Na_2O \cdot 2SiO_2$, $K_2O \cdot 2SiO_2$ and $Cs_2O \cdot 2SiO_2$ glasses, respectively, showing
that it decreases with increasing cation mass or ionic radius.

BOND STRENGTH OF NETWORK MODIFIERS

The network modifiers, such as alkali ions, are usually
considered not to take a part of network structure but to occupy
the interstices of network structures. If the network structure
remains almost unchanged when network modifiers are substituted
from one kind to another, the change in internal energy arises
mainly from the difference in bond energy between two kinds of network modifiers. This is the basis of our approach that the interatomic bond strength of network formers can be obtained. The
interatomic bonding nature of network modifiers is generally considered ionic rather than covalent, so that their bond strength
may be approximately represented by the Born potential [27]. The
simple Born potential is given by

$$U = - \frac{q_1 q_2}{r} + \frac{b}{r^n} \qquad (4)$$

where r is the separation of ions, q_1 and q_2 are the effective ionic charges of anion and cation, respectively. The minimum in this potential energy occurs at a separation r_0, where the first derivative equals zero.

$$\left(\frac{dU}{dr}\right)_{r=r_0} = \frac{q_1 q_2}{r_0^2} - \frac{nb}{r_0^{n-1}} = 0 \tag{5}$$

from Eqs (4) and (5), the potential energy at r_0 can be approximated by

$$U = - \frac{q_1 q_2}{r_0} \left(1 - \frac{1}{n}\right) \tag{6}$$

For most of the ionic solids, n is about the same ($9 \sim 12$), so that U is approximately proportional to $-(q_1 q_2 / r_0)$. The similar conclusion may be obtained even if the Born-Mayer potential is used, because the contribution of the repulsive term is about 10% for almost all ionic compounds. The force constant for a simple harmonic oscillator at the equilibrium separation is the second derivative of U with respect to r and thus,

$$f = \left(\frac{d^2 U}{dr^2}\right)_{r=r_0} \quad \propto \ - \frac{q_1 q_2}{r_0^3} \tag{7}$$

When anharmonicity is neglected, the simple harmonic oscillator relationship between the force constant and frequency at the minimum of the potential well is

$$\omega_0 = 2\pi \left(\frac{f}{m}\right)^{1/2} \tag{8}$$

where m is the mass. Consequently,

$$\omega_0 \propto \left(\frac{q_2}{m r_0^3}\right)^{1/2} \tag{9}$$

This frequency ω_0 can be correlated with the characteristic temperature as well as with the far infrared peak position ν_E in the following.

By the definition of $\theta_E(K) = \frac{h}{k}\, \omega_0(sec^{-1}) = 1.44\nu(cm^{-1})$ and Eq. (8), $\theta_E \propto \left(\frac{q_2}{m r_0^3}\right)^{1/2}$ can be obtained, where h and k is the Planck and Boltzmann constant, respectively. In Fig. 5 the values of θ_E for various alkali ions listed in Table II are plotted as a function of

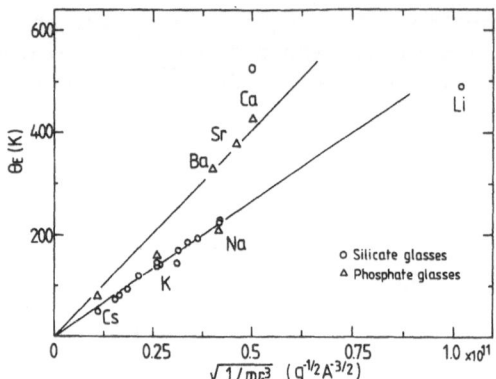

Fig. 5   Relationship between θ_E and $(1/mr_0^3)^{1/2}$ for alkali and
         alkaline earth ions.

$(\dfrac{1}{mr_0^3})^{1/2}$, the values of which for alkali-oxygen bonds were calculated from the sum of ionic radii of alkali and oxygen ions and the
atomic weight of alkali ions. A linear relationship between two
quantities can be clearly seen. The similar relationship holds for
the glasses containing alkaline earth oxides as the network modifiers.
As shown in Fig. 5, the values of θ_E for CaO, SrO and BaO fit on
another straight line, whose slope is about $\sqrt{2}$ times of that for
alkali oxides in Fig. 5, showing that the effective charge of alkaline earth ions is about 2. This is in accordance with the fact
that the effective charge, q_2, of these ions is about 2.

    In Fig. 6, the values of band maxima ν_E for various alkaline
metal-oxygen bonds in silicate glasses and phosphate glasses listed

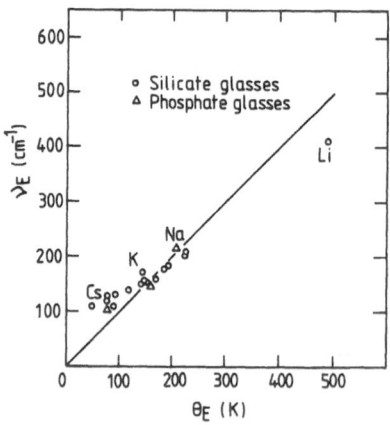

Fig. 6   Relationship between ν_E and θ_E.

in Table IV are plotted as a function of θ_E. A good fit on one straight line implies that the values of θ_E obtained for various kinds of alkali containing glasses by heat capacity measurements are in accordance with those of vibrational modes of alkali ions in far infrared absorption spectra. According to the definition of θ_E, the proportional constant for the relationship between θ_E and ω_0 should be 1.44, which is slightly larger than the experimental value of about 1. This discrepancy may arise from the simplified treatment of alkali-oxygen bonds by means of an Einstein function, but the real cause is not known at this time.

Table IV. Characteristic temperatures θ_E and far infrared band maxima in alkali silicate and phosphate glasses

Glass composition			θ_E(K)	ν_E(cm^{-1})	Glass composition			θ_E(K)	ν_E(cm^{-1})
Na$_2$O	-K$_2$O	-SiO$_2$			Li$_2$O	-SiO$_2$			
33.0	0	67.0	230	210	33.0	67.0		490	484
22.2	11.1	66.7	195	185	15.0	85.0		-	480
16.7	16.7	66.6	185	179					
11.1	22.2	66.7	170	160	Li$_2$O	-P$_2$O$_5$			
0	33.0	67.0	145	152	50.0	50.0		-	414
Na$_2$O-Cs$_2$O	-SiO$_2$				Na$_2$O	-K$_2$O	-P$_2$O$_5$		
33.0	0	66.7	230	210	50.0	0	50.0	210	215
20.0	13.3	66.7	145	170	25.0	25.0	50.0	-	179
16.7	16.7	66.6	140	155	0	50.0	50.0	163	147
6.3	27.0	66.7	80	130					
0	33.0	67.0	50	112	K$_2$O	-Cs$_2$O	-P$_2$O$_5$		
					50.0	0	50.0	163	147
K$_2$O	-Cs$_2$O	-SiO$_2$			25.0	25.0	50.0	-	130
33.0	0	67.0	145	152	0	50.0	50.0	80	102
23.3	10.0	66.7	120	138					
16.7	16.7	66.6	95	130					
10.0	23.3	66.7	75	120					
0	33.0	67.0	50	112					
Na$_2$O	-K$_2$O	-SiO$_2$							
14.0	0	86.0	230	208					
7.0	7.0	86.0	-	178					
0	15.0	85.0	150	150					
Na$_2$O-Cs$_2$O	-SiO$_2$								
14.0	0	86.0	230	208					
7.0	7.0	86.0	-	148					
0	15.0	85.0	90	108					
K$_2$O	-Cs$_2$O	-SiO$_2$							
15.0	0	85.0	150	150					
7.0	8.0	85.0	-	128					
0	15.0	85.0	90	108					

The comparison in θ_E between the glassy and the crystalline states shown in Table I and II indicates that θ_E for a glass is slightly lower than that for the crystal having the same composition. This means that the bond strength of alkali-oxygen bonds is weaker in glass structure than that in crystal structure, which may be attributable to the difference in average bond length between these two states.

BOND STRENGTH OF NETWORK FORMER

Since θ_1 is assigned to reflect the bond strength of network formers, the relative values of θ_1 for B-0, Si-0, Ge-0 and P-0 bonds may be discussed in a similar way as in the case of network modifiers, although an uncertainty in potential form for these oxide bonds is much larger than that for alkali metal-oxygen bonds. As the first approximation, it was assumed that the Born or Born-Mayer potential also holds for these oxides, and the value of $(q_1q_2/mr_0^3)^{1/2}$ for these bonds were calculated from the interatomic distance based on their ionic radii, the mean average mass of these ions and the charges of ions. The results are plotted in Fig. 7 against θ_1 for B_2O_3, SiO_2, P_2O_5 and GeO_2 glass. A linear relationship was observed like in the case of network modifying ions in Fig. 5. In order to compare the values of θ_1 with the infrared absorption bands, the dispersive nature of elastic continuum should be taken into con- sideration. According to the dispersion theory of crystal lattice, the zero-point energy of the nondispersive chain of Debye and that of the dispersive chain of Born and Karman are related to each other as follows:

$$\int_0^{\nu_{max}} \frac{h\nu}{2} \cdot \frac{3N}{\nu_{max}} \, d\nu = \int_0^{(\nu_{max})disp} \frac{h\nu}{2} \cdot \frac{6}{\pi} \cdot N \cdot \frac{d\nu}{\sqrt{\nu_{max}^2 - \nu^2}} \quad (10)$$

By integrating this equation, the characteristic temperature θ_1 equivalent to θ_{disp} for the Born and Karman chain can be obtained as follows.

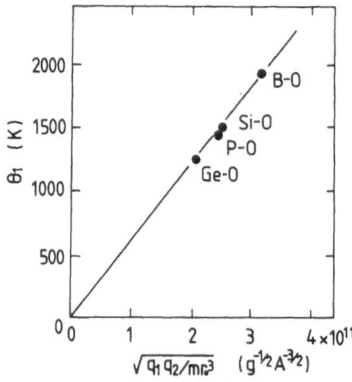

Fig. 7   Relationship between θ_1 and resonance frequency of cation- oxygen bonds, $(q_1q_2/mr_0^3)^{1/2}$.

$$\frac{3}{4} Nh\nu_{max} = \frac{3}{\pi} Nh(\nu_{max})_{disp} \tag{11}$$

and

$$\frac{(\theta_1)_{eq}}{(\theta)_{disp}} = \frac{\nu_{max}}{(\nu_{max})_{disp}} = \frac{4}{\pi} \tag{12}$$

The maximum frequency $(\nu_{max})_{disp}$ can be calculated from the longitudinal and transverse maximum frequencies using the following equation.

$$\frac{3}{(\nu_{max})_{disp}} = \frac{2}{\nu_{max}^{trans}} + \frac{1}{\nu_{max}^{longi}} \tag{13}$$

Thus, when the average frequencies of the absorption band associated with the stretching and bending vibrations are adopted as ν_{max}^{longi} and ν_{max}^{trans} , respectively, the characteristic temperature $(\theta_1)_{eq}$ can be determined from the infrared spectroscopic data as follows.

$$(\theta_1)_{eq} = \frac{4}{\pi} (\theta_1)_{disp} = \frac{4}{\pi} (1.44\nu_{max})$$

$$= 1.44 \cdot \frac{4}{\pi} \cdot \frac{3\nu_{max}^{trans} \cdot \nu_{max}^{longi}}{(2\nu_{max}^{longi} + \nu_{max}^{trans})} \tag{14}$$

The infrared spectra for SiO_2, GeO_2, B_2O_3 and P_2O_5 glasses have been studied widely [28]. According to their studies, the average of the maximum stretching frequency band was at 1098 cm^{-1}, 880 cm^{-1}, 1390 cm^{-1} and 1170 cm $^{-1}$ for Si-O, Ge-O, B-O and P-O bond, respectively and the maximum bending frequency band was at 780 cm^{-1}, 590 cm^{-1}, 740 cm^{-1}, 700 cm^{-1}, respectively, although these values may vary slightly because of the broadness of these bands. The values of $(\theta_1^{Si-O})_{eq}$, $(\theta_1^{Ge-O})_{eq}$, $(\theta_1^{B-O})_{eq}$ and $(\theta_1^{P-O})_{eq}$ calculated by Eq. (14) were listed in Table V, together with θ_1 determined from heat capacity measurements. Clearly a good agreement can be seen between them except for B_2O_3 glass. This result indicates that the characteristic temperature θ_1 calculated on the basis of the present three-band theory corresponds well to the value calculated spectroscopically. As for B_2O_3 glass, its Raman spectrum is strikingly different from the infrared spectrum and consists of broad bands in the regions 1250 $\sim$ 1550 cm^{-1}, and a strong sharp band at 807 cm^{-1} [29]. When this band is used as ν_{max}^{trans}, the value of $(\theta_1)_{eq}$ = 1720K is obtained and approaches to θ_1, but the difference between them remains unknown.

Table V. The values of θ_1 determined spectroscopically and from
the heat capacity data for SiO_2, GeO_2, B_2O_3 and P_2O_5
glasses.

glass	θ_1	$(\theta_1)_{eq.}$	$\nu_{max}^{longi.}$	$\nu_{max}^{trans.}$
SiO_2	1580	1584	1098	780
GeO_2	1250	1216	880	590
B_2O_3	1920	1608	1390	740
B_2O_3	1920	1767	1550*	807*
P_2O_5	1500	1483	1170	700

* Taken from Raman spectra in Ref.29

EFFECT OF NETWORK MODIFIER ON BOND STRENGTH OF NETWORK FORMER

In Fig. 8, the change in θ_1 with Na_2O content for various
oxide glasses was shown as a function of the atomic ratio of
(Network Modifying Cation/Network Forming Cation). In the case of
silicate glasses, θ_1 decreases sharply with increasing (Na/Si)
ratio when (Na/Si) ratio is small, but it becomes almost constant
at high (Na/Si) ratio up to 2. This initial decrease in θ_1 was
attributed to the appearance of nonbridging oxygens in glass
structure, or the breakdown of continuous three-dimensional SiO_2
network, which causes the change in effective charge of oxygen
ions and/or interatomic distance in Si-O chains. The effect of
this change on θ_1 becomes less as the number of nonbridging oxygens
become more and may approach to a certain value for the case of
(Na/Si)=2, where only two oxygens of SiO_4 tetrahedra are bridging
oxygens in average.

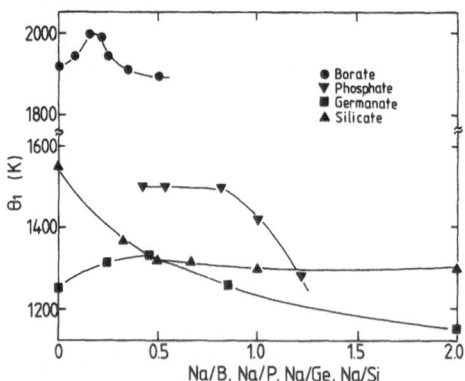

Fig. 8 Change in θ_1 with Na_2O content for binary silicate, germanate,
phosphate and borate glasses.

This result is consistent with the behavior of infrared spectra. As shown in Fig. 9 after Ferraro and Manghanani [30], an increase of the alkali oxide content results in a decrease monotonically in the peak position of the Si-O stretching bonds within the tetrahedra. In all cases, the frequency decreases with an increase of Na_2O, exception in the case of the bending vibration frequency ($\sim$ 460 cm^{-1}), which reflects little increase. This weakening effect of alkali ions on Si-O bonds depends on the kind of alkali ions. When Na is substituted by K or Cs in the composition of $M_2O \cdot 2SiO_2$ glasses, ν_{max} of Si-O stretching vibration decreases in the order of Na, K and Cs, as described in Section IV. This observation is consistent with the fact that the characteristic temperature θ_1 decreases in the order Na>K>Cs for the glasses with the same alkali content.

The effect of Na_2O content on θ_1 for oxide glasses other than silicate glasses is complicated. The increase in θ_1 for borate and germanate glasses is attributable to the change in coordination number. The amount of Na_2O giving the maximum value of θ_1 corresponds reasonably well with that for other physical properties.

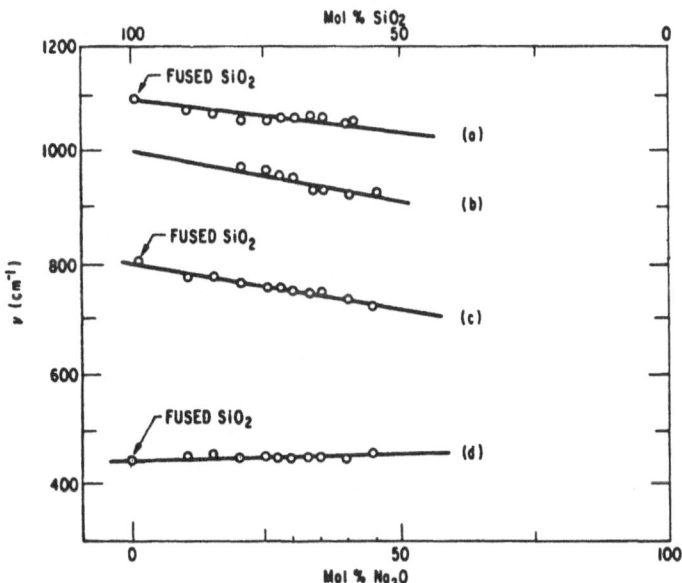

Fig. 9 Band frequencies versus amount of Na_2O content in sodium silicate glasses. (a) Si-O-Si stretching mode, (b) terminal Si-O stretching mode, (c) O-Si-O bending mode, (d) Si-O-Si bending mode, from Ref. [30].

As for phosphate glasses [31], not all oxygens take the bridging
positions but some take the double bonded states. When Na_2O is
added to phosphate glass, these double bonded oxygens remain unchanged
but the bridging oxygens change into nonbridging oxygens. Up to the
metaphosphate composition, or (Na/P)=1, continuous P-O chains can be
visualized, but the breakdown of continuous one-dimensional P-O
chains takes place when (Na/P) ratio exceeds 1. This can be seen
in Fig. 8.   Up to near (Na/P)=1, θ_1 remains about the same as in
the case of silicate glasses, but θ_1 decreases sharply beyond
(Na/P)=1.

The effects of mixed alkali ions on characteristic temperatures,
θ_1 and θ_E are shown in Fig. 10. The values of θ_E for mixed ions
fall on a line drawn between the values of two end members, indica-
ting that the bond strength of mixed M-O bonds is the average of
two kinds of M-O bonds. On the other hand, θ_1 shows a positive
deviation from the linear additivity of two end members, indicating
that the bond strength of Si-O bonds becomes stronger by mixing two
kinds of alkali ions.   The maximum deviation of about 7% is observed
for Na-Cs glasses.

As shown in Section IV, the band maxima of far infrared
absorption of mixed alkali glasses show little negative deviation
from the linear additivity of two end members, indicating that the
Si-O chains are tightened slightly by mixing two kinds of alkali
ions. This behavior is consistent with the change in θ_1.

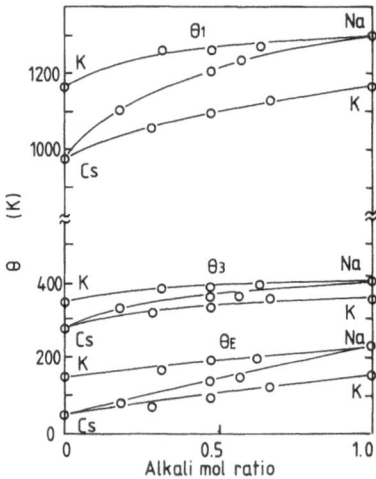

Fig. 10   Characteristic temperatures for mixed alkali disilicate
          glasses.

DEGREE OF DISORDER FOR GLASSY STATE

According to the theory of lattice dynamics, the maximum vibrational frequency of a discrete lattice depends on the size of vibrational units [32]. Since the characteristic temperature θ_3 is proportional to the maximum vibrational frequency of the three-dimensional continuum, a low value of θ_3 means that the average size of the vibrational units is large. Thus, the comparison of the values of θ_3 for various glasses and crystals gives an information about the size of repeated assemblies of atoms or ions. As shown in Table III, θ_3 for Na_2O-SiO_2 glass is close to that for Na_2SiO_3 crystal. Thus, it is considered that the average size of the repeated units is similar for both states. In other words, the medium range structure of $Na_2O \cdot SiO_2$ glass is similar to that of Na_2SiO_3 crystal. On the other hand, the value of θ_3 for fused silica is much smaller than that of any crystalline SiO_2, indicating that fused silica has a medium range structure different from SiO_2 crystal. Qualitatively speaking, these results suggest that fused silica has a more disordered or random structure than $Na_2O \cdot SiO_2$ glass. In order to discuss this medium range structure more quantitatively, the estimation of the size of repeated units is necessary. This may be done if the relationship between θ_1 and θ_3 becomes known. At this moment, an explicit relationship is difficult to derive, and so some reasonable assumptions have to be incorporated. In the previous study, the following simple assumption was applied; the minimum wavelength of vibrational modes responsible for θ_3 is comparable to the average size of the repeated units and it is multiple of the minimum wavelength for θ_1, which is related with the interatomic distance. This assumption may be easily visualized from the model shown in Fig. 1(c). The choice of the interatomic distance affects the size. As described in the previous section, θ_1 decreases with increasing amount of network formers, resulting probably from the changes in effective charge of oxygen ions and interatomic distance. If the former effect is dominant, the Si-O bond length remains almost constant and the bond length for fused silica may be used to estimate the size of repeated units. On the other hand, if the latter effect is dominant, the change in bond length has to be estimated. This may be done by taking a linear relationship between θ_1 and $(q_1 q_2 / m r_0^3)^{1/2}$, which gives about 10% change in bond length when θ_1 varies from 1550K to 1300K. In Fig. 11, the estimated size of repeated units for glasses and crystals in Na_2O-SiO_2 system is shown as a function of Na_2O content. A constant bond length of 1.6 Å, which is the average of Si-O bond length for various silicate compounds, was used. The size may increase about 10% when the change in bond length is assumed to take place. The estimated size of repeated units of about 17 Å for fused silica is about twice that of SiO_2 crystals and is comparable to the size of characteristic order of 15-30 Å indicated by Phillips [33] for covalent noncrystalline solids. The interaction of alkali ions

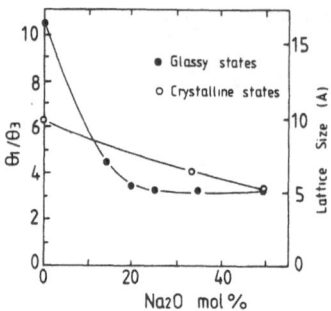

Fig. 11   The estimated lattice size of Na_2O-SiO_2 composition.

to fused silica decreases the size to about 5 Å up to the amount
of 20 mol% Na_2O, beyond which it remains almost constant. This
size is comparable to the size found from θ_1/θ_3 for Na_2SiO_3 or
$Na_2Si_2O_5$ crystal. It is also close to the minimum crystallographic
cell dimension of silicates as shown in Table III, as well as the
mean free path of phonons for silicate glasses [34]. Such dependence of the size of repeated units on the amount of network
modifiers indicates that the bridging angle of Si-O-Si bonds
varies widely for fused silica but becomes narrow with addition
of network modifiers and approaches to that of the corresponding
crystals. This tendency is in accordance with the fact that the
difference in density between the glassy and crystalline states
of SiO_2 is quite large but it becomes small with addition of network modifiers. This result shows that a large number of atoms
should be used as the system size for the molecular dynamic simulations for fused silica than Na_2O-SiO_2 glasses. However, the
development of computer will enable us to construct the glass
model containing a large number of atoms by using the interatomic
potential. In that case, the results obtained here certainly are
useful in overcoming some limitations of this simulation.

REFERENCES

1.   W. Huckel, J. Prakt. Chem. 5:105 (1957).
2.   N. Soga, J. de Physique 43:C9-557 (1982).
3.   K. Hirao and N. Soga, Rev. Sci. Instr., 54:1538 (1983).
4.   A. Einstein, Ann. de. Phys. 22:180 (1907).
5.   P. J. W. Debye, Ann. Phys. (Leipzig), 39:789 (1912).
6.   V. V. Tarasov, Zh. Fiz. Khim., 24:111 (1950).
7.   E. F. Westrum, Proc. IV Congres du Verre, Paris, 396 (1956).
8.   G. K. White, and J. A. Birch, Phys. Chem. Glasses, 6:85 (1965).
9.   K. Hirao, N. Soga, and M. Kunugi, J. Am. Ceram. Soc. 62:570
        (1979).
10. K. Clusius, J. Goldman, and A. Perlick, Z. Naturforsch, 4a:117
        (1949).
11. S. F. Giauque, and R. C. Archibald, J. Am. Chem. Soc. 59:561
        (1937).

12. W. Desorbo, J. Chem. Phys. 21:1144 (1953).
13. C. T. Anderson, J. Am. Chem. Soc., 52:2296 (1930).
14. V. V. Tarasov, Zh. Fiz. Khim., 29:198 (1955).
15. W. Desorbo, and W. W. Tyler, J. Chem. Phys. 21:42 (1958).
16. K. Hirao, and N. Soga, J. Ceram. Soc. Japan, 90:390 (1982).
17. K. Hirao, and N. Soga, J. Ceram. Soc. Japan, 90:476 (1982).
18. C. T. Anderson, J. Am. Chem. Soc., 58:568 (1963).
19. K. K. Kelley, J. Am. Chem. Soc., 61:471 (1939).
20. R. P. Beyer, J. Chem. Engineering Data, 24:171 (1979).
21. K. K. Kelley, J. A., Chem. Soc., 65:339 (1943).
22. S. Cristensen, Z. Physik Chem., B25:273 (1934).
23. E. G. King, J. Am. Chem. Soc., 80:1799 (1958).
24. P. A. Soboleva, Rus. J. Phys. Chem., 44:1667 (1970).
25. G. Gremier, J. Am. Chem. Soc., 78:6226 (1956).
26. B. N. Nelson, and G. J. Exarhos, J. Chem. Phys. 71:2739 (1979).
27. M. Born, and K. Huang, Dynamic Theory of Crystal Lattice, (Oxford Univ., 1954) 24.
28. J. Wong, and C. A. Angell, Glass Structure by Spectroscopy, (Dekker, New York, 1976) 429.
29. M. C. Tobin, and T. Baak, J. Opt. Soc. Am., 60:368 (1970).
30. J. R. Ferraro, and M. H. Manghnani, J. Appl. Phys. 43:4595 (1972).
31. N. Soga, K. Hirao, M. Matsuno, and R. Ota, submitted to Soviet J. Glass Phys. Chem.
32. O. L. Anderson, J. Phys. Chem. Solids, 12:41 (1960).
33. J. C. Phillips, J. Non-Cryst. Solids, 43:37 (1981).
34. W. D. Kingery, H. K. Bowen, and D. R. Uhlmann, Introduction to Ceramics, (John Wiley, New York, 1976) 627.

SEMIEMPIRICAL MOLECULAR ORBITAL STUDIES OF INTRINSIC DEFECTS IN a-SiO$_2$

Arthur H. Edwards

U.S. Army ET&D Laboratory, Electronic Materials Div.
Device Physics and Analysis Branch
Fort. Monmouth, NJ 07703

and

W. Beall Fowler

Dept. of Physics and Sherman Fairchild Laboratory
Lehigh University
Bethlehem, NJ 18015

ABSTRACT

The intrinsic point defects in a-SiO$_2$ have recently received considerable attention. Careful use of paramagnetic resonance, coupled with annealing and optical studies, has led to unambiguous identification of three fundamental defects. These are the E' center, the superoxide radical and the nonbridging oxygen hole center (NBOHC). Theoretical studies of the first two defects have led to greater understanding of experiment through inclusion of atomic relaxation. Two models exist for the NBOHC. One, by Skuja and Silin, invokes a Jahn-Teller splitting to explain 2 eV optical transitions. The other, by Griscom, is an extension of a model devised for alkali silicate glasses and involves pairs of oxygens, one of which is adjacent to a proton. Using MOPN, a semiempirical spin-unrestricted molecular orbital program, we have done molecular orbital studies of both NBOHC models. Our results support the Griscom model and not the Skuja-Silin model. These results, coupled with our earlier calculations on the E' and superoxide defects, allow us to address defect formation and transformation processes in a logical way. In particular, these results are consistent with our speculations on the sequential creation of NBOHC, superoxide precursor, and superoxide radical by hole trapping.

I. INTRODUCTION

Over the past twenty years a great deal of research has been carried out on various aspects of amorphous silicon dioxide. While there had been interest in this material as a model for other glass systems, the advent of MOS technology and of fiber optics communication considerably heightened the activity in this field. Defects in the glass structure have been studied with great intensity. At first blush, a discussion of defects in an amorphous material may seem to be a semantic rather than a scientific exercise. However, the existence of localized charge traps in MOS oxides [1] as well as in bulk fused silica [2-4] blunts this objection.

In this paper we present the theory of a fairly narrow class of defects. We consider only those that, in at least one charge state, give rise to a paramagnetic resonance signal, or that are associated with an optical absorption band distinct from those seen in perfect crystalline quartz. At this point we should elucidate some subtleties of this restriction. First, it is well known that all forms of amorphous SiO_2 (fused quartz, thermal and deposited MOS oxides) contain substantial amounts of water [5]. Only when water is responsible for either charge trapping or a spin signal do we consider it part of a defect. That is to say, OH groups incorporated into the glass network are considered an intrinsic part of the system. Second, there are features of the vibrational spectra of $a\text{-}SiO_2$ that are not observed in crystalline quartz [6-8]. As some of the added features are apparently due to local network topology (i.e., the presence of three- and four-membered rings [6,7]), rather than point defects [8], we do not appeal to these in the discussion of our results. Finally, we restrict our attention to intrinsic defects, namely, those arising from the intrinsic constituents of the glass. This excludes discussion of impurity atoms other than hydrogen.

In this chapter, then, we present the results of recent calculations of the electronic structure and conformation of several intrinsic defects in amorphous silicon dioxide. These calculations were performed for several purposes. First, and most importantly, they are used to test microscopic models generated predominantly from electronic spin resonance [9], but from optical spectroscopy as well [10]. Second, they are used to gain insight into the defect charge states that are not spin active, or as yet are not associated with an optical absorption band. Third, they are used to explore models for defect formation and transformation. The results of these explorations are used to formulate further experiments.

The chapter is organized in the following manner: in section II we present an overview of the essential experimental results.

In section III we discuss the theoretical technique only briefly, as this has been dealt with at length elsewhere [11-13]. In section IV we present our results for the various defects, as well as our models for defect transformation and annealing. We conclude in section V.

II. EXPERIMENTAL BACKGROUND

As mentioned above, the most useful data on point defects in SiO_2 have been obtained through electron spin resonance studies. Optimally, this technique yields detailed microscopic information about the defect wave function, including the local symmetry of the defect, spin densities on the various chemical species, and orbital decomposition of the spin density on each atom (i.e., fractional s and p character). While this technique is quite naturally adapted for use in crystals, it can be a similarly powerful technique in glasses, provided that appropriate angular averaging is performed [14]. We should mention that there are crystalline analogues for some of the most important glass defects, so that wherever possible we will draw from these results as well.

In Fig. 1 we show models of the three spin active intrinsic defects in a-SiO_2 that have been observed to date. These models are more elaborate than the spin resonance data can unambiguously determine. We have included the added detail, to be elucidated below, on the basis of indirect experimental evidence, of results obtained in crystalline SiO_2, and of the results of our calculations.

The E_1' center, Fig. 1a, is the most thoroughly studied of the three intrinsic defects [2,3,15-21]. It was first observed over twenty years ago by Weeks and Nelson [2] in neutron irradiated, crystalline quartz. Silsbee [17] analyzed the ^{29}Si hyperfine and determined that the spin spends 80% of its time on a single silicon atom in an $sp^{2.7}$ orbital. This result was at first puzzling. It was assumed that the primary damage created by neutron irradiation was oxygen vacancies. If an oxygen vacancy were to trap an electron or a hole, one would expect that the spin would be almost evenly shared by both of the nearest neighbor silicon atoms. This puzzle was resolved when Feigl, Fowler, and Yip [15,16] developed a model wherein, upon trapping a hole, an oxygen vacancy would undergo an asymmetric, pseudo Jahn-Teller [22,23] relaxation. As illustrated in Fig. 1a, the silicon atom on which the unpaired electron resides moves slightly into the vacancy, while the other silicon atom relaxes back into the plane of its three back-bonded oxygen atoms. Yip and Fowler [24] performed some detailed LCDAO calculations that strongly support this model. Recently, Edwards and Fowler [13] and Rudra and Fowler [24] have repeated these calculations using MOPN [13,25] and have obtained very good agreement with the Yip-Fowler results, and with experiment [26].

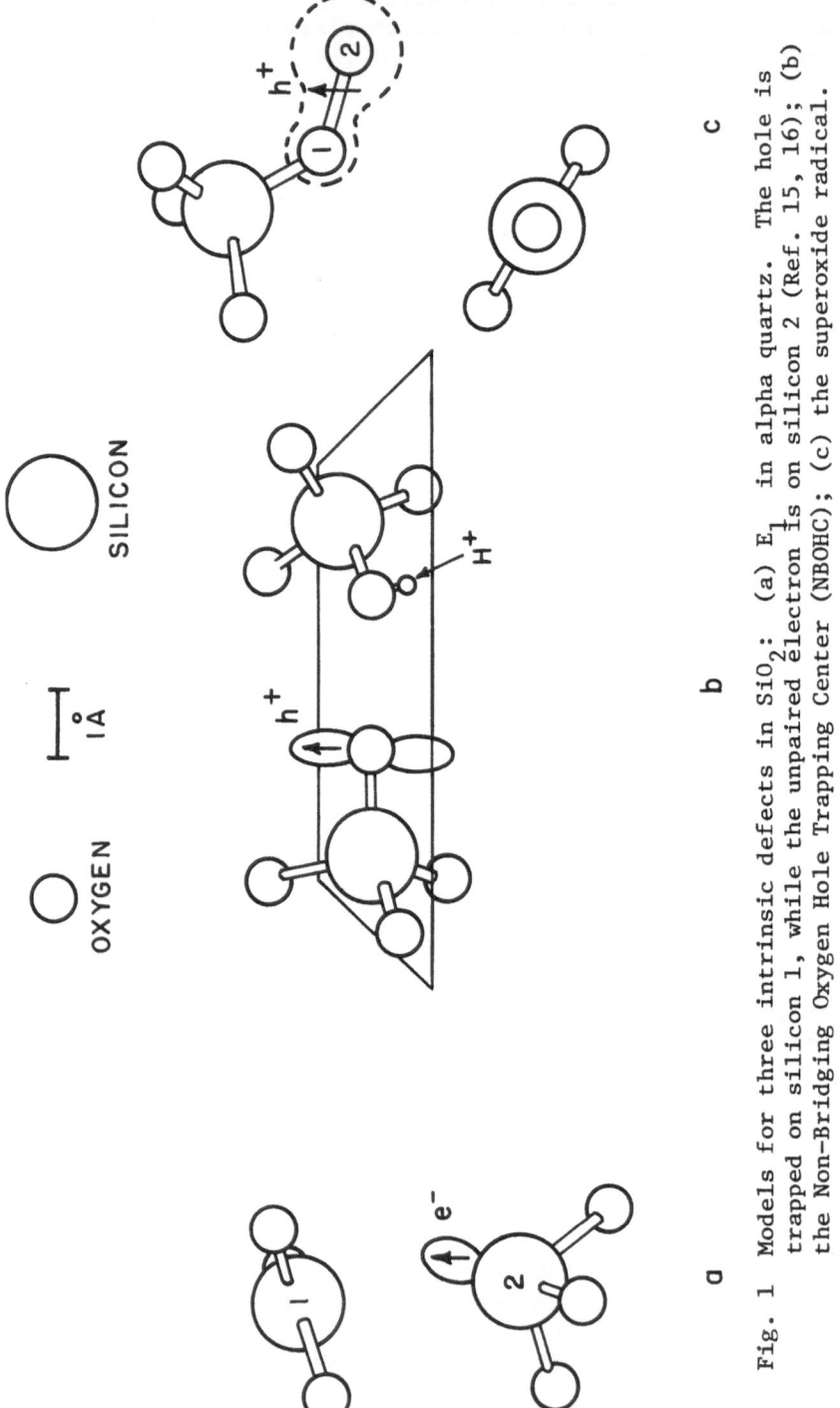

Fig. 1 Models for three intrinsic defects in SiO_2: (a) E_1' in alpha quartz. The hole is trapped on silicon 1, while the unpaired electron is on silicon 2 (Ref. 15, 16); (b) the Non-Bridging Oxygen Hole Trapping Center (NBOHC); (c) the superoxide radical.

Griscom and çoworkers have developed a remarkably detailed picture of the E_1' center in a-SiO$_2$ [3]. Isotopic enrichment (both ^{17}O and ^{29}Si) combined with extensive computer simulation yielded the following picture:

1. As in the crystalline case, the E_1' spin resonance signal arises from an unpaired electron in a dangling sp^3 orbital primarily on a single silicon atom. The strong hyperfine lines (a 420 G doublet) were unambiguously demonstrated to arise from interaction with a ^{29}Si nucleus, the one on which the spin is localized [27], rather than from interaction with a proton, as had been previously proposed [28].

2. In contrast to the crystalline case, the weak hyperfine structure of the glass E_1' center does not arise from interaction with other ^{29}Si nuclei, but from interaction with protons.

3. While the hyperfine and g tensors established only the lower half of Fig. 1a, the description of the E' center as an oxygen vacancy is strongly supported by the similarity of the g tensors in both the crystalline and the amorphous material, combined with the calculations of Gobsch et al. [29]. Hence we include the added structure of the relaxed silicon in the upper half of the figure.

The remaining defects in Fig. 1 are the two oxygen associated hole centers (OHC's) [4,30,31]. Fig. 1b is the Non-Bridging Oxygen Hole trapping Center (NBOHC). This is the dominant OHC in wet silica (~ 1200 ppm OH). The spin resonance data, including both ^{17}O and ^{29}Si isotopic enrichment studies, indicate that the spin is 100% localized in an oxygen lone pair 2p state, as indicated on the left side of Fig. 1b. The other half of this defect is postulated on the basis of thermodynamic arguments favoring pairing of O-H groups in glasses, and on the splitting of the two lone pair states on the spin active oxygen atom, obtained from the observed g tensor (this will be discussed at length below). We note that there is a competing model for the NBOHC in which the nonbridging oxygen atom faces a large void [10]. In this model, Skuja and Silin envision a large dynamic Jahn-Teller distortion in which weak bonding occurs between the NBO and the other three back-bonded oxygen atoms. We have performed calculations on both of these models and will discuss the results in section IV.

In Fig. 1c, we show the OHC that predominates in dry silica (<10 ppm OH), the superoxide (or peroxy) radical defect. The ESR spectrum of this defect is very similar to that of O$_2$, although, as the figure indicates, the spin is unequally shared between the two oxygen atoms. ^{17}O and ^{29}Si hyperfine data indicate that the hole spends 24% of its time on the inner oxygen atom (closest to the silicon) and 76% of its time on the outer oxygen atom [30,31].

Further, the ^{29}Si data indicates that the O_2 fragment is attached
to a single silicon atom [30]. As for the NBOHC, there is no
compelling ESR data that dictates the structure that the O_2-
fragment faces. However, we will give arguments below that the
bottom half of Fig. 1c is quite reasonable.

III. THEORETICAL TECHNIQUE

    The calculations discussed in this chapter were all performed
using MOPN [12]. This is an unrestricted Hartre-Fock version of
MINDO/3 [11], a semiempirical molecular orbital program written
by M. J. S. Dewar and coworkers. Central to this method is the
Neglect of Differential Overlap (NDO) approximation,

$$\phi_{m_A}(r) \; \phi_{n_B}(r) = 0, \; m_A \neq n_B \tag{1}$$

where A and B denote different atoms, and m_A and n_B denote sets of
atomic quantum numbers. This constraint eliminates all three and
four center integrals while greatly reducing the number of two
center integrals. If strictly enforced, the NDO approximation
yields no term splitting. A slightly less restrictive approximation, Intermediate NDO (INDO), retains all one-center, two
electron integrals, thus ameliorating this deficiency. In Modified
INDO (MINDO)/3, the surviving one and two center integrals are
evaluated either by using semiempirical formulae, or by fitting
to experimental heats of formation, bond lengths, bond angles,
dipole moments and ionization potentials. We note that the
unrestricted Hartree-Fock formalism automatically yields values
for spin density in various atomic orbitals, thus facilitating
comparison with hyperfine tensors derived from spin resonance
studies. We have successfully applied this technique to bulk
SiO_2 and to various defects observed in both amorphous SiO_2 and
crystalline quartz [13,25].

    We applied MINDO/3 to finite clusters of atoms whose geometries
match proposed models for defects in amorphous SiO_2. These
clusters were terminated with hydrogen atoms to lower the energy
of the surface states, and hence prevent admixture with the defect
states. Justification of modeling an infinite solid with a cluster,
and of this termination scheme warrants some discussion.

    It is well known that many features of the electronic
structure of the various crystalline polymorphs and of amorphous
SiO_2 are strikingly similar, despite large variations in long
range order. These features include the SiK_β and $L_{II,III}$ X-ray
emission spectra [32], the intrinsic band gap, and the density of
states in the valence bands [33]. The obvious implication is that
local order, on the scale of the SiO_4 tetrahedron, is preserved,

and that any model of the glass system that includes this local
order stands a good chance of reproducing these experimental
results. This has in fact been the case. There is a modest
literature of cluster calculations [13,34-38] wherein the simple
SiO_4 tetrahedron is successfully used to explain the optical
experiments discussed above. Furthermore, recent calculations on
the SiO_4 tetrahedron as well as on the Si_2OH_6 and the $Si_2O_7H_6$
clusters wherein the geometry was optimized with respect to total
energy have reproduced the Si-O bond length surprisingly well
(1.61 Å experiment vs. 1.62-1.67 Å calculation [13,39]). In summary,
our experience with hydrogen-terminated atomic clusters gives us
confidence that this is a decent representation of the solid. For
defect calculations we feel that the justification for cluster
termination is even stronger than for representing the perfect
infinite solid because, as spin resonance has shown, the defect
wave functions are strongly localized.

IV. CALCULATIONS

     In this section, we will discuss calculations for a hierarchy
of defects and defect precursors, shown schematically in Fig. 2.
In these figures, the largest spheres represent silicon atoms,
the smaller spheres represent oxygen atoms, and the smallest
spheres represent hydrogen atoms. These are, in fact, the clusters
we used for most of our calculations. Not shown are the terminating
hydrogen atoms. In Figs. 2b and 2d we show the NBOHC and the superoxide radical, respectively, while in 2a and 2c we show their proposed
precursors. Before we discuss our results for each individual defect,
we should mention a few features common to all the calculations.

     The position of the six outer oxygen atoms were chosen with
consideration of our previous results [25] for the superoxide
radical [31]. In studying that defect, we performed a series of
calculations in which the two sets of three outer oxygen atoms were
fixed at increasingly large separations. In each calculation,
the four internal atoms (two silicon and two oxygen) were allowed
to search for the minimum in total energy. The purpose was to
find the minimum cage size for which the following criteria hold:

     1. In the neutral charge state, a two atom superoxide bridge
is favored over a single oxygen atom.

     2. In the positive charge state, the superoxide bridge breaks,
and the two oxygen atoms attach to a single silicon, as shown in
Fig. 2d.

     We used the cage geometry described above for most calculations.
The only exception was the Skuja-Silin NBOHC, for which the right
hand side of Fig. 2b is absent. We should also point out that
we performed calculations on both the Skuja-Silin and Griscom

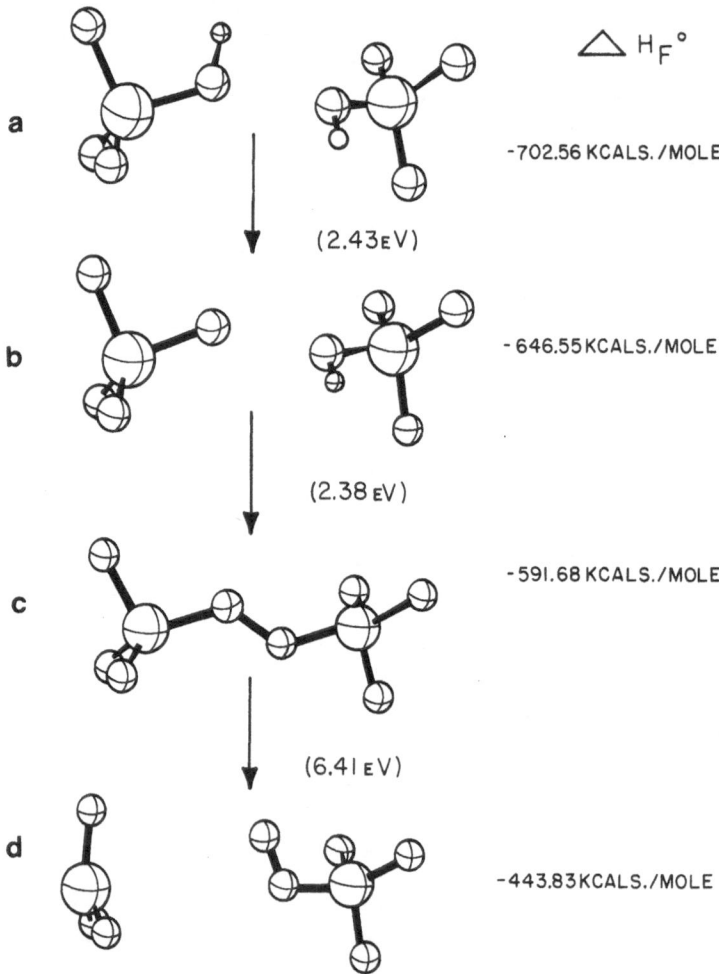

$\triangle H_F^\circ$

a   -702.56 KCALS./MOLE

(2.43EV)

b   -646.55 KCALS./MOLE

(2.38 EV)

c   -591.68 KCALS./MOLE

(6.41 EV)

d   -443.83 KCALS./MOLE

Fig. 2   Hierarchy of defects considered in this chapter: (a) The
          presumed NBOHC precursor; (b) the Griscom model for the
          NBOHC (note: the Skuja–Silin model excludes the right hand
          side of the figure); (c) the superoxide bridge, the presumed
          superoxide radical precursor; (d) the superoxide radical.
          The terminating hydrogen atoms are not shown. Computed heats
          of formation ΔH_f° are shown for each, and in parentheses
          the differences in heats of formation from one defect to the
          next.

models using extended clusters. In these clusters, the terminating
hydrogen atoms were replaced by $Si(OH)_3$ groups. In all of the
small cluster calculations, all atoms inside the oxygen cage were
given complete freedom to search for a minimum in total energy.
In the extended cluster calculations, we froze the outer shell of
silicon atoms and their terminating O-H groups, allowing the six
oxygens that comprise the small cage to relax. As will be shown
below, this added freedom had relatively little effect on our results.

A.   Skuja-Silin NBOHC

        The simplest model that might account for Griscom's EPR
data for a nonbridging oxygen is the left hand side of Fig. 2b.
This model was indirectly offered by Skuja and Silin [10]. In
their paper, they were concerned with optical properties, not with
the spin resonance data; however, the charge state they considered
was in fact the neutral, spin active state. As noted in section
II, they envisioned a dynamic Jahn-Teller effect: specifically,
the nonbridging oxygen atom would form a weak bond with one of
the other three oxygen atoms that surround the nearest neighbor
silicon atom. In the ground vibronic state, the unpaired electron
would reside on these oxygen ligand atoms. This prediction,
however, is not consistent with Griscom's spin resonance data.

        We felt that it would be instructive to pursue this model
for two reasons. First, while the dynamic Jahn-Teller model could
not be reconciled with EPR, there was no a priori reason to
discount the possibility that some fraction of NBOHC's are well
represented by this model. Second, even if we could show that
there were no circumstances under which we would expect this large
Jahn-Teller distortion, the nonbridging oxygen may still face a
void. We were then still hopeful that we might be able to
give some theoretical basis- either spin densities, or orbital
splittings- for distinguishing between this model and Griscom's
model, Fig. 2b.

        The geometrical results of our calculations for both large
and small clusters are shown in Table 1. Note that there is no
bonding with the three ligand oxygen atoms in either calculation.
We should point out that we made special effort to encourage this
bonding. In all initial geoemtries, the nonbridging oxygen was
placed close to one of the ligand oxygen atoms. Further, in the
large cluster calculations, special care was taken to ensure that
the lone pair orbital on the ligand pointed toward the nonbridging oxygen, thus maximizing the orbital overlap. Because the
lone pair direction is perpendicular to the Si-O-Si plane, we
can effect the desired hybridization by manipulating the outer
silicon atoms. In all calculations, the predominant distortion
was found to be a breathing motion between the nonbridging oxygen
and its nearest neighbor silicon atom. We note that there was a

Table I.   Computed geometrical parameters for the equilibrium
           configuration of the isolated NBOHC. All distances are
           in angstroms, and all angles are in degrees. O* denotes
           the nonbridging oxygen atom, while O denotes the three
           back-bonded oxygen atoms. Note that the Si-O-Si angle is
           very close to the value 144° found in alpha quartz.

	Small cluster	Large cluster
R_{Si-O*}	1.675	1.678
R_{Si-O}	1.65	1.65
R_{O*-O}	2.67 (2)	2.743 (2)
	2.65 (1)	2.741 (1)
∢ O*-Si-O	106.93 (2)	110.5 (2)
	105.75 (1)	110.1 (1)

$$144.9° < ∢\ Si\text{-}O\text{-}Si < 145.6°$$

small Jahn-Teller distortion that lowered the symmetry from C_3 to
C_s, as dictated by the Jahn-Teller theorem. This distortion cannot,
however, be construed to be oxygen-oxygen bonding, as the minimum
distance between oxygen atoms is 2.73 Å in the large cluster,
and 2.65 Å in the small cluster. We feel that these results
clearly rule out the Skuja-Silin dynamic Jahn-Teller model.

     We now turn to the problem of reconciling these calculations
with Griscom's ESR results. As discussed in section II, the ^{17}O
hyperfine data indicate that the spin is completely localized on
a single oxygen atom, in a lone pair 2p orbital. Our calculations
agree quite well with this picture. For both clusters, more than
99 percent of the spin density is localized in the 2p states.
Further, analysis of the molecular orbital associated with the spin
wave function indicates that it is indeed a lone pair state, i.e.,
that the lobe is perpendicular to the Si-O bond (90.07° for the
large cluster and 89.76° for the small cluster). These results
alone are entirely consistent with the spin resonance data, and
hence the isolated nonbridging oxygen is still a viable model
for the NBOHC.

B. Griscom NBOHC

We now turn to the more elaborate model for the nonbridging
oxygen atom, wherein it faces an OH group. As discussed above, we
used the same oxygen cage as we did for the superoxide radical.
Holding the six outer oxygen atoms fixed, we allowed the five
internal atoms to relax to the minimum energy configuration. The
geometrical results of this calculation are summarized in Table II.

Table II. Geometrical parameters for the equilibrium configuration
           of the Griscom NBOHC. All distances are in angstroms,
           and all angles are in degrees. O* denotes the non-
           bridging oxygen atom, while O(B1) etc. denote the three
           back-bonded oxygen atoms. O denotes the oxygen associa-
           ted with the central OH group.

	Small cluster	Large cluster
R_{Si-O*}	1.645	1.643
$R_{Si-O(B)}$	1.62	1.62
R_{O*-O}	2.21	2.14
R_{O-H}	0.95	0.95
$R_{O*-O(B1)}$	2.75	2.80
$R_{O*-O(B2)}$	2.74	2.83
$R_{O*-O(B3)}$	2.41	2.43
⋟ O*-Si-O(B1)	114.5	115.9
⋟ O*-Si-O(B2)	113.7	118.2
⋟ O*-Si-O(B3)	95.5	95.5

$$143.0° < \text{⋟ Si-O-Si} < 146.1°$$

Several features bear comment. First, the distance between
the nonbridging oxygen and its nearest neighbor silicon is slightly
smaller here (1.64 Å) than in the isolated NBO model (1.67 Å).
Second, the SiO₄ tetrahedron is more distorted. This is initially
surprising, because the Jahn-Teller theorem does not apply. However,

the distortion is a size effect: it turns out that the nonbridging
oxygen is repelled from the opposing OH group, rather than being
attracted to the back-bonded oxygen atom. The last noteworthy
geometrical feature is the distance between the hydrogen atom
associated with the central OH group and the nonbridging oxygen
atom. This distance, 2.4 Å, compares favorably with Griscom's
prediction [4] of the minimum distance, 2.5 Å, based on the absence
of proton hyperfine splitting in the ESR spectrum. The hydrogen
atom is also positioned such that, given the OH bond length, the
distance between it and the NBO is maximized.

The electronic structure of the Griscom NBOHC is similar to
that of the isolated NBO model. That is to say, the spin is again
completely localized on the NBO, and the spin wave function has
lone pair 2p character. Because of the presence of the other side
of this defect (i.e., the SiO_4H_4 tetrahedron), we must be more
careful in defining the term "lone pair". Unlike the isolated NBO
case, where there were two almost degenerate lone pair states
(both perpendicular to the Si–O bond), here the only lone pair
state is perpendicular to the Si–O*–O plane, where O* denotes the
oxygen on which the hole resides. It is interesting to note that
this true lone pair orbital is not the spin wave function. The
hole actually resides in the slightly anti-bonding state that is
in the Si–O*–O plane. The anti-bonding interaction is between O*
and the nearest of the three back-bonded oxygen atoms. Note that
this is somewhat different from Fig. 1, where the O*–O–H plane
determined the character of the lone pair state. Our results
indicate that the hydrogen atom has very little influence on the
nature of the hole state. This is consistent with the absence of
hydrogen admixture in the spin wave function, implicit in the lack
of ^{1}H hyperfine interaction.

As noted above, we have also performed calculations on this
center using an extended cluster. The character of the defect is
preserved in great detail, as illustrated in Table II.

C. Negatively Charged Non-Bridging Oxygen Atoms

Thus far, we have given no calculated parameters that would
delineate between the two proposed models for the NBOHC. Experimentally, it is known from the g tensor that this defect is
not axially symmetric. In a simple one electron picture, this
lack of axial symmetry results in a splitting of the lone pair
states. The magnitude of this splitting, derived from Griscom's
spin resonance data, is 0.5 eV [41]. If we could compare this to the
splittings obtained in our calculations, we could perhaps favor
one of the two models. However, in an unrestricted Hartree-Fock
formalism this comparison is not straightforward, as the splitting
is strongly dependent on occupation. Because one of the lone pair
states is singly occupied, and the other is doubly occupied, the

splitting is much larger than one-electron theory would predict.
To overcome this difficulty, we considered the negative charge
state, wherein both lone pair states are doubly occupied. We
should mention that, for this comparison of splittings, we held
all atoms fixed in the equilibrium geometry for the neutral state.
The comparison is summarized in Table III. Note that, while the

Table III.   Lone-pair level splittings and electrical levels
             (negative charge state) for the isolated and the Griscom
             NBOHC's. All energies are in eV's. The electrical
             levels are calculated relative to the conduction band
             minimum (see Eq. (2)). Note that there is negligible
             change in the electrical level in going from the small
             to the large cluster, indicating that the defect wave
             function is strongly localized.

	Isolated NBO		Griscom
	Small Cluster	Large Cluster	Small cluster
Level Splitting	0.001	0.0008	0.26
Electrical Level (0/-1)	1.42	1.44	1.38

Jahn-Teller distortion of the isolated NBO does split the lone pair
levels, this splitting is computed to be far too small to account
for the observed g tensor. However, the splitting evidenced in the
Griscom NBO, derived partly from distortion and partly from the
presence of the other half of the defect, is substantially larger,
and is of the same order of magnitude as experiment. Thus, within
the accuracy of our approach, we believe the Griscom model is favored
over the isolated NBO model.

    It is important to note that both models for the NBOHC turn
out to be viable electron traps as well. The approximate electrical
levels for both models of the NBOHC are shown in Table III. These
were calculated according to equation (2).

$$E_{defect} = E^- - E^o - \chi \tag{2}$$

Here, χ is the experimental value for the electron affinity (1.2 eV),
and E^- and E^o are the self-consistent total energies calculated

using MOPN for the negative and neutral charge states, respectively.
We note that these centers have the feature that they are spin-
inactive when they are occupied by an electron, and thus might be
candidates for the unidentified electron traps in SiO_2. This
defect is half of the valence alternation pair model proposed by
Lucovsky [42,43], i.e., the C_1^- defect.

If our calculations for the negatively charged NBO are even
qualitatively correct, then, noting that Poindexter et al. [44] and
others [45,46] have observed the three principal intrinsic defects
(see Fig. 1) in MOS samples, we suggest the following experiment.
On a p type silicon substrate, grow a thick (1000-2000 Å) steam
oxide and cover with a metal gate. Subject this sample to strong
ionizing radiation and measure the spin resonance signal. Following
this, subject the sample to a series of avalanche electron injec-
tions. After each injection cycle, compare the growth of negative
charge with the NBOHC spin signal. It is our prediction that these
signals will be strongly anti-correlated. As the negative charge
builds up, the spin signal should decline.

D.   Defect Transformation Schemes

As mentioned above, we have also performed calculations on
the presumed precursors of the two OHC's. These are depicted in
Figs. 2a and 2c. The geometries of these centers were also optimized with respect to total energy. Also shown in Fig. 2 are
the heats of formation for each of the four defects [47]. We note
that these heats of formation form a hierarchy, with the NBOHC
precursor lowest in energy and the superoxide radical highest in
energy. We also note that each successive center can be obtained
from the previous one by trapping a single hole. (The transformation from 2a to 2b involves the concomitant shedding of a
proton, as does the transformation from 2b to 2c). We assert
that this is a very reasonable scenario for successive transformation between defects in the presence of radiation. As mentioned
above, the NBOHC precursor is not merely a convenient assumption,
but is thermodynamically favored over dispersed, single OH groups [48].
Furthermore, Griscom and coworkers have studied the evolution of
atomic hydrogen during radiation by monitoring the ^{1}H spin resonance
at low temperature [49]. They have found that the amount of free
atomic hydrogen generated is greater than the number of NBOHC's, in
at least qualitative agreement with the scheme of Fig. 2. Finally,
we note that the growth of superoxide radicals during radiation
almost certainly involves the superoxide bridge (Fig. 2c). This
mechanism is completely different from our proposed diffusion
mechanism [25], appropriate to post-radiation thermal annealing,
in which neutral O_2 diffuses through the material and becomes
trapped at E' centers [50]. In the radiation environment, the
ambient temperature is far too low to account for any appreciable
diffusion.

V. SUMMARY

We believe that the calculations presented here have served several purposes. First, they have reinforced the principal features of microscopic models generated by spin resonance. These include the symmetry of the defect (both the NBOHC and the superoxide radical), as well as the nature of the wave function. We should also point out that the calculations have put the Skuja-Silin model in serious doubt; in fact, we believe they render the model completely inappropriate. Second, they have given the first strong support to at least half of the VAP model for defects, the C_1^- center [51]. We have also elucidated an experiment that promises to give direct observation of these centers through the annihilation of NBOHC's. Finally, they have allowed us to present a complete and plausible picture of defect creation and transformation during the radiation process.

VI. ACKNOWLEDGMENTS

The authors thank Drs. M. D. Newton, G. V. Gibbs, J. A. Weil, E. H. Poindexter, R. Pfeffer, L. Trombetta, G. G. DeLeo, D. L. Griscom, F. J. Feigl, and Mr. J. K. Rudra and P. J. Caplan for valuable conversations. We also thank the staff of the Lehigh University Computing Center for help with the computational aspects of this work. The Quantum Chemistry Program Exchange supplied computer programs used in this work. This research was supported in part by the Office of Naval Research under Contract No. N00014-76-C-0125.

REFERENCES

1. D. J. DiMaria, in The Physics of SiO₂ and its Interfaces, ed. S. T. Pantelides, Pergamon, New York, 160:references therein, (1978).
2. R. A. Weeks and C. M. Nelson, J. Am. Ceram. Soc. 43:399 (1960).
3. D. L. Griscom, Phys. Rev. B22:1729 (1980).
4. M. Stapelbroek, D. L. Griscom, E. J. Friebele and G. H. Sigel, Jr., J. Non-Cryst. Solids 32:313 (1979).
5. The water content of fused silica is measured in ppm OH. The dryest fused silica, suprasil W1, contains 5-10 ppm OH, while suprasil 1 contains ~ 1200 ppm OH. Recent SIMS studies of thermal MOS oxides indicate there is at least 5-10 ppm OH in the oxide for even the dryest growing conditions. (R. Gale, F. J. Feigl, C. W. Magee, and D. R. Young, J. Appl. Phys. 54:6938 (1983)).
6. F. L. Galeener and M. F. Thorpe, Phys. Rev. B28:5802 (1983).
7. A. E. Geissberger and F. L. Galeener, Phys. Rev. B28:3266 (1983).
8. J. C. Phillips, in Solid State Physics, ed. F. Seitz and D. Turnbull, Academic Press, New York, 37:93 (1982).
9. As examples we cite references 3 and 4.
10. L. N. Skuja and A. R. Silin, Phys. Stat. Sol. (a) 70:43 (1982).

11. R. C. Bingham, M. J. S. Dewar, and D. H. Lo, J. Am. Chem. Soc. 97:1285 (1975).

12. P. Bischof, J. Am. Chem. Soc. 98:6844 (1976).

13. A. H. Edwards and W. B. Fowler, J. Phys. Chem. Sol. 46:841 (1985).

14. P. C. Taylor, J. F. Baugher and H. M. Kriz, Chem. Rev. 75:205 (1975).

15. F. J. Feigl, W. B. Fowler, and K. L. Yip, Sol. State. Comm. 14:225 (1974).

16. K. L. Yip and W. B. Fowler, Phys. Rev. B11:2327 (1975).

17. R. H. Silsbee, J. Appl. Phys. 32:1459 (1961).

18. D. L. Griscom, E. J. Friebele and G. H. Sigel, Jr., Sol. State Comm. 15:479 (1974).

19. D. L. Griscom, Nucl. Inst. and Methods B1:481 (1984).

20. M. G. Jani, R. B. Bossoli, and L. E. Halliburton, Phys. Rev. B27:2285 (1983).

21. At this conference, we have heard about three spectroscopically different E' centers in a-SiO$_2$. In this chapter, we will only consider the E' γ center.

22. H. A. Jahn and E. Teller, Proc. Roy. Soc. London A161:220 (1937).

23. F. S. Ham, Phys. Rev. B8:2926 (1973).

24. J. K. Rudra and W. B. Fowler, Bull. Am. Phys. Soc. 30:369 (1985).

25. A. H. Edwards and W. B. Fowler, Phys. Rev. B26:6649 (1982).

26. Recently, Jani, Bossoli and Halliburton have performed an extensive study on the formation mechanisms and the spin Hamiltonian parameters of the E_1' center in alpha quartz (Ref. 20). In this study they proposed a new model, based on an analysis of the angular dependence of the weak hyperfine lines. The two sets of lines are attributed to two of the next nearest neighbor silicon atoms, each bonded to one of the back-bonded oxygen atoms in the lower half of Fig. 1a. Further, the absence of a hyperfine signal from the third silicon atom on this side of the defect was attributed to a second oxygen vacancy. This is in stark contrast to the Feigl-Fowler-Yip model, wherein the two sets of weak hyperfine lines are attributed to the silicon atom in the upper half of Fig. 1a, and to another unspecified silicon atom. The recent calculations of Rudra and Fowler (Ref. 24) have rendered the double vacancy model extremely doubtful.

27. Using ENDOR, Jani et al. (Ref. 20) have recently determined that the strong hyperfine lines in alpha quartz are also due to interaction with a ^{29}Si nucleus, rather than a proton.

28. A. V. Shendrick and D. M. Yudin, Phys. Stat. Sol. (b) 85:343 (1978).

29. G. Gobsch, H. Haberlandt, H. J. Weckner, and J. Reinhold, Phys. Stat. Solidi 90:309 (1978).

30. D. L. Griscom and E. J. Friebele, Phys. Rev. B24:4896 (1981).

31. E. J. Friebele, D. L. Griscom, M. Stapelbroek, and R. A. Weeks, Phys. Rev. Lett. 42:1346 (1979).

32. O. A. Ershov, D. A. Goganov, and A. P. Lukirskii, Sov. Phys. Solid State 7:1903 (1966).
33. D. L. Griscom, J. Non-Cryst. Solids 24:155 (1977).
34. K. L. Yip and W. B. Fowler, Phys. Rev. B10:1400 (1974).
35. J. A. Tossel, J. Phys. Chem. Sol. 34:307 (1973).
36. S. J. Louisnathan and G. V. Gibbs, Am. Mineral. 57:1614 (1972).
37. J. A. Tossel, J. Am. Chem. Soc. 97:4840 (1975).
38. G. A. D. Collins, D. W. Cruickshank, and A. Breeze, J. Chem. Faraday Trans. II 68:1189 (1972).
39. G. V. Gibbs, Amer. Mineral. 67:421 (1982).
40. J. M. Baker and P. T. Robinson, Sol. State Comm. 48:551 (1983).
41. D. L. Griscom, private communication.
42. G. Lucovsky, Philos. Mag. B39:513 (1979); B39:531 (1979); 41:457 (1980).
43. G. Lucovsky, J. Non-Cryst. Solids 35-36:825 (1980).
44. E. H. Poindexter, P. J. Caplan, R. L. Pfeffer, A. H. Edwards, and W. Muller-Warmuth, Bull. Am. Phys. Soc. 29:368 (1984).
45. B. I. Vikhrev, N. N. Gerasimenko, and G. P. Lebedev, Soviet Physics: Microelectronics 6:71 (1977).
46. K. L. Brower, P. M. Lenahan, and P. V. Dressendorfer, Appl. Phys. Lett. 41:251 (1982).
47. The heat of formation is defined in MINDO/3 as the difference between the total self consistent energy, and the sum of the isolated, single atom energies.
48. S. Urnes, Trans. Brit. Ceram. Soc. 60:85 (1961).
49. D. L. Griscom, M. Stapelbroek, and E. J. Friebele, J. Chem. Phys. 78:1638 (1983).
50. R. Pfeffer and D. L. Griscom, to be published.
51. This defect has been investigated previously, using a one-electron, tight-binding technique (E. P. O'Reilly and J. Robertson, Phys. Rev. B27:3780 (1983)). However, electron-electron interactions (for the negative state) and electron lattice interactions were not included.

ELECTRONIC PROPERTIES OF SiO$_2$ AND MODELS FOR ITS POINT DEFECTS

W. Beall Fowler

Department of Physics and Sherman Fairchild Laboratory
Lehigh University
Bethlehem, PA 18015

ABSTRACT

Research on silicon dioxide has developed in several distinct, nearly independent patterns. The electronic band structure of crystalline SiO$_2$ is well established, and progress is being made on amorphous SiO$_2$. Thus the transport of electrons and holes may be understood and mechanisms for defect creation by ionizing radiation may be considered. Theoretical work on defects has centered on molecular cluster models for the defects, with considerable success as measured by comparison with various spectroscopic measurements (primarily electron spin resonance). And some of the most interesting recent work has involved the study of transient defects and development of models for them. This paper summarizes our understanding in all of these areas, with the intent of emphasizing the connections between them.

I. INTRODUCTION

Considerable work has been carried out within the past decade in attempting to understand the electronic properties of crystalline and amorphous SiO$_2$ and how defects are produced and transformed among one another by various physical, chemical, and thermal stimuli. Much of this work has been the subject of other reviews [1-13]. In the present paper we will attempt to provide a coherent account of aspects of this work, but where appropriate we will refer to these other reviews rather than repeat their contents.

Historically, there have been several logical frameworks used in attempting to treat this topic. These frameworks have included the following:

(1) The band-structure approach: Most easily applied to perfect crystalline SiO_2, this yields the Bloch states, valence and
conduction bands, effective masses, optical absorption spectra, and
other properties which could be related to the bulk electronic
structure. Theoretical techniques include the various methods
developed and applied to other solids over the years [4-23].

(2) The quantum-chemistry approach: As generally carried out,
this involves approximating the extended system by a cluster of atoms
with suitable boundary conditions, and then applying one of various
techniques used to calculate the electronic properties of molecules.
This is especially useful in treating defects, and by calculating
total energy vs atomic position information on the defect structure
can be obtained along with other electronic properties such as wave
functions, spin densities, and energies [24-33].

(3) Vacancy-interstitial models for defects: Such models tend
to be in the mainstream of solid-state activity, and if one allows
interstitials (e.g. oxygen) to be incorporated into peroxy linkages
and vacancies to exhibit large relaxations, it is possible to understand considerable experimental data within this framework.

(4) Valence-alternation pair models for defects: The notion
that in amorphous materials nearby oppositely-charged defects can
readily form has been less successful in SiO_2 than in some other
glassy systems, but it has been used to analyze certain transport
properties in sputtered films [34-36].

(5) Ionic or covalent models and notation: Because SiO_2 is
regarded as mixed ionic-covalent [37-40], both extremes have been
used to provide conceptual frameworks as well as defect notations.
Several analyses based on ionic concepts have been successful,
without, however, "proving" the ionic nature of the Si-O bond.
Indeed, given the ambiguities in defining ionicities and in appor-
tioning charge, one should be careful not to over-interpret any
such results.

Defect notation can easily generate confusion. For example,
in an ionic notation, a germanium which substitutes for silicon and
traps an electron is Ge^{+3}; in a covalent notation it is Ge^{-1}.

This list could certainly be extended; in fact, the controversy
over the structure of amorphous SiO_2 has not yet been mentioned [41],
nor have the several schemes used to explain Si-O-Si bond angles by
means of steric effects [42,43]. The point is, there are many ways
to attempt to understand SiO_2. Many of these approaches have provided valuable insights, but it has not always been clear just how
they might be interrelated; for example, how might a knowledge of
the electronic band structure of α-quartz (or idealized β-cristobalite)
be used to help evaluate defect models or to predict mechanisms for

defect production? What is the relationship, if any, between the results of <u>ab initio</u> quantum chemistry calculations on clusters of Si and O atoms and electronic transport in amorphous SiO_2?

While detailed answers to most of these questions are not yet known, we are at a point where at least plausible comments and speculations can be made which can help guide further investigations. That is the main purpose of this paper. It is divided into the following parts: Section II treats the theory of electronic properties of bulk SiO_2, both crystalline and amorphous, both band and exciton states. Comparisons with experiment are made and some still-unanswered questions are presented.

Section III summarizes the models for <u>some</u> of the better-established defects (mostly either intrinsic or dressed with hydrogen) and their status. Section IV discusses possible defect production mechanisms, and in particular how knowledge of the bulk electronic properties may be useful in understanding defect properties.

One of the most significant aspects of current SiO_2 research is directed towards the properties of thin SiO_2 films in the MOS electronic device configuration. While most, if not all, of the contents of this paper have some bearing on understanding phenomena in the MOS system, a discussion of this topic is beyond the scope of this paper.

II. ELECTRONIC PROPERTIES OF BULK SiO_2

A. Electronic Band Structure of Crystalline SiO_2 - Theory

Over the past decade many theoretical calculations and analyses of the electronic band structure of crystalline SiO_2 have been carried out. Although some work has been done on amorphous SiO_2 ($a-SiO_2$), there has been considerably less activity than in the crystalline case because of the difficulties in structurally modelling $a-SiO_2$ and in carrying out electronic calculations for such a non-periodic system.

These calculations on extended SiO_2 followed (and paralleled) calculations on cluster models. Quantum-chemical calculations on clusters have been useful not only in revealing the physics of SiO_2 but also in providing comparisons of various approximations. Reilly's analysis [44] based on a Si-O-Si fragment yielded a first understanding of the properties of the valence bands as consisting of non-bonding oxygen states at the top, silicon-oxygen bonding states at lower energies.

It is generally possible to compare the discrete valence-band states of the cluster calculations with valence bands of similar

nature derived from calculations of the extended solid. One must
be careful with respect to symmetry, however, in cases where selection rules may be important; e.g., an O-centered cluster will have
different symmetry from a Si-centered cluster, and in the solid
translational symmetry can play an additional role.

Most of the interest in crystalline SiO_2 has focused on α-quartz,
since this is the most common form and has important technological
applications. Several other forms have also been investigated; calculations on at least two of these are scientifically useful,
idealized β-cristalobite and stishovite.

Idealized β-cristobalite can be thought of as an expanded
silicon lattice with an oxygen atom inserted midway between each
near-neighbor silicon pair. It differs from true cristobalite
structures [45,46] in that its Si-O-Si bonds are all linear; it shares
with all other crystalline forms (except stishovite) four-fold Si and
two-fold O coordination. Because of its relative simplicity it has
served as a useful prototype system for which to evaluate band-
structure effects in SiO_2.

α-quartz has the same coordination but a bent Si-O-Si bond
(144°) and an involved topology. Figure 1 shows schematically a
fragment of α-quartz.

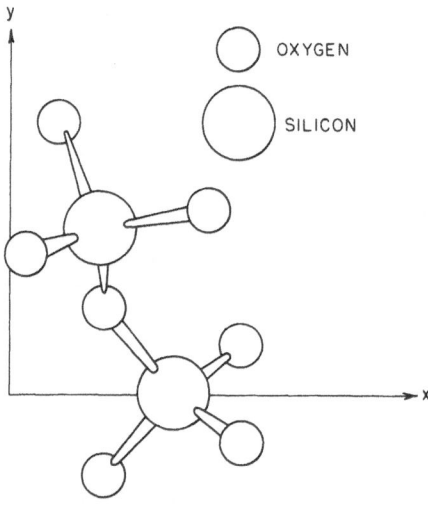

Fig. 1 Fragment of right-handed α-quartz.

Stishovite is a tetragonal form of SiO_2 having the rutile
structure. Each silicon has six oxygen neighbors, each oxygen three
silicon neighbors. While this coordination is unusual for SiO_2,
it is pertinent here since it is conceivable that it might be found
occasionally in a-SiO_2.

Idealized β-cristobalite was the first form to be treated in detail. Its band structure as obtained by Schneider and Fowler [14] is shown in Fig. 2. Features of its band structure include the following:

(1) The valence bands are oxygen-rich. This is not surprising; even with an assumption of zero ionicity, there are 8 valence p electrons associated with the two oxygens vs 4 electrons associated with the silicon. Partial ionic bonding favors oxygen population even more. If, for example, each oxygen has a net charge of -1, there will be 10 oxygen p electrons vs 2 valance electrons associated with the silicon.

(2) The valence bands are divided into two non-overlapping regions (in addition to a much lower oxygen 2s band). The highest region is derived primarily from non-bonding oxygen 2p states; these pπ-like orbitals exhibit only weak bonding effects. This region is about 2 eV wide, and $\sim$ 2 eV below there is a wider (4-5 eV) band formed from bonding oxygen 2p orbitals and Si sp^3 hybrids. This description of the valence bands is consistent with simple chemical arguments and with the results of quantum-chemical calculations on atomic clusters. Good agreement with these computed bands has been obtained with two versions of tight-binding fits: one used only oxygen 2p functions [14] while the other used "bond orbitals" [47] associated with silicon-oxygen bonds.

(3) The highest valence band has a maximum at $\vec{k}$ = 0 and is very flat, predicting a large effective mass for holes.

(4) The lowest conduction band is free-electron-like, with and s-like minimum at $\vec{k}$ = 0 (Γ_1).

(5) The lowest energy optical absorption is predicted to be a <u>direct</u> <u>forbidden</u> transition ($\Gamma_{25} \rightarrow \Gamma_1$).

Many of these features are preserved in α-quartz, although the bands appear considerably more complex. Fig. 3 shows the recent results of Calabrese [21] which are partially parametrized to fit experiment. One should also note the earlier, widely-cited self-consistent bands of Chelikowsky and Schluter [15], and a paper of Calabrese and Fowler [16] which contains a discussion of comparisons between calculations and between theory and experiment. The computed gap within the valence bands of α-quartz appears to be somewhat smaller ($\sim$ 1 eV) than in β-cristobalite. It is not clear whether the valence-band maximum is at $\vec{k}$ = 0 or at some other point in the Brillouin zone; most calculations have predicted the latter, but only by differences in energy of order tenths of an eV. Either way, the uppermost valence band is very flat and the onset of optical absorption is forbidden; the transition at $\vec{k}$ = 0 is direct forbidden, while a transition starting away from $\vec{k}$ = 0 would be indirect.

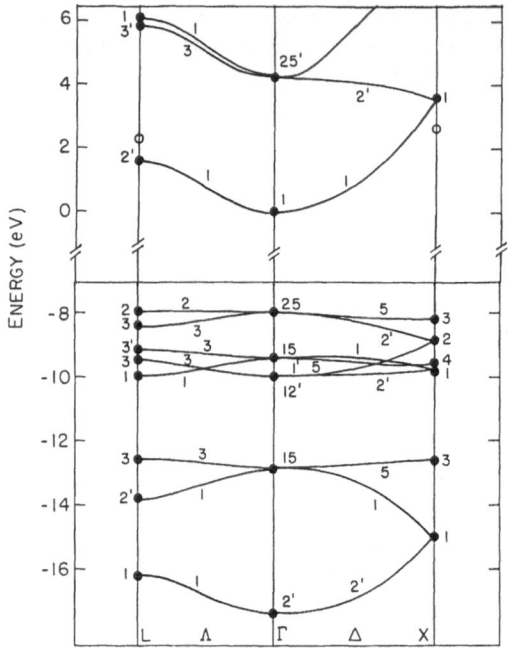

Fig. 2   Band structure of idealized β-cristobalite (Ref. 14).

It is not surprising that the bands in stishovite are rather
different from those in β-cristobalite and α-quartz, given the
different short-range structures. Fig. 4 shows the 2p valence bands
as obtained by Rudra and Fowler [23] using a simple parametrized
tight-binding method. Although these bands are oxygen-rich, there is
no gap within them; the bonding and non-bonding parts are not clearly delineated. The maximum of the valence bands is at $\vec{k}$ = 0, and
the topmost band is much less flat than in either of the other
two cases discussed here. The lowest conduction band is again
s-like, with a minimum at $\vec{k}$ = 0 [22]. The optical absorption edge
is predicted to be direct forbidden.

It is of considerable interest that the lowest-energy optical
transitions at $\vec{k}$ = 0 are in all three cases predicted to be forbidden. This property can be understood by considering the nature
of the highest valence-band states at $\vec{k}$ = 0. These states are
derived from oxygen 2pπ orbitals, and the highest-energy states are
those for which these orbitals are the least bonding, or have the
least charge between oxygen atoms. This condition leads to phase
relations between the orbitals on different oxygen atoms which are
shown in Fig. 5.

Fig. 5a shows (by arrows) the orientation of the p functions
on the oxygens for the highest valence state of a hypothetical

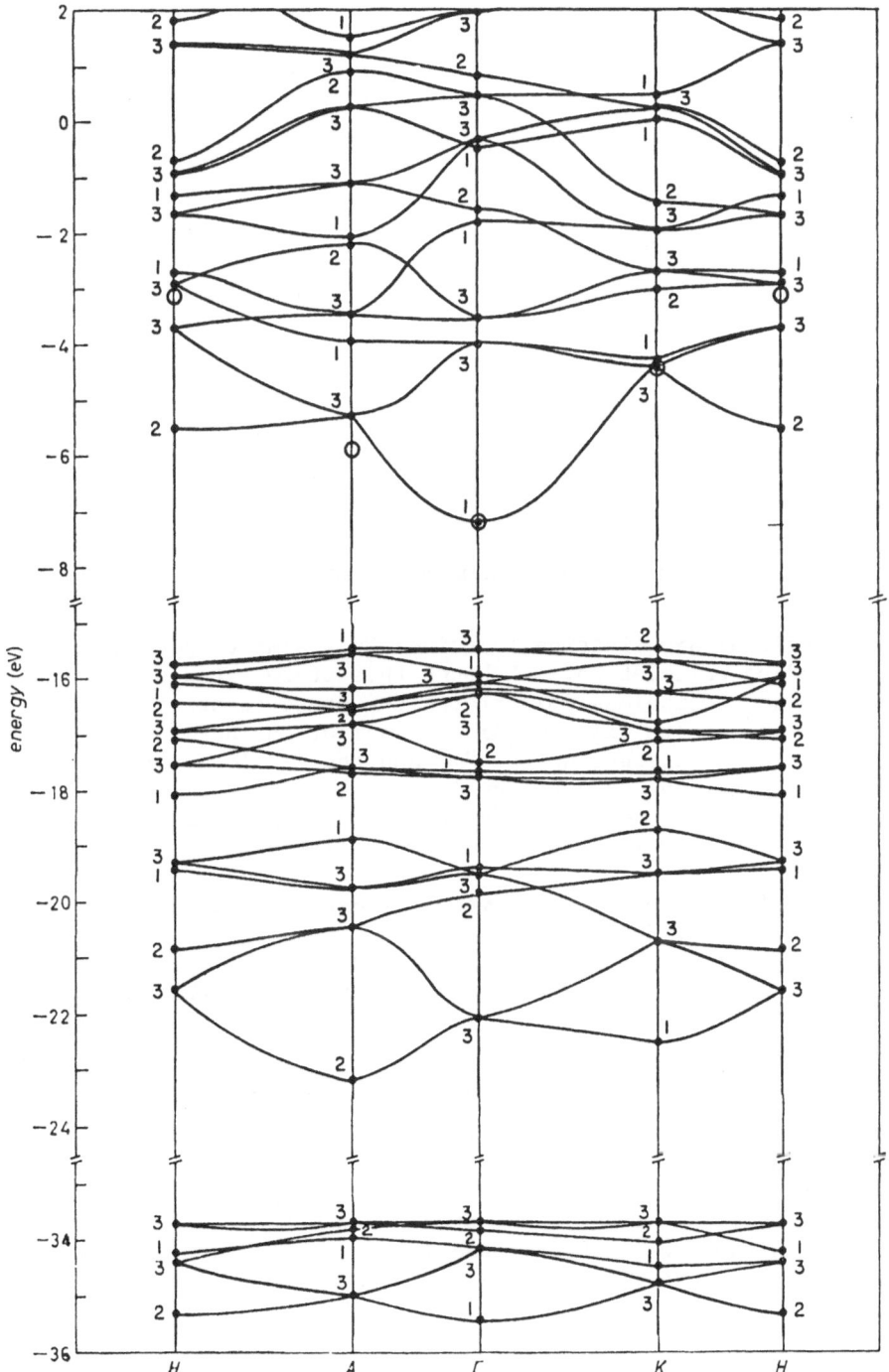

Fig. 3 Band structure of α-quartz (Ref. 21).

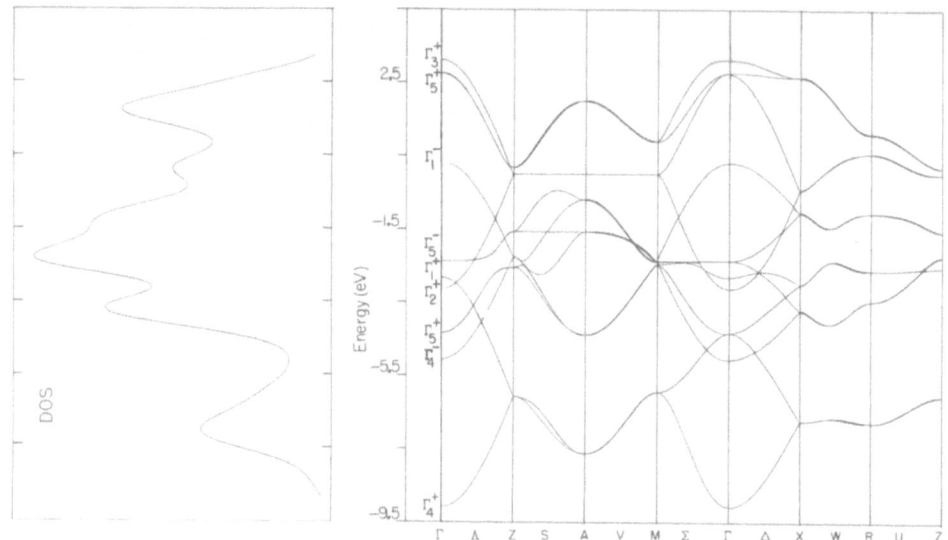

Fig. 4   Valence band structure of stishovite (Ref. 23).

two-dimensional SiO_2 lattice, obtained by a simple nearest-neighbor
tight-binding calculation. Although this state turns out not to be
at $\vec{k} = 0$, it nonetheless illustrates the phase relations in that the
region between atoms will contain the sum of the head and the tail
of two arrows, representing the "+" and the "−" lobes of two p
functions. Fig. 5b shows the analogous situation in idealized
β-cristobalite for the four oxygens surrounding one silicon. Again,
"+" and "−" lobes will have maximum cancellation.

     A similar argument can be made for α–quartz, although not,
strictly speaking, from group theory. This principle holds for
stishovite, as well, although the extra degrees of freedom and
complex bands make it somewhat harder to judge (in the absence of
detailed calculations) which of several valence states will in fact
be highest.

     It is these phase relations which lead to the forbiddenness
of the direct transitions. The lowest conduction state is Γ_1, which
can be thought of in the tight-binding sense as being formed from
linear combinations of s functions on oxygens and silicons, all with
constant phase. Whereas in the atomic sense one would expect that
these transitions would be allowed (p → s), it is the relative
phases of the oxygen 2p orbitals in the valence-band wave functions
which cause the electric dipole transition matrix elements between
valence- and conduction-band states to be zero.

     Since these selection rules do not derive from atomic ones
(in particular, parity of atomic wave functions), one must also look

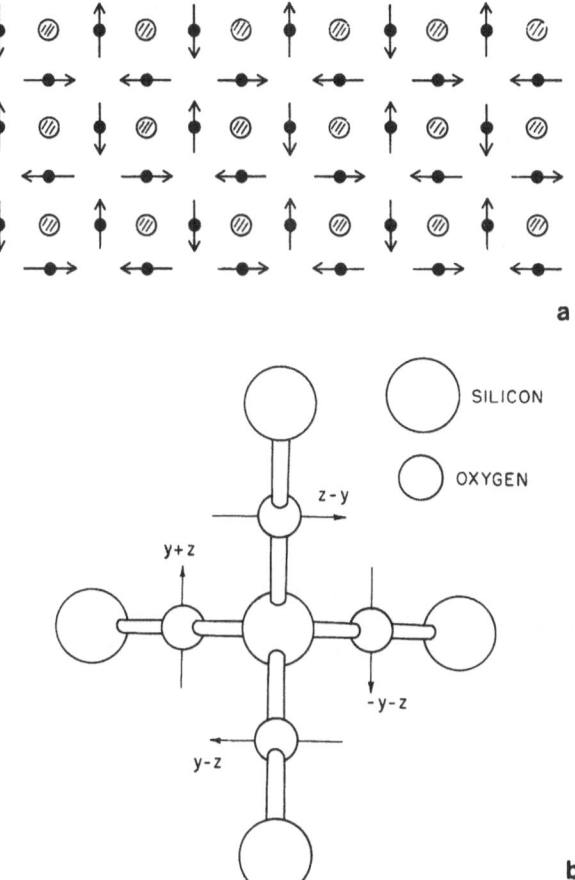

Fig. 5   2p wave function phases at the valence-band maximum for
a) hypothetical two-dimensional SiO$_2$; b) idealized beta
crystobalite.

in more detail at the selection rules for transitions involving more
than one photon. And indeed, it turns out (by group-theoretical
arguments) [48] that for β-cristobalite the $\Gamma_{25} \to \Gamma_1$ transition is
forbidden for two photons as well. It is however, allowed for
three photons.

B. Electronic Band Structure of Crystalline SiO$_2$ - Experiment

Briefly stated, most of the theoretical results presented
in Sec. A for 4:2 crystalline SiO$_2$ are consistent with experiment.
Extensive comparisons of theory with experiment have appeared elsewhere, and we will simply summarize some of the conclusions here:

(1)   Conduction electrons have high mobilities and behave as band-like particles, with a mobility at room temperature in a-SiO$_2$ of $\underset{\sim}{} 20$ cm^2/V sec [49,50].

(2)   Holes have low mobility and move as small polarons by means of a hopping mechanism. At 200 K the hole mobility in a-SiO$_2$ is less than 10^{-6} cm^2/V sec [49,50].
(Note that most of the experiments related to items (1) and (2) have been carried out on amorphous rather than crystalline SiO$_2$.)

(3)   The X-ray emission and photoemission spectra are generally consistent with the theoretical description of the bands [51-54].

(4)   The valence bands are oxygen-rich; in fact, an ionic model provides a good fit to Compton scattering [39] and to most features of X-ray emission. (However, one must be cautioned not to over-interpret these results.)

(5)   The optical reflectivity spectrum of α-quartz [55,56], shown in Fig. 6, is consistent with a forbidden band gap (no strong excitons) at $\underset{\sim}{} 9$ eV, followed at $\underset{\sim}{} 12$ eV by an allowed gap which has a strong n = 1 exciton at 10.4 eV [18]. Exciton effects may be noted in the higher energy spectra as well.

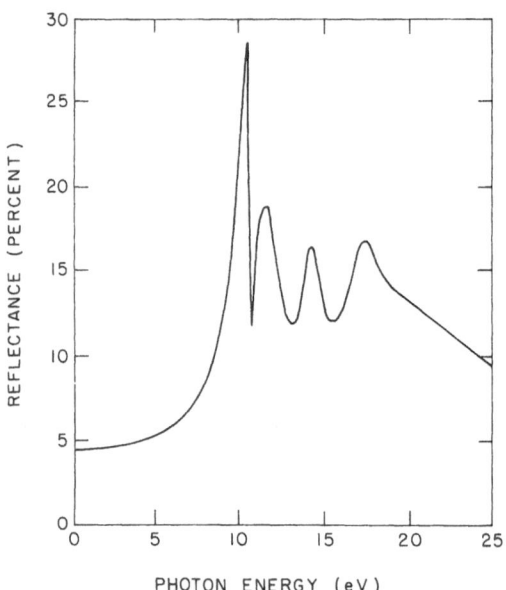

Fig. 6   Optical reflectivity of alpha quartz (Ref. 55).

C. Excitons in Crystalline SiO₂

There are in general three elements involved in the point symmetry of an exciton: the hole, the electron, and the envelope function which describes the relative electron-hole orbital [57]. Since envelope functions may have various symmetries, one expects both allowed and forbidden exciton transitions. If only the lowest-energy (n = 1) direct exciton is considered, one expects that the envelope function will be of Γ_1 symmetry (s-like), by analogy with the hydrogen atom. If this is the case, the symmetry of the exciton is determined solely by that of the electron and hole, and if the electron has Γ_1 symmetry as in the cases which we are discussing, the exciton point symmetry is just that of the hole. The only additional item is that since the exciton is made from two "particles" one must consider its spin, which will lead to overall singlet or triplet symmetry in the absence of spin-orbit admixture.

In the discussion of excitons we may use some well-established ideas [57-62] to make the following points. Transport of excitation energy involves matrix elements of the form

$$\langle \psi_g(\vec{r}'-\vec{R})\psi_e(\vec{r})|e^2/|\vec{r}-\vec{r}'||\psi_g(\vec{r})\psi_e(\vec{r}'-\vec{R})\rangle$$

$$= \langle \phi_1(\vec{r})|e^2/|\vec{r}-\vec{r}'||\phi(\vec{r}'-\vec{R})\rangle \qquad (1)$$

and

$$\langle \psi_g(\vec{r}'-\vec{R})\psi_e(\vec{r})|e^2/|\vec{r}-\vec{r}'||\psi_e(\vec{r}-\vec{R})\psi_g(\vec{r}')\rangle \qquad (2)$$

Here, $\phi(\vec{r}-\vec{R})$ is a state of the system in which the atom at point $\vec{R}$ is excited and all other atoms are in their ground state. This may also be written as a Slater determinant which when expanded yields products of one-electron functions such as

$$\psi_e(\vec{r}-\vec{R})\psi_g(\vec{r})\psi_g(\vec{r}-\vec{R}')\psi_g(\vec{r}-\vec{R}'') \,\text{---} \qquad (3)$$

where "e" is for excited, "g" for ground state.

The first of these matrix elements (Eq. 1) is a direct term, the second (Eq. 2) and exchange term. The first does not require wave function overlap: it resembles a classical electromagnetic energy transfer, and if $e^2/|\vec{r}-\vec{r}'|$ is expanded it involves multipole-multipole interactions. The second (exchange) term does require wave function overlap and so is presumably of rather short range.

Dexter [60] argued in a slightly different context that if the direct matrix element is of dipole-dipole character, the transfer rate for nearby atoms will be faster than vibrational rates and one may expect the excitation to be transferred before any lattice relaxation occurs. This would result in a "proper" exciton; it represents the idealized case in that the excitation may be written as a Bloch function with only electron dynamics involved. If the direct transfer is not of dipole-dipole character, then the exchange term will be larger than the direct, but this term will probably be small enough that some lattice relaxation will occur before transfer takes place.

The foregoing arguments were made in the context of excitation transfer involving impurities, and the exciton situation is slightly more involved in that the excitation must be localized (i.e., a wave packet must be formed) before significant lattice relaxation can be expected to occur. Nevertheless, this does occur, even in alkali halides where the transfer mechanism is dipole-dipole [63]. Haken and Reineker [61] have pointed out the importance of temperature and argued that cases exist (e.g., in anthracene) in which at low temperature the exciton moves as a Bloch function while at higher temperature its motion is best described by a hopping process.

Applying the above logic to SiO_2 suggests the following:

(1) The 10.4 eV exciton, which is allowed, would be transferred via a dipole-dipole mechanism. However, this exciton is degenerate with free electron-hole states and so would likely decay to a free hole and electron before moving very far.

(2) The forbidden excitons associated with the 9 eV band gap would be transferred via an octupole-octupole mechanism. This transfer rate is small enough (and falls off as R^{-9}) that the exchange process would certainly predominate, but this process in turn would be small enough (by Dexter's arguments) that lattice relaxation or polaron effects would come about. Even if the exciton-lattice interaction were too small to lead to exciton self-trapping, the effective mass of the exciton would be larger since lattice distortion must move with the excitation. Note that because the exciton is forbidden, it would have along lifetime against radiative decay.

(3) However, should the lattice relaxation suggested in (2) be such as to localize the hole (and electron) on one oxygen, the situation would change dramatically. Although localization would require admixture of states throughout the uppermost valence band and would "cost" an energy $\sim 1/2$ the band width, or ~ 1 eV, it seems possible that more than 1 eV could be gained through lattice relaxation and/or chemical rebonding, thus facilitating such relaxation.

A p-like hole and an s-like electron localized on <u>one</u> oxygen atom would have a radiative recombination rate which is electric-dipole allowed. It might be spin-forbidden, however (triplet -- singlet). The excitation transfer rate would be small because although facilitated by a dipole-dipole process, the large lattice relaxation (and resulting removal of degeneracy among different oxygen sites) would strongly quench the transfer process. We will return to this topic later in connection with defect production.

D. Band Structure and Excitons in Amorphous SiO₂

Can any of these concepts be carried over to amorphous SiO₂? The answer must be "yes", with various qualifications depending on one's concept of the structure of a-SiO₂. At present, two proposed structures of a-SiO₂ are attracting particular attention. The random-network model [64,65] has long been accepted as defining a valid idealized structure for a-SiO₂, to which defects could be introduced such as broken bonds, vacancies, and interstitials. In this model all silicons have 4 oxygen neighbors in approximately tetrahedral coordination, and all oxygens have 2 silicon neighbors. The noncrystalline nature derives from the flexibility of the Si-O-Si bond angle, which allows long-range order to be destroyed by means of various bonding topologies.

Recently Phillips [41] has proposed a model of a-SiO₂ as consisting of regions of β-cristobalite with oxygen-rich interfaces between regions. Convincing (and strongly held) arguments have been made on obth sides of the Phillips vs random network model issue, and the matter is still not settled.

At least two further points should be made here:

(1) Although it is often assumed that a-SiO₂ will have the same form regardless of preparation, this may well not be true. In particular, one should question [11] whether the structure of bulk a-SiO₂ prepared from the melt is identical to that of a-SiO₂ formed by oxidizing silicon.

(2) Even in the bulk there exist at least two forms of a-SiO₂ ("normal" and "densified") which appear to have rather different microscopic properties [66] related to structure. And recent arguments have been made [67] that bulk a-SiO₂ prepared from the melt contains high concentrations of oxygen vacancies.

With reference to the present discussion, in all of the models of a-SiO₂ most of the silicons have 4 oxygen neighbors, most of the oxygens have 2 silicon neighbors. So to the extent that the electronic properties are dominated by this short-range order, we anticipate that all of these models would lead to similar basic predictions as regards electronic properties, with refinements depending on the differences in the models.

Relatively few calculations which would be analogous to the band structure calculations done on crystalline SiO_2 have been carried out on a-SiO_2. The most comprehensive of these seems to be that due to Ching [68], who obtained the electronic structure for a "perfect" random-network model of a-SiO_2. Ching presented a rather thorough analysis of the effect of Si-O bond length and Si-O-Si angle on the electronic energies. The valence-conduction band gap is predicted to be $\stackrel{\sim}{\sim}$ 1 eV smaller for a-SiO_2 than for α-quartz. The edges of the valence bands (in particular, the top) are predicted to be more localized in a-SiO_2, but the lowest conduction states are delocalized in both cases. While no information was provided in Ching's paper as to relative phases in the upper valence-band wave functions, we expect that our previous arguments would be valid here and that the wave function associated with the highest valence state would have very little density between nearest oxygen atoms. Thus we anticipate that the optical absorption edge would be "quasi-forbidden" and that the n = 1 exciton associated with this edge would be weak.

As implied earlier, amorphous and crystalline SiO_2 share many similar electronic properties. This includes electron and hole transport [49,50], X-ray emission and photoemission [51-54], and optical absorption [55,56]. This indicates that such properties are strongly dependent on short-range order and weakly dependent on long-range order, provided that the number of defects is small.

E. Some Open Questions

Some problems related to the band structure and exciton properties still exist. They include the following:

(1) It does not seem possible to fully reproduce the Si $L_{2,3}$ X-ray emission without considering Si 3d-like orbitals within the upper non-bonding valence bands [39]. A relatively small (perhaps 15 to 20%) admixture may, however, be sufficient.

(2) The lowest conduction band at $\vec{k}$ = 0 is s-like, but whether it is mostly like Si 3s, O 2s, or O 3s is a subject of discussion [19]. Given the large extent of the 3s functions and their overlap, it may be difficult to provide a totally meaningful answer.

(3) The $\stackrel{\sim}{\sim}$ 1 eV gap within the valence bands of α-quartz probably does not exist in a-SiO_2. Ching's calculation [68] suggests a "pseudo-gap" in that the gap states are strongly localized. Defects such as three-fold coordinated oxygen are likely to provide states in that region, as well. This will be considered again in Sec. IV.

(4) Experimental evidence is accumulating that excitons can self-trap in SiO_2, and that one result of this self-trapping is the

creation (in at least a transient sense) of defects. This will be dealt with in Sec. IV.

III. MODELS FOR POINT DEFECTS IN SiO₂

Fig. 7 shows models for 8 "established" intrinsic defects in SiO₂, while Fig. 8 shows models for 3 "speculative" intrinsic defects. Because hydrogen is such a ubiquitous part of SiO₂, we consider H-related defects to be intrinsic. Not all will agree on the classification of established vs speculative, and some of the "established" defects have not been observed but fall into a logical relationship with observed defects. It should be noted that some of these defects are appropriate to α-quartz, some to a-SiO₂, and some to both structures.

A. Established intrinsic defects in SiO₂

Perhaps the best identified defect of all is the E_4' center [69], a H atom replacing an O atom in α-quartz. It is paramagnetic, with 3 electrons (one each from the H and its neighboring silicons). The H atom moves in a shallow double potential well, which means that its equilibrium position and excursions from equilibrium are strong functions of temperature, and this is reflected in the observed electron-spin-resonance spectrum. The unpaired spin has sizeable amplitude on both silicons and on the hydrogen. These experimental features have been accurately reproduced by two independent theoretical calculations [69-71].

The E_1' center has been observed in α-quartz, in bulk glasses and thin films, and at SiO₂ surfaces [72-76]. Its strongest, indisputable feature is a spin 1/2 ESR signal whose g-tensor yields a model of an sp^3 electron spending most of its time on one Si, "pointing" towards an oxygen vacancy. The current model [77,78], first suggested 10 years ago for the E_1' center in α-quartz, envisions that the other near-neighbor Si of the vacancy has trapped a hole and has relaxed away from the oxygen vacancy into the plane defined by its 3 remaining oxygen neighbors.

This model, also predicted theoretically [78], explains why the hyperfine interaction is so large with only <u>one</u> silicon; a spontaneous symmetry-breaking displacement has removed the near-degeneracy of the two silicons adjacent to the positively-charged vacancy, yielding a lower total energy when the electron is localized on one of them. The model is also consistent with two weak silicon hyperfine signals, and although their origin has not been totally clear the recent assignments of Jani <u>et al.</u> seem convincing [79]. However, the conjecture of Jani <u>et al.</u> that two oxygen vacancies are involved in the defect is probably incorrect. And it should be noted that while in α-quartz the weak hyperfine signals have been definitely attributed to Si, in SiO₂ glass a similar signal has been identified with a proton [80].

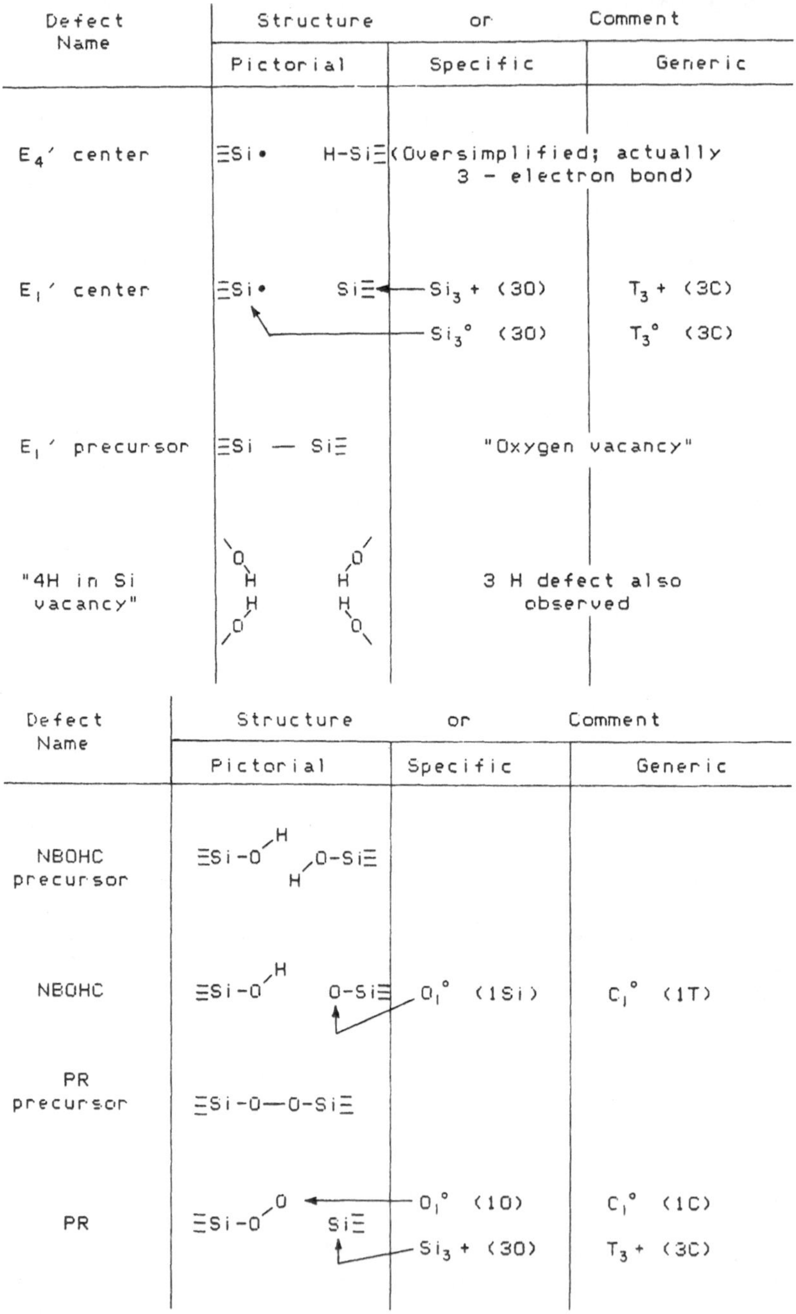

Defect Name	Structure	or	Comment
	Pictorial	Specific	Generic
E_4' center	$\equiv Si\bullet$ $H-Si\equiv$	(Oversimplified; actually 3 − electron bond)	
E_1' center	$\equiv Si\bullet$ $Si\equiv$ ←	$Si_3 +$ (30) Si_3° (30)	$T_3 +$ (3C) T_3° (3C)
E_1' precursor	$\equiv Si - Si\equiv$	"Oxygen vacancy"	
"4H in Si vacancy"	(pictorial of O, H, H, O structure)	3 H defect also observed	

Defect Name	Structure	or	Comment
	Pictorial	Specific	Generic
NBOHC precursor	$\equiv Si-O\overset{H}{}\,\,\overset{H}{}O-Si\equiv$		
NBOHC	$\equiv Si-O\overset{H}{}$ $O-Si\equiv$	O_1° (1Si)	C_1° (1T)
PR precursor	$\equiv Si-O-O-Si\equiv$		
PR	$\equiv Si-O\overset{O}{}$ $Si\equiv$ ←	O_1° (10) $Si_3 +$ (30)	C_1° (1C) $T_3 +$ (3C)

Fig. 7   Models for 8 "established" intrinsic defects in SiO_2.

Defect Name	Structure	or	Comment
	Pictorial	Specific	Generic
SPR	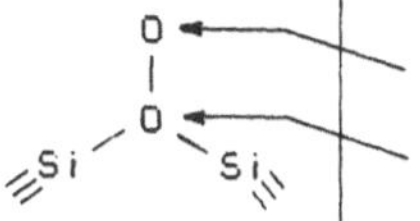 O↓ O (≡Si Si⋰)	O_1^+ (1O) $O_3^°$ (1O,2Si)	C_1^+ (1C) $C_3^°$ (1C,2T)
3-fold oxygen+	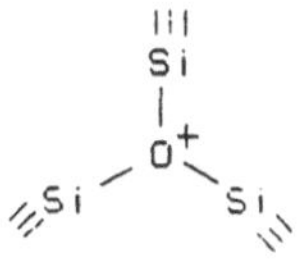 Si ‖ O⁺ (≡Si Si⋰)	O_3^+ (3Si)	C_3^+ (3T)
2-fold silicon	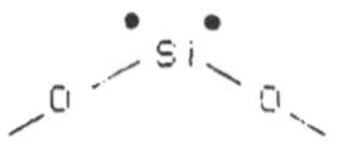 Si (O Si O)	$Si_2^°$ (2O)	$T_2^°$ (2C)

Fig. 8 Models for 3 "speculative" intrinsic defects in SiO₂.

While it is possible that in a-SiO$_2$ or near a surface the E_1' electron projects into a void, that is, the right half of the pictorial does not exist or is far away, it is probably more accurate to assume the same model as in α-quartz unless compelling reasons indicate otherwise. One reason for this statement is that there is growing indirect evidence for an E_1' precursor, an "oxygen vacancy", whose structures in α-quartz and in a-SiO$_2$ are similar and which upon trapping a hole becomes an E_1' center.

It was argued several years ago by O'Reilly and Robertson [81] that an "oxygen vacancy" which arose during the thermal preparation of bulk a-SiO$_2$ would in fact resemble a Si-Si bond: in effect, in the melt there would be nothing to keep the two silicons apart and so they would be found at a separation close to the Si-Si distance in crystalline Si, 2.35 Å. This argument, which Robertson [82] subsequently incorporated into a more extensive model of intrinsic defect chemistry in a-SiO$_2$, immediately raises a question: if the neutral E_1' precursor is essentially an oxygen vacancy of zero size, will it trap a hole? And if it does, will it relax asym-etrically into the E_1' center observed experimentally? One might expect, for example, that when the hole is trapped the remaining electron would be shared by both silicons, which because of confinement of their neighboring atoms would be unable to relax asymmetrically.

Recent calculations by Rudra [83] on the neutral O vacancy in α-quartz indicate that even with structural constraints imposed by the crystalline environment the silicons will approach to $\sim$ 2.5 Å apart, not much larger than the 2.35 Å mentioned above. So even in quartz Robertson's suggestion that the O vacancy is better thought of as a Si-Si bond seems to be valid. But we know already that when this entity in quartz traps a hole, it relaxes asymmetrically into the E_1' center. So we therefore assert that it is likely that the same thing happens in a-SiO$_2$. By studying just how the atoms relax, Rudra found that it is the flexibility of the Si-O-Si bond which allows sufficient freedom of motion, even when the silicons are initially close together, for the silicons to move apart.

Other E'-like centers have been observed in α-quartz and in a-SiO$_2$. The E_2' center [84,85] in α-quartz resembles the E_1' center, only the spin is attached to the "other" silicon (the "long" bond, as opposed to the "short" bond for E_1'). And there is a nearby hydrogen, but probably not within the vacancy. E -type centers associated with germanium [86] and aluminum [87] impurities have also been studied.

Griscom [88] has recently identified so-called E_α' and E_β' centers in a-SiO$_2$, E_α' is a transient defect which will be discussed in Sec. IV, while E_β' involves a hydrogen; it may be the analog of the E_2' center in α-quartz.

The next 4 "established" defects involve holes trapped on oxygens, and their suggested precursors. Since these defects are the subject of another article in this volume [89,90], they are mentioned here only briefly and for the sake of completeness. Two of these defects are observed by spin resonance in a-SiO$_2$ - the NBOHC (non-bridging oxygen hole center) [91,92] and the PR (peroxy radical). In each case the left half of the pictorial is established experimentally while the right half is assumed (with some indirect evidence). Neither precursor is observed directly but their assumed existence is part of a logical sequence of processes of defect chemistry, discussed more fully elsewhere [89,90] and summarized below.

One can envision a sequence of hole trapping events, starting with the NBOHC precursor as shown. The first hole trapped causes a proton to be released, leaving the system neutral and paramagnetic, the NBOHC. The second hole again leads to a proton release, so the system is still neutral; now the two oxygens bond together and form the PR precursor. Finally, a third hole is trapped, forming the positively charged, paramagnetic PR.

The final "established" defect, studied by EPR, involves 3 or 4 H atoms trapped at a Si atom vacancy [93,94]. This may be a key defect in unlocking several aspects of defect chemistry in SiO$_2$.

It illustrates that H can substitute for either O or Si in SiO_2.
This model has in fact been used as a basis for understanding the
hydrolysis of α-quartz [95]. And it should be noted that this
defect may be looked at as two adjacent NBOHC precursors, thus
providing a natural way for such precursors to exist.

One other defect which has been studied in detail and which
seems reasonably well established should be mentioned here: this
is the oxygen divacancy center studied by Bossoli, Jani, and
Halliburton [96,97]. The "divacancy" center is a spin 1 defect with
the most of the spin density located on the two silicons which define the perimeter of the defect.

B. Speculative intrinsic defects in SiO_2

We now turn to the "speculative" defects shown in Fig. 8.
The SPR (small peroxy radical) has been predicted to exist, possibly
in a-SiO_2 but primarily in α-quartz, but to date no experimental
evidence for it has been produced. Edwards and Fowler [90] argued
that such a defect would be produced in α-quartz if an O_2 molecule
were trapped by an E_1 center. It would have a net charge of +1 and
a spin resonance signal similar to that of the PR observed in a-SiO_2,
but with hyperfine interactions of comparable strengths with two
silicons rather than just one.

Three-fold coordinated oxygen(+) has been suggested to occur [81]
in a-SiO_2, as part of valance alternation pair models of defects.
It would presumably occur only in the +1 charge state, which is
diamagnetic; were it to trap an electron it would rebond so that
the oxygen would become two-fold coordinated. No experimental
evidence for its existence is presently available.

Two-fold coordinated silicon is a model suggested by Skuja
et al. [98] to explain luminescence polarization data associated
with the so-called B_2 center in a-SiO_2. The data yield information
about the symmetry of the defect (C_{2v}), and it is argued that such
a model is the most reasonable one which would have that symmetry.
This intriguing suggestion has not yet been tested theoretically.

C. Extrinsic defects and the "missing" trapped electrons

A major puzzle in understanding defects in SiO_2 is how
and where electrons are trapped. All of the paramagnetic defects
discussed in the previous sections involve trapped holes. At
present there are no intrinsic defects which have been convincingly
associated with trapped electrons, and ESR signals associated with
trapped electrons are typically absent after irradiation while
trapped-hole ESR signals are abundant. While one might speculate
that some sort of "negative U" interactions are causing diamagnetic
electron pairs to form, no firm candidates for such defects exist.

The closest candidate for non-magnetic electron trapping comes
from a study some years ago by Lorenze and Feigl [99,100] on
α-quartz containing Al, Ge, and alkali (Na) impurities. Normally
one finds two defects associated with these impurities, Ge
substituted for Si, and Al substituted for Si with a nearby charge-
compensating alkali ion. Upon irradiation at room temperature Ge
traps an electron and Al traps a hole and releases an alkali which
then migrates to a Ge^{-1} (germanium plus trapped electron) [101,102].
Lorenze and Feigl discovered and studied a tendency for the alkalis
to aggregate in the region of the germanium in such a way that the
region remained neutral. They observed by EPR one- and three-alkali
defects and inferred by kinetic analysis the existence of two-alkali
diamagnetic defects. However, such defects have not been observed
in the absence of germainium, so alkali clustering cannot be a
complete explanation of the fate of the missing electrons, even in
cases where enough alkalis are present. Another possibility could
involve the formation of hydrogen molecules [103].

Aluminum is a nearly universal impurity in α-quartz, and spin
resonance and optical properties of holes trapped at Al have been
widely studied and thoroughly reviewed [49,50,104]. Relevant to
the present article is the nature of the trapped hole; at low
temperature it is located in a pπ orbital on one nearest-neighbor
oxygen. At higher temperatures the hole undergoes thermally
activated hops to other nearest-neighbor oxygens. These facts alone
suggest that the model of hole trapsport discussed earlier is
correct; holes tend to self-trap in pπ orbitals on a single oxygen
and behave as small polarons in their motion through the material
[49,50].

This model has been corroborated by ab initio cluster
calculations [105]. Thus properties of a point defect, and its
analysis by a cluster model, in fact lead to valuable insight
concerning hole transport in SiO_2.

One might expect that excitons would self trap in a similar
manner; however, as we will see in the next section, although
exciton self-trapping apparently does exist, the self-trapped
exciton is not at all like a self-trapped hole plus a benign,
weakly-bound electron.

While a number of other extrinsic defects in SiO_2 have been
studied, discussion of them is beyond the scope of this article.

IV. DEFECT PRODUCTION IN SiO_2 BY IONIZING RADIATION

A. Permanent Defects

Although ionizing radiation can lead to transport of
electrons, holes, alkalis, and hydrogen in bulk SiO_2, creation of

stable structural defects such as vacancies and interstitials by
radiolysis is an inefficient process. This conclusion and others
are summarized in recent articles which should be consulted for
further details [2.10].

Creation of E' centers in α-quartz by X-irradiation has been
studied for some time, and it has been found that the number of E'
centers saturates with radiation dose [106]. The simplest explanation for this is the ionization of pre-existing vacancies which
ceases as all the vacancies become ionized. In general one should
also consider alternative explanations, such as vacancy creation
by radiation which saturates because of a radiation-induced back
reaction; this, however, is probably not occurring in α-quartz.
And particularly in the case of E' centers one should not neglect
the importance of electron trapping by other defects as a possible
limiting process. A recent paper by Jani and Halliburton [97]
summarizes some of the still-puzzling aspects of E_1-center
creation in α-quartz by ionizing radiation.

There is evidence for defect creation by ionization in α-quartz
by means of electrons whose source is an electron microscope [107-111].
Some of this clearly takes place near structural defects or water
impurities, but a portion appears not to be related to pre-existing
defects [107-109]. It has been suggested that in the high electron
flux of the electron microscope double ionization may be the source
of the latter type of defect creation [112].

Defect creation in a-SiO₂ by ionizing radiation has long been
studied and attempts have been made to explain it by the breaking of "strained bonds [113-115]. Such may occur, although
calculations by Edwards and Fowler [90] suggest that a rather large
amount of strain ($\approx$ 50%) may be required for an ionized bond to
break. It should also be noted that when an oxygen pπ electron is
removed from the upper valance bands there does not exist a "broken
bond" in the literal sense, but simply a missing non-bonding
electron. A hole in the lower "bonding" valance bands may be more
effective in leading to damage, and we have argued [10] that in
α-quartz and even, perhaps, in a-SiO₂, such a hole might have a
sufficiently long lifetime in the bonding band to allow relaxation
to occur.

A possible alternative to the bond-breaking process is the
interatomic Auger process which Knotek and Feibelman [116,117] have
shown to be effective in the desorption of ionic materials near the
surface. This is a Coulomb effect: an anion becomes positively
charged and thus is likely to be ejected. Just how important this
mechanism is in the bulk is not known, in particular how often a
positively charged oxygen would escape from its normal site before
conduction electrons recombine with it.

Griscom and Friebele's recent studies of E' center creation and thermal annealing in a-SiO_2 have led them to suggest the existence of three varieties of E' centers, depending upon how they are created and the amount of hydrogen in the material [88,118]. 100 keV X-rays, which do not cause knock-on damage, create "E_α'" centers in dry a-SiO_2. These bleach below room temperature and have a large orthorhombic component in their g matrix as seen in EPR. It is suggested by Griscom and Friebele that E_α' centers are formed by radiolysis but that the displaced oxygen is nearby, perhaps forming a peroxy bridge.

E_β' centers are also created by X-rays, in high-OH a-SiO_2. These are suggested to occur by a reaction of a pre-existing Si^+ with a mobile H atom, which then leads to a 3-coordinated Si^0 (an E' center) plus a proton which then diffuses away. Whether such a reaction is energetically feasible remains to be seen.

E_γ' centers are created by γ rays which are sufficiently energetic to lead to knock-on damage (along with considerable ionization). These are assumed to resemble the E_1 center in α-quartz; they are thermally stable to ~ 800 K, presumably because the interstitial oxygen is far away.

B.   Transient Defects

In contrast with the difficulty in generating permanent defects in SiO_2 by radiolysis, it appears that transient defects can be readily generated by ionizing radiation. Some ten years ago Griscom, Sigel and co-workers [3,4] reported on the short-lived optical absorption and luminescence found in both quartz and a-SiO_2 during 500 keV pulsed electron irradiation. The similarity of these transient optical spectra with those of certain permanent defects, particularly the E' center, led to the suggestion that short-lived oxygen vacancies were being created which resembled E' centers. This argument fits nicely with the discussion of the previous section. In this context the E_α' centers are similar to transient E' centers, but they are stable in certain temperature ranges.

More recent work seems to be corroborating the transient vacancy picture while suggesting the location of the transient interstitial. Tanimura, Tanaka, and Itoh [112] were able to measure a transient expansion of the quartz crystal during the excitation process which was consistent with vacancy formation. They also correlated the time dependence of decay with that of absorption and luminescence. Very recently Hayes et al. [119] have carried out ODMR (optical detection of magnetic resonance) on a component of the well-known blue luminescence which peaks at 2.6 eV at low temperature. They conclude that a triplet state is involved in the luminescence and obtain principal axes of that triplet. While other interpretations may be forthcoming, their results are consistent with a transient vacancy occurring adjacent to a transient peroxy linkage. This would be a self-trapped exciton with an enormous relaxation and, indeed, a very large Stokes' shift ($\sim$ 9–10 eV absorption, 2.6 eV emission).

C. Discussion

While the recent experiments cited above on transient defects have been carried out on α-quartz, a similar situation may well occur in a-SiO₂. This process does not appear to lead to permanent defect formation in α-quartz; rather, the luminescence occurs with high efficiency after which the sytem returns to the original perfect α-quartz structure. It may sometimes lead to permanent defect formation (the E_α' center) in a-SiO₂.

While most of the experiments related to this conclusion have utilized high-energy electrons or X-rays, at least one example exists where band-gap excitation has been used. Trukhin and Plaudis [120] found that at 77K the excitation spectrum for the 2.6 eV luminescence peaks at 10.2 eV, and suggested that the luminescence is due to the decay of a self-trapped exciton. Thus it appears that there is nothing pathological about X-ray or electron excitation, although further studies with band-gap excitation would be desirable.

It is indeed remarkable that such a large atomic relaxation can accompany the existence of a triplet exciton in α-quartz, and the physics (or chemistry) of the relaxation mechanism remains to be explored. This slightly resembles the creation of the intimate F center – H center pair in alkali halides. It should be noted that the assignment of Hayes et al. differs in detail from earlier ones in that it involves a neutral oxygen vacancy rather than a E' center. This creates two difficulties; first, there is no evidence that the neutral oxygen vacancy should have an optical absorption band near the position of the E' center, as is observed. Second, as discussed in Sec. IIIA, the neutral oxygen vacancy is expected to lead to contraction, not expansion, of the material, in apparent contradiction with the experimental results of Tanimura et al. [112].

As in the alkali halides, the trapped electron seems critical to the relaxation process: a self-trapped hole by itself does not lead to defect formation. Just how the electron participates is not clear; Toyozawa [121] suggested the existence of an adiabatic instability in the case of the alkali halides, but there is evidence that this may not be appropriate [122].

Although no calculations have yet been reported which bear on the stability of this self-trapped exciton, it is reasonable to assume that the system has found the lowest energy atomic arrangement consistent with its being an excited electronic state; in a configuration-coordinate picture it is at the minimum of the

excited-state curve. Apparently there are no low-energy cross-
overs to the ground-state configuration coordinate since radiative
de-excitation occurs.

It seems clear that we are at a very early stage in understanding this process, and there is little doubt but that the
next few years will see considerable progress in delineating the
mechanisms involved.

V.   ACKNOWLEDGMENTS

Discussions with Frank J. Feigl, David L. Griscom, Arthur H.
Edwards, and Jayanta K. Rudra were helpful in developing parts of
this work. This research was supported by the U.S. Navy Office
of Naval Research, Contract No. N00014-81-0005.

REFERENCES

1. D. L. Griscom, J. Non-Cryst. Solids 24:155 (1977).
2. E. J. Friebele and D. L. Griscom, in Treatise on Materials
      Science, Vol. 17, Academic Press, New York, (1979).
3. D. L. Griscom, Proc. 32nd Freq. Control Symp., Atlantic City,
      NJ, 98 (Electronic Industries Assoc. Washington, DC, (1979).
4. G. H. Sigel, Jr., J. Non-Cryst. Solids 13:372 (1973-74).
5. G. N. Greaves, Phil. Mag. B 37:447 (1978).
6. The Physics of SiO$_2$ and its Interfaces, S. T. Pantelides, ed.
      Pergamon Press, New York (1978).
7. The Physics of MOS Insulators, G. Locovsky, S. T. Pantelides,
      and F. L. Galeener, eds., Pergamon Press, New York (1980).
8. D. L. Griscom, Ref. 6, p. 232.
9. Radiation Effects in Insulators, G. W. Arnold and J. A. Borders,
      eds., North-Holland (1984). Reprinted from Nuclear Instru-
      ments and Methods in Phys. Research B1(1984).
10. W. Beall Fowler, Semicond. Insul. 5:583 (1983); Rad. Effects
      72:27 (1983).
11. D. L. Griscom, Symposium on Glass Science and Technology,
      Problems and Prospects for 2004 (1984).
12. J. Robertson, Phys. Chem. Glasses 23:1 (1982).
13. J. A. Weil, Phys. Chem. Minerals 10:149 (1984).
14. P. M. Schneider and W. Beall Fowler, Phys. Rev. Lett. 36:425
      (1976); Phys. Rev. B 18:7122 (1978).
15. J. R. Chelikowsky and M. Schluter, Phys. Rev. B 15:4020 (1977).
16. E. Calabrese and W. Beall Fowler, Phys. Rev. B 18:2888 (1978).
17. R. N. Nucho and A. Madhukar, Phys. Rev. B 21:1576 (1980).
18. R. B. Laughlin, Phys. Rev. B 22:3021 (1980).
19. R. B. Laughlin, J. D. Joannopoulos, and D. J. Chadi, Phys. Rev.
      B 20:5228 (1979).
20. S. Ciraci and I. P. Batra, Phys. Rev. B 15:4923 (1977).
21. E. Calabrese, Nuovo Cimento 3D:361 (1984).
22. W. Y. Ching (present volume).

23. J. K. Rudra and W. B. Fowler, Phys. Rev. B 28:1061 (1983).
24. Structure and Bonding in Crystals, M. O'Keeffe and A. Navrotsky, eds., Academic Press, New York, (1981).
25. A. R. Ruffa, Phys. Rev. Lett. 25:650 (1970).
26. A. J. Bennett and L. M. Roth, J. Phys. Chem. Solids 32:1251 (1971).
27. K. L. Yip and W. B. Fowler, Phys. Rev. B 10:1400 (1974).
28. K. L. Yip and W. B. Fowler, Phys. Rev. B 11:2327 (1975).
29. J. A. Tossell, J. Phys. Chem. Solids 34:307 (1973).
30. J. A. Tossell, J. Am. Chem. Soc. 97:4840 (1975).
31. R. Meier and T. -K. Ha, Phys. Chem. Minerals 6:37 (1980).
32. M. D. Newton and G. V. Gibbs, Phys. Chem. Minerals 6:221 (1980).
33. C. A. Ernst, A. L. Allrod, M. A. Ratner, M. D. Newton, G. V. Gibbs, J. W. Moskowitz, and S. Topid, Chem. Phys. Lett. 81:424 (1981).
34. G. Lucovsky, Phil. Mag. B 41:457 (1980).
35. J. Robertson, Phys. Chem. Glasses 23:1 (1982).
36. K. Shimakawa and A. Kondo, Phys. Rev. B 27:1136 (1983).
37. R. F. Stewart, M. A. Whitehead, and G. Connay, Am. Min. 65:324 (1980).
38. M. Rosenberg, F. Martoni, W. A. Reed, and P. Eisenberger, Phys. Rev. B 18:844 (1978).
39. W. Beall Fowler, J. Phys. Chem. Solids 42:623 (1981).
40. M. O'Keeffe, in Ref. 24, p. 299.
41. J. C. Phillips, Solid State Physics 37:93 (1982).
42. M. O'Keeffe and B. G. Hyde, Acta Cryst. B 34:27 (1978).
43. Y. T. Thathachari and W. A. Tiller, J. Appl. Phys. 53:8615 (1982).
44. M. H. Reilly, J. Phys. Chem. Solids 31:1041 (1970).
45. R. W. G. Wyckoff, Crystal Structure, Wiley (1963).
46. H. D. Megaw, Crystal Structures: A Working Approach, Saunders, (1973).
47. S. T. Pantelides and W. A. Harrison, Phys. Rev. B 13:2667 (1976).
48. W. B. Fowler, unpublished.
49. R. C. Hughes, Phys. Rev. Lett. 30:1333 (1973).
50. R. C. Hughes, Rad. Effects 26:225 (1975); Phys. Rev. B 15:2012 (1977).
51. G. Wiech, Soft X-ray Band Spectra, D. J. Fabian, ed., Academic Press (1968).
52. G. Klein and H. -U. Chun, Phys. Stat. Sol. B 49:167 (1962).
53. G. Weich, E. Zopf, H. -U. Chun, and R. Bruckner, J. Non-Cryst. Solids 21:251 (1976).
54. B. Fischer, R. A. Pollak, T. H. DiStefano, and W. D. Grobman, Phys. Rev. B 15:3193 (1977).
55. H. R. Philipp, Solid State Commun. 4:73 (1966); J. Phys. Chem. Solids 32:1935 (1971).
56. M. Rossinelli and M. A. Bosch, Phys. Rev. B 25:6482 (1982).
57. R. S. Knox, Theory of Excitons, Academic Press (1963).
58. R. S. Knox, Phys. Chem. Solids 9:238, 265 (1959).
59. W. R. Heller and A. Marcus, Phys. Rev. 84:809 (1951).
60. D. L. Dexter, J. Chem. Phys. 21:836 (1953).

61. H. Haken and P. Reineker, Z. Physik 249:253 (1972).
62. R. S. Knox and K. J. Teegarden, Physics of Color Centers,
    W. B. Fowler, ed., Academic Press (1968).
63. N. F. Mott and A. M. Stoneham, J. Phys. C 10:3391 (1977).
64. R. L. Mozzi and B. E. Warren, J. Appl. Crystallogr. 2:164 (1969).
65. S. K. Mitra, Phil. Mag. B 45:529 (1982).
66. M. Grimsditch, Phys. Rev. Lett. 52:2379 (1984).
67. A. R. Silin, P. J. Bray, and J. C. Mikkelsen, Jr., J. Non-Cryst.
    Solids 64:185 (1984).
68. W. Y. Ching, Phys. Rev. Lett. 46:607 (1981); Phys. Rev. B 26:6622
    (1982).
69. J. Isoya, J. A. Weil, and L. E. Halliburton, J. Chem. Phys. 74:5436
    (1981).
70. A. H. Edwards, Ph.D. dissertation, Lehigh University (1981).
71. A. H. Edwards and W. Beall Fowler, J. Phys. Chem. Solids 46:841 (1985).
72. R. A. Weeks, J. Appl. Phys. 27:1376 (1956).
73. R. A. Silsbee, J. Appl. Phys. 32:1459 (1961).
74. P. M. Lenahan and P. V. Dressendorfer, J. Appl. Phys. 55:3495
    (1984).
75. G. Hochstrasser and J. F. Antonini, Surf. Sci. 32:644 (1972).
76. J. Arends, A. J. Dekker, and W. G. Perdole, Phys. Stat. Sol.
    3:2275 (1963).
77. F. J. Feigl, W. Beall Fowler, and K. L. Yip, Solid State Commun.
    14:225 (1974).
78. K. L. Yip and W. Beall Fowler, Phys. Rev. B 11:2327 (1975).
79. M. G. Jani, R. B. Bossoli, and L. E. Halliburton, Phys. Rev. B
    27:2285 (1983).
80. D. L. Griscom, Phys. Rev. B 22:4192 (1980).
81. Eoin P. O'Reilly and John Robertson, Phys. Rev. B 27:3780 (1983).
82. J. Robertson, J. Phys. C 17:L221 (1984).
83. J. K. Rudra, unpublished.
84. R. A. Weeks, Phys. Rev. 130:570 (1963).
85. J. G. Castle, D. W. Feldman, P. G. Klemens, and R. A. Weeks,
    Phys. Rev. 130:577 (1963).
86. F. J. Feigl and J. H. Anderson, J. Phys. Chem. Solids 31:575
    (1970).
87. K. L. Brower, Phys. Rev. Lett. 41:879 (1978).
88. D. L. Griscom, Ref. 9, p. 481.
89. A. H. Edwards and W. Beall Fowler, present volume.
90. A. H. Edwards and W. Beall Fowler, Phys. Rev. B 26:6649 (1982).
91. M. Stapelbroek, D. L. Griscom, E. J. Friebele, and G. H. Sigel, Jr.,
    J. Non-Cryst. Solids 32:313 (1979).
92. D. L. Griscom and E. J. Friebele, Phys. Rev. B 24:4896 (1981).
93. R. H. D. Nuttall and J. A. Weil, Solid State Commun. 33:99 (1980).
94. J. A. Weil, to be published.
95. A. C. McLaren, R. F. Cook, S. T. Hyde, and R. C. Tobin, Phys.
    Chem. Minerals 9:79 (1983).
96. R. B. Bossoli, M. G. Jani, and L. E. Halliburton, Solid State
    Commun. 44:213 (1982).
97. M. G. Jani and L. E. Halliburton, J. Appl. Phys. 56:942 (1984).

98. L. N. Skuja, A. N. Streletsky, and A. B. Pakovich, Solid State
       Commun. 50:1069 (1984).
99. R. V. Lorenze and F. J. Feigl, Phys. Rev. B 8:4833 (1973).
100. R. V. Lorenze, Ph.D. dissertation, Lehigh University (1972).
101. J. H. Mackey, Jr., J. Chem. Phys. 39:74 (1963).
102. J. A. Weil, J. Chem. Phys. 55:4685 (1971).
103. D. L. Griscom, to be published.
104. J. A. Weil, Rad. Effects 26:261 (1975).
105. M. J. Moboquette, J. A. Weil, and P. G. Mezey, Can. J. Phys.
       62:21 (1984).
106. R. W. Ditchburn, E. W. J. Mitchell, E. G. S. Paige, J. F.
       Custers, H. B. Dyer, and C. D. Clark, Defects in Crystalline
       Solids, Physical Society of London, p. 92 (1955).
107. L. W. Hobbs and M. R. Pascucci, J. Phys. C 41:237 (1980).
108. M. R. Pascucci, J. L. Hutchinson, and L. W. Hobbs, The
       Scientific Basis for Nuclear Waste Management, S. V. Topp, ed.,
       Elsevier, p. 689 (1982).
109. M. R. Pascucci, J. L. Hutchinson, and L. W. Hobbs, Radiation
       Effects 74:219 (1983).
110. D. Cherns, J. L. Hutchinson, M. L. Jenkins, P. B. Hirsch, and
       S. White, Nature 287:314 (1980).
111. C. B. Carter and D. L. Kohlstedt, Phys. Chem. Minerals 7:110
       (1981).
112. K. Tanimura, T. Tanaka, and N. Itoh, Phys. Rev. Lett. 51:423
       (1983).
113. C. M. Nelson and J. H. Crawford, Jr., J. Phys. Chem. Solids
       13:296 (1960).
114. G. W. Arnold and W. D. Compton, Phys. Rev. 116:802 (1959).
115. C. W. Gwyn, J. Appl. Phys. 40:4886 (1969).
116. M. L. Knotek and P. J. Feibelman, Phys. Rev. Lett. 40:964 (1978);
       P. J. Feibelman and M. L. Knotek, Phys. Rev. B 18:6531 (1978).
117. M. L. Knotek, Semicond. Insul. 5:361 (1983).
118. D. L. Griscom and E. J. Friebele, Rad. Effects 65:63 (1982).
119. W. Hayes, M. J. Kane, O. Salminen, R. L. Wood, and S. P. Doherty,
       J. Phys. C 17:2943 (1984).
120. A. N. Trukhin and A. E. Plaudis Sov. Phys. Solid State 21:644
       (1979).
121. Y. Toyozawa, J. Phys. Soc. Japan 44:482 (1978).
122. K. Tanimura and N. Itoh, Semicond. Insul. 5:473 (1983).

RAMAN STRUCTURAL CORRELATIONS FROM STRESS-MODIFIED AND BOMBARDED

VITREOUS SILICA

G. E. Walrafen and M. S. Hokmabadi

Chemistry Department, Howard University
Washington, DC 20059

I. INTRODUCTION

This chapter deals with angle and length correlations related
to microscopic structures in vitreous silica. These correlations
result from changes in the vitreous silica structure produced by
stress modification and particle bombardment -- they were determined
by Raman spectroscopy of samples in optical fiber as well as bulk
form. In this work, stress modification refers to the following
conditions: (1) large reversible tensile stress, optical fiber,
(2) irreversible uniaxial compaction, $\geq$ 90 kbar, bulk, (3) reversible
hydrostatic compression at room temperature, and at elevated
temperature, bulk and (4) large reversible torsion, optical fiber.
Particle bombardment refers to: (5) intense neutron bombardment,
bulk, and (6) prolonged bombardment by alpha particles, mostly bulk.

The angle and length correlations refer to the following three
microscopic quantities: (i) the mean equilibrium dihedral or tilt
angle, $\bar{\delta}$, between adjoining SiO_4 tetrahedra that share a common
oxygen atom, (ii) the mean equilibrium Si-O-Si bridging angle, $\bar{\phi}$,
formed between adjoining SiO_4 tetrahedra, 144° for ordinary fused
silica, and (iii) the mean equilibrium Si-O bond length, $\bar{r}$.

Specifically, the correlation as deduced from Raman spectra is
that a decrease in $\bar{\phi}$ produces a corresponding increase in both $\bar{r}$
and $\bar{\delta}$, or vice versa. Stated alternatively, both $\bar{r}$ and $\bar{\delta}$ are
negatively correlated with $\bar{\phi}$. These changes in $\bar{\phi}$, $\bar{r}$, and $\bar{\delta}$ can be
visualized by considering a hypothetical single structure whose
angles and distances are the average values for the entire silica
network.

185

Consider the nonlinear structure $O_3Si-O-SiO_3$ formed when two SiO_4 tetrahedra share an oxygen atom. When the nonlinear Si-O-Si bridging angle of this structure, ϕ, decreases, repulsion between the two nearest Si atoms, as well as between the nonbridging oxygen atoms, i.e., the oxygen atoms of the SiO_3 units, increases, or vice versa. Two effects then occur which counteract this repulsion and lower the energy, as follows. The first of these effects is an increase in the mean equilibrium Si-O bond distance, $\bar{r}$. This increase in $\bar{r}$ occurs because it counteracts the repulsion by moving all atoms in the structure farther apart. The second effect, which also occurs to counteract the repulsion, involves counter rotations of the SiO_3 units around the O-Si bonds of the nonlinear Si-O-Si bridge. Here $\bar{\delta}$ increases as the second-neighbor oxygen atoms of the two SiO_3 units move farther apart. Hence, the repulsion is lowered even more. The basic feature of the $O_3Si-O-SiO_3$ structure with average equilibrium $\bar{\phi}$, $\bar{r}$, and $\bar{\delta}$ values, is that it can be used as an approximation for the complicated cooperative changes that must occur in the entire silica network, yet its steric-like repulsive effects are easily visualized.

Of course, in obtaining the preceding structural correlations, it is necessary to connect specific Raman spectral changes with changes in $\bar{\phi}$, $\bar{r}$, and $\bar{\delta}$. The relationships, briefly, are as follows: (j) a Raman peak from vitreous silica near 800 cm^{-1} is thought to refer to the bending vibration of the nonlinear Si-O-Si bridge. A decrease in the mean equilibrium Si-O-Si bridging angle, $\bar{\phi}$, is thus thought to correspond to an increase in the frequency of the nominal 800 cm^{-1} peak. (jj) Two Raman lines from vitreous silica near 1060 and 1200 cm^{-1} are thought to refer to Si-O stretching. A decrease in frequency for these lines is considered to occur when the Si-O bond distance, $\bar{r}$, increases. (jjj) An increase in $\bar{\delta}$ is inferred to correspond to an intensity decrease of a contour component peaking in the vicinity of 200-300 cm^{-1}, as explained subsequently in detail. The broad 200-300 cm^{-1} component is part of a much broader contour extending from 0-450 cm^{-1}, and the intensity decrease of the 200-300 cm^{-1} component is seen visually in the spectra as an asymmetric low-frequency narrowing of the 0-450 cm^{-1} contour.

The reasons behind the Raman spectral changes relating to changes in $\bar{\phi}$, and in the negatively correlated quantities, $\bar{r}$ and $\bar{\delta}$, are presented next. Then the Raman data corresponding to the stress-modification and particle-bombardment conditions, (1)-(6), will be presented in that order. However, a subordinate theme develops as the Raman data are presented. This refers to two "defect" Raman lines near 490 and 600 cm^{-1}. Changes in the 490 and 600 cm^{-1} features from vitreous silica will be detailed for conditions (1)-(6), but no structural correlations will be made for them. The structural correlations that are presented in this chapter refer solely to angles and distances of the microscopic structures of the main equilibrium network structure of vitreous silica.

Relationship Between Raman Spectral Changes and Silica Microstructures

The 800 cm^{-1} Raman line from vitreous silica is thought to refer to bending (deformation vibration) of the nonlinear Si-O-Si bridge [1]. When the mean equilibrium Si-O-Si bridging angle, $\overline{\phi}$, decreases, repulsion between the two nearest-neighbor Si atoms, and between the second-neighbor O atoms (between O atoms of the SiO_3 group) increases. Hence, an increase in the frequency of the nominal 800 cm^{-1} feature, means that the bending or deformation force constant has increased because the repulsive forces have increased, i.e., an increase of Raman frequency at $\sim$ 800 cm^{-1} means that the mean equilibrium Si-O-Si bridging angle, $\overline{\phi}$, has decreased.

The Raman lines from vitreous silica near 1060 and 1200 cm^{-1} arise from Si-O stretching [2]. A decrease in the frequencies of these lines means that the Si-O stretching force constant has decreased [3]. But it is well known [4] that an increase in bond length decreases the bond stretching force constant, which in turn decreases the vibrational frequency. Thus, a decrease in Raman frequency of the nominal 1060 and 1200 cm^{-1} features means that the mean equilibrium Si-O bond distance has increased.

Thus far, it is evident from Raman considerations that a decrease in $\overline{\phi}$ corresponds to an increase in $\overline{r}$, i.e., negative correlation. Phenomenologically, this means that the 800 cm^{-1} peak position moves upward in frequency when the 1060 and 1200 cm^{-1} peak positions move downward. Clearly, only the additional correlation between $\overline{\phi}$ and $\overline{\delta}$ remains to be established.

When the 800 cm^{-1} peak position moves upward in frequency and the 1060 and 1200 cm^{-1} positions move downward, it has been observed in every case that a low-frequency sharpening of the 0-450 cm^{-1} Raman contour occurs, and vice versa. No exception to this observation has yet been found for vitreous silicas whose densities are large, i.e., roughly 2.2-2.7 gm cm^{-3}. A relationship between the low-frequency Raman contour sharpening (or broadening), and the mean equilibrium dihedral or tilt angle, $\overline{\delta}$, is presented below.

It has been shown previously by Gaussian analysis of the Raman contour from fused silica, [5] that a component centered roughly near 235 cm^{-1} (the 200-300 cm^{-1} component mentioned before) makes a very significant contribution (76% of the most intense component). Similar Gaussian decompositions of Raman contours shown subsequently here, indicate either that low-frequency narrowing or broadening of the 0-450 cm^{-1} contour can be associated with a decrease or an increase, respectively, in the integrated intensity of the very broad component nominally at 235 cm^{-1}. Further, because of the low frequency involved, it is not unreasonable to associate the 235 cm^{-1} component with torsional oscillations of the bent $O_3Si-O-SiO_3$

structure, that is, the torsional oscillation in which the O₃Si groups rotate <u>counter</u> to each other. However, X-ray data from alpha-quartz under pressure lead to much more compelling conclusions.

Jorgensen [6] has conducted precise X-ray measurements of the Si-O-Si bridging angle, $\bar{\phi}$, of the dihedral or tilt angle, $\bar{\delta}$, and of the Si-O bond distance, $\bar{r}$, for crystalline quartz at a series of pressures to 28.2 kbar. It is shown subsequently that these X-ray results for alpha-quartz parallel all of the present Raman results for vitreous silica so closely, that it seems virtually certain that they provide a good model for the effects of various stresses on vitreous silica, as well as for alpha-quartz.

Jorgensen observed that $\bar{\phi}$ decreased from 144.1° to 138.7°, as $\bar{\delta}$ increased from 15.9° to 20.4° (and he observed the same general trend for the alpha-quartz form of crystalline GeO₂). This large negative correlation was the principal effect noted by Jorgensen. The negative correlation between $\bar{\phi}$ and $\bar{r}$ for vitreous silica has been established from previous Raman considerations, hence, only the correlation between $\bar{\phi}$ and $\bar{\delta}$ remains. Because the negative correlation between $\bar{\phi}$ and $\bar{\delta}$ is so obvious for alpha-quartz (and the corresponding alpha-quartz form of GeO₂), this correlation is now considered to apply to vitreous silica as well. However, the only remaining Raman effect that can be reasonably associated with an increase of $\bar{\delta}$ is the low-frequency sharpening of the 0-450 cm⁻¹ Raman contour. This effect occurs simultaneously with an increase of the Raman frequency at 800 cm⁻¹, which in turn refers to a decrease of $\bar{\phi}$. <u>Hence, the sharpening of the 0-450 cm⁻¹ Raman con-</u> <u>tour from vitreous silica is assigned here to an increase in the</u> <u>mean equilibrium dihedral or tilt angle, $\bar{\delta}$, and vice versa.</u>

Once the three connections between, (1) the 0-450 cm⁻¹ Raman contour sharpening and an increase of $\bar{\delta}$, (2) the increase in the 800 cm⁻¹ peak position with decrease of $\bar{\phi}$, and (3) the decrease of the 1060 and 1200 cm⁻¹ peak positions with increase of $\bar{r}$ have been made, it can be shown that all of Jorgensen's results for alpha-quartz scale in direction and even roughly in magnitude with the corresponding Raman effects for fused silica.

The smallest of the Raman effects in terms of percentage change refers to changes in the positions of the 1060 and 1200 cm⁻¹ peaks. These frequency changes refer to changes in the Si-O bond length, $\bar{r}$. In regard to this, Jorgensen observed no significant change in the average Si-O bond length, $\bar{r}$, for alpha-quartz up to 28.2 kbar. However, two of his distance values at 21.1 and 23.8 kbar are noticeably larger than the 1 atm. value, namely, 1.610 and 1.612 Å versus 1.607 Å, respectively. An increase in the Si-O bond distance of 0.003-0.005 Å, can probably be seen by Raman spec-troscopy, and would correspond to a <u>decrease</u> in frequency, in agreement with an increase in $\bar{\delta}$ and a decrease in $\bar{\phi}$. Such a small distance increase might be negligible in the X-ray method.

As an example, the Raman OH stretching frequency for ice VII has been found to change by about 140 cm^{-1} for a change of 0.1 Å in the O-O distance [4]. Because a change of 5 cm^{-1} is readily detectable in the Raman frequency, it is evident that a change of 0.004 Å can be detected in the O-O distance for ice VII. Similar sensitivity would be expected from Raman methods for other systems.

In the case of the reversible hydrostatic compression of fused silica to ∿ 10 kbar (to be detailed subsequently) a downward frequency shift (∿ 0.5%) was observed for the 1060 cm^{-1} peak due to Si-O stretching. An upward frequency shift of ∿ 0.8% was observed for the 800 cm^{-1} peak due to Si-O-Si bending. And, an intensity decrease of the 0-450 cm^{-1} Raman contour of about 12% was observed. For comparison, Jorgensen's data suggest a possible very slight increase in $\bar{r}$ (which would correspond to a small frequency decrease), and he observed a 3.7% decrease in $\bar{\phi}$ (corresponding to a frequency increase), and a 28% increase in $\bar{\delta}$ (corresponding to loss of Raman intensity below 450 cm^{-1}). Clearly, there is a close parallel between the X-ray results for alpha-quartz, and the Raman results for fused silica, both under reversible hydrostatic pressure, and this parallel refers both to magnitude as well as to the sign of the effect. Hence, it seems reasonable to use Jorgensen's negative $\bar{\phi}$ versus $\bar{\delta}$ correlation, which is the largest effect reported for alpha-quartz, for vitreous silica, as well.

The Raman effects (changes) for vitreous silica that are considered to correspond to changes in $\bar{\phi}$, $\bar{\delta}$, and $\bar{r}$ are presented in Table I. Again, it should be strongly stressed that only the negative $\bar{\phi}$ versus $\bar{\delta}$ correlation is derived from the X-ray data for alpha-quartz. The negative correlation between $\bar{\phi}$ and $\bar{r}$ comes from well-established Raman criteria.

Large Reversible Tensile Stress

A Raman investigation of the effects of tensile stress to 33 kbar on pure fused silica optical fibers was reported recently [7]. Raman spectra derived from that work are shown in Fig. 1, (a) stressed to 33 kbar, (b) unstressed.

The Raman spectra of Fig. 1 were transferred to horizontal baselines using a baseline template as described in Ref. 7. They were then decomposed into Gaussian components using a duPont 310 analog computer as described in Ref. 5. In the region from 0-450 cm^{-1}, five Gaussian components were employed. However, only the three components A, B, and C, of interest in the present work (two components used to fit the asymmetric 60 cm^{-1} peak are not shown) appear in the figure, for simplicity. The sharp 490 cm^{-1} component was also not included in the analysis, as shown by the dashed nearly linear baseline in this region. In addition, a second dashed line between about 200-400 cm^{-1} shown in the (a) spectrum, is the (b) spectral intensity normalized at the 450 cm^{-1} peak position.

Table I.   Raman changes thought to correspond to changes in $\bar{\phi}$, $\bar{r}$, and $\bar{\delta}$ for vitreous silica.

INCREASE OF FREQUENCY FOR THE NOMINAL 800 cm^{-1} PEAK.	DECREASE OF THE MEAN EQUILIBRIUM Si-O-Si BRIDGING ANGLE, $\bar{\Phi}$.
DECREASES IN THE FREQUENCIES OF THE TWO PEAKS NOMINALLY AT 1060 cm^{-1} AND 1200 cm^{-1}	INCREASE IN THE MEAN EQUILIBRIUM Si-O BOND DISTANCE, $\bar{r}$.
LOSS OF RAMAN INTENSITY BETWEEN 0-450 cm^{-1} DUE TO INTENSITY LOSS OF COMPONENT CENTERED BETWEEN 200-300 cm^{-1}, *e.g.*, THE BROAD 235 cm^{-1} COMPONENT IN THE CASE OF TENSILE STRESS	INCREASE IN THE MEAN EQUILIBRIUM DIHEDRAL ANGLE, $\bar{\delta}$.

Comparisons between (a) and (b) of Fig. 1 indicate that the 490 cm^{-1} "defect" peak increases in intensity with increasing tensile stress, as detailed previously [7]. Also, the fact that the normalized (b) intensity, dashed line, between about 200-400 cm^{-1} lies below the solid line, indicates that an intensity increase occurred in this region with increasing tensile stress. This increase in intensity is evident in the Gaussian contour decomposition

The three Gaussian components shown in Fig. 1, designated A, B, and C in order of decreasing Raman frequency, are centered near 475, 375, and 235 cm^{-1}. The integrated intensities of components A, B, and C were determined, and the total integrated intensity in the region from 0-900 cm^{-1} was also determined for purposes of quantitative comparison. Here, both the 490 and 600 cm^{-1} "defect" components were removed as shown, for example, by the baseline below the 490 cm^{-1} component (a) and (b). Ratios of integrated component intensities and of total contour intensities are presented in Table II.

In Table II the ratios $(I_C/I_T)/(I^o_C/I^o_T)$ and $(I_C/I_{A+B})/(I^o_C/I^o_{A+B})$ are listed. The superscript (^{o}) refers to a condition of zero tensile stress. A slight increase above unity is observed for both ratios, which indicates that the intensity increase between 200-400

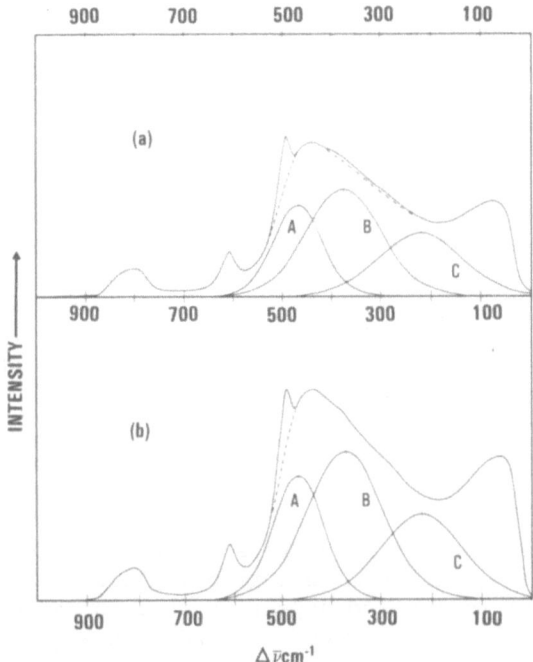

Fig. 1 Raman spectra from vitreous silica optical fibers under a
tensile stress of 33 kbar, (a), and unstressed, (b).

Table II Raman integrated intensity ratios for vitreous silica
under various conditions. The superscript (^{o}) refers
to the normal or unaffected condition, e.g., unstressed.

CONDITION	$\dfrac{I_C/I_T}{I^o_C/I^o_T}$	$\dfrac{I_C/I_{A+B}}{I^o_C/I^o_{A+B}}$
REVERSIBLE TENSILE STRESS, 33 KBAR	1.01	1.06
IRREVERSIBLE UNIAXIAL COMPACTION, 90 KBAR	0.87	0.88
REVERSIBLE HYDROSTATIC COMPRESSION, 10 KBAR	0.85	0.77

cm^{-1} shown by the dashed line in (a) arises primarily from the
broad C component centered near 235 cm^{-1}. Here, it should be
remembered that the normalized (b) intensity is also only slightly
smaller than the (a) intensity.

The two important effects of tensile stress on the Raman spectrum of fused silica are thus that the 490 cm^{-1} "defect" intensity
and the $\sim$ 235 cm^{-1} component intensity both <u>increase</u> with increasing
tensile stress. In terms of the correlation of Table I, the increase
of the $\sim$ 235 cm^{-1} component C intensity means that the mean equilibrium dihedral angle, $\bar{\delta}$, <u>decreases</u> with increasing tensile stress.
Also, as seen from the work of Jorgensen, it is apparent that the
effect of uniaxial tensile stress on $\bar{\delta}$ is just the opposite of that
due to pressure, i.e., the effects of elongation and compaction on
$\bar{\delta}$ are opposite to each other.

Irreversible Uniaxial Compaction

Bridgman [8] was the first to show that application of pressure
in excess of 90 kbar at room temperature produces an irreversibly
compacted form of silica whose density is much larger than that of
ordinary fused silica, but whose structure is amorphous, as determined from the lack of X-ray diffraction. A Raman spectrum corresponding to fused silica subjected to prolonged uniaxial compaction
at 90 kbar is shown in Fig. 2(a). This spectrum, and the uncompacted
spectrum, Fig. 2(b), were derived from Ref. 9 in the same manner as
described for Fig. 1.

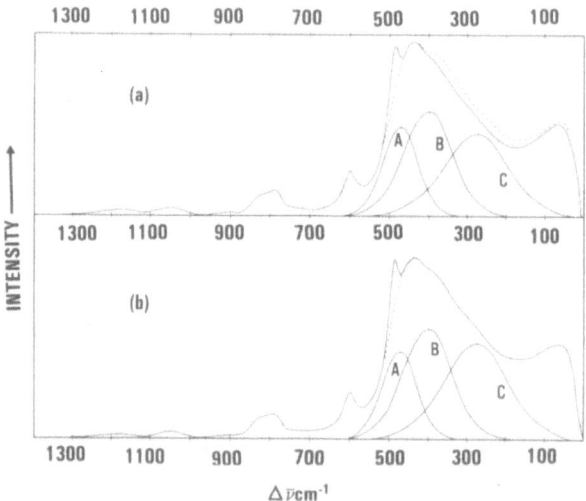

Fig. 2   Raman spectra from irreversibly compacted vitreous silica,
90 kbar, (a), and the parent uncompacted material, (b).

In Fig. 2 the dashed line in (a) between about 25-450 cm^{-1}, which refers to the normalized uncompacted spectrum, occurs <u>above</u> the solid line corresponding to the (a) spectrum. Hence, the compacted spectrum shows a loss of Raman intensity between 25-450 cm^{-1}. The three Gaussian components, A, B, and C, are also shown in the figure. Changes in component C, centered at $\sim$ 260 cm^{-1}, relative to components A plus B, and relative to the total contour intensity from 0-900 cm^{-1} are given in Table II. The ratios shown in Table II indicate that the loss of Raman intensity between $\sim$ 25-450 cm^{-1} must be attributed primarily to component C.

The effect of irreversible uniaxial compaction on component C is just the opposite of that of tensile stress. Hence, from this observation, and from Table I, it is evident that δ <u>increases</u> upon compaction.

No large changes in the 490 and 600 cm^{-1} defect intensities were apparent in the case of irreversible compaction, although normalized difference Raman spectra, shown subsequently, indicate a slight decrease in the 490 cm^{-1} intensity.

Reversible Hydrostatic Compression

A. Room Temperature

    Raman spectra derived from Ref. 10 corresponding to fused silica compressed under a reversible hydrostatic pressure of 10 kbar, and at 1 atm. are shown in Fig. 3, (a) and (b), respectively.

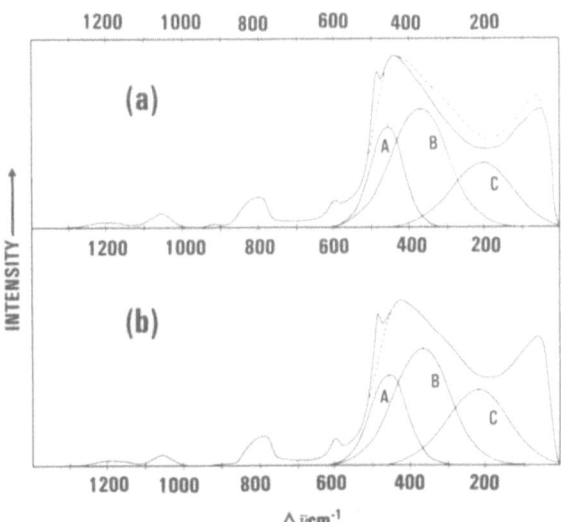

Fig. 3 Raman spectra from vitreous silica under reversible hydrostatic compression, 10 kbar, (a), and no compression, (b).

The dashed line of Fig. 3(a), like that of Fig. 2(a), occurs above
the solid line in the region from $\sim$ 50-450 cm^{-1}. This observation
indicates that a reversible hydrostatic pressure of 10 kbar decreases
the low-frequency Raman intensity. The corresponding effect on
component C is also listed in Table II. Here, component C is
centered near $\sim$ 220 cm^{-1}, and it clearly decreases in intensity
under pressure.

A normalized difference Raman spectrum calculated from the original spectra from which Fig. 3 were obtained, i.e., Fig. 4, is
shown in Fig. 5. Negative cross-hatched areas refer to the loss
of low-frequency Raman intensity below 450 cm^{-1} and at the 600 cm^{-1}
"defect" peak. In addition, an upward shift of $\sim$ 6 cm^{-1} was observed
for the 800 cm^{-1} Si-O-Si bending peak, as shown by the derivative-
like feature centered near 800 cm^{-1}. And a qualitatively opposite
effect near 1060 cm^{-1} indicates that a downward frequency shift,
determined to be $\sim$ 3-5 cm^{-1}, occurred for the Si-O stretching peak.
These effects indicate that reversible hydrostatic compression
decreases ϕ and increases both $\bar{r}$ and $\bar{\delta}$, as seen from Table I.

The decrease in the integrated intensity of the 600 cm^{-1} "defect"
peak with pressure rise is quantitatively shown in Fig. 6, see also,
Fig. 7. This decrease corresponds to about 27% in 10 kbar. However,
no significant change in the 490 cm^{-1} "defect" intensity was observed.

B. Elevated Temperature

McMillan et al. [11] have recently reported Raman spectra
corresponding to ambient fused silica (previously) subjected to

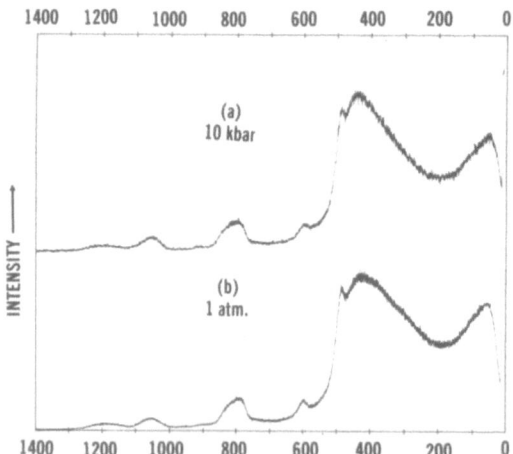

Fig. 4 Original Raman spectra from vitreous silica under reversible
        hydrostatic compression, 10 kbar, (a), and no compression,
        (b), from which Fig. 3 and Fig. 5 resulted.

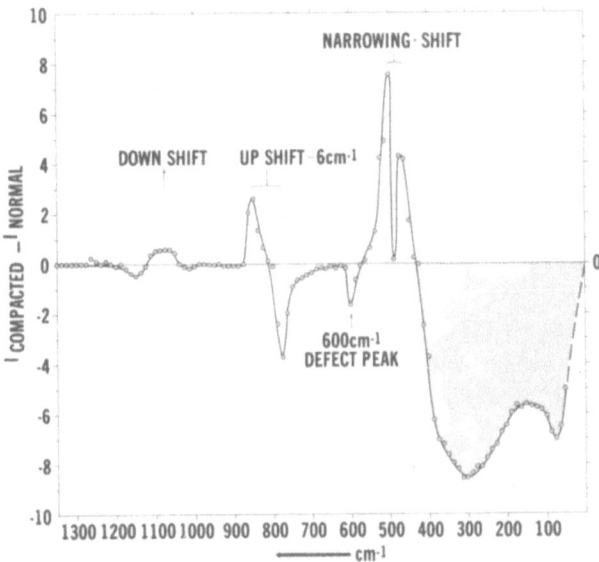

Fig. 5   Difference Raman spectrum, I compacted – I normal, obtained
from Fig. 4.   Cross-hatched areas refer to intensity loss
under hydrostatic pressure.

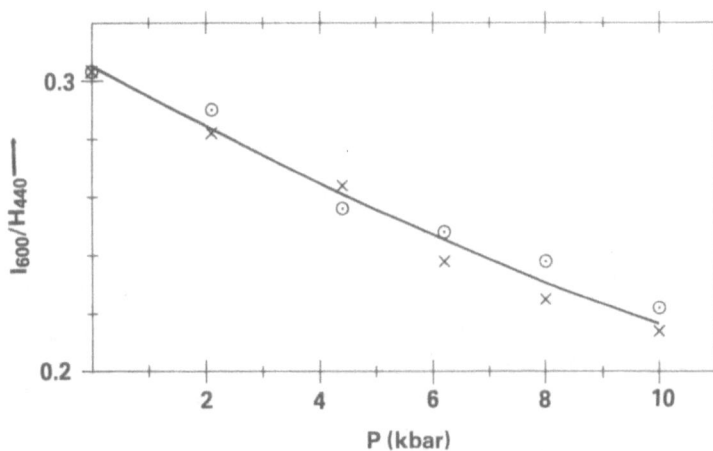

Fig. 6   Integrated Raman intensity at ~ 600 cm^{-1}, I_{600}, relative to
peak height at 440 cm^{-1}, H_{440}, for vitreous silica under
reversible hydrostatic compression.   (o) and (x) refer to
repeat determinations of I_{600} and H_{440}, but not to repeated
spectra.

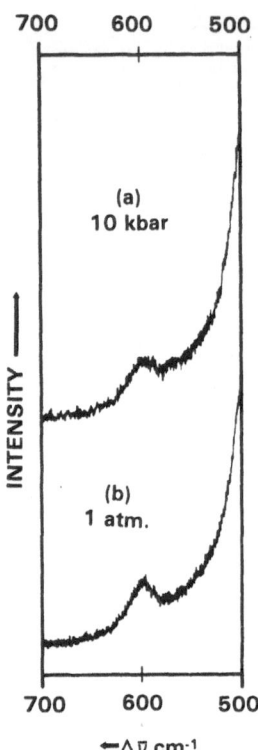

Fig. 7   Raman spectra showing effect of reversible hydrostatic
compression upon $\sim$ 600 cm^{-1} intensity, see Fig. 6.

39.5 kbar at 530°C. They observed a marked loss of Raman intensity
below 450 cm^{-1}, an upward shift of the 800 cm^{-1} peak position, and
a downward shift of the 1060 and 1200 cm^{-1} bands to 1020 and 1150
cm^{-1}, respectively.

     From the previous correlation, Table I, it is evident that the
McMillan results indicate that $\bar{\phi}$ decreases as both $\bar{r}$ and $\bar{\delta}$ increase,
i.e., the effects are qualitatively the same as at 10 kbar and
25°C, although they are quantitatively larger, as expected. The
high temperature of 530°C may be very effective kinetically in
modifying the silica structure, compared for example to the effect
of irreversible compaction at 90 kbar and 25°C. Hence, in view
of the fact that $\bar{\delta}$ may increase in both cases along with the
correlated changes in $\bar{\phi}$ and $\bar{r}$, it seems very doubtful that the
differences in mechanism claimed by McMillan et al. are correct,
unless such differences refer to the rate of the structural
transformation in vitreous silica.

Raman results similar to those described here have also been reported by Mochizuki and Kawai [12] and by Seifert et al. [13].

Large Reversible Torsion

Raman spectra corresponding to large reversible torsion, and uniaxial irreversible compaction are contrasted in Fig. 8(b) and (d), Ref. 3. The normalized difference spectra, (a) minus (b), and (c) minus (d), indicate that loss of Raman intensity below 450 cm^{-1} occurs for reversible torsion. as well as for irreversible compaction. A decrease of intensity at the 490 cm^{-1} "defect" peak position is also suggested by the (c) minus (d) difference spectrum for the irreversible compaction case, as mentioned before.

From the previous correlation, Table I, it is apparent that large reversible torsion, as well as uniaxial irreversible compaction, and reversible hydrostatic compression, involve an increase in δ. This suggests that reversible torsion is a densification process, and that an increase of $\bar{\delta}$ occurs with densification. Tensile stress, on the other hand, involves a decrease in $\bar{\delta}$. Here, elongation occurs.

Intense Neutron Bombardment

Raman spectra from fused silica after intense neutron bombardment have been reported by Bates et al. [14]. Intense neutron bombardment, like reversible and irreversible compression, is a densification process.

From Raman spectra, Bates et al. observed a strong low-frequency intensity loss below 450 cm^{-1}, a ∿ 12 cm^{-1} upward frequency shift at the 800 cm^{-1} peak, and decreases in frequency of the 1060 and 1200 cm^{-1} peaks amounting to 22 and 25 cm^{-1}, respectively. Obviously, neutron densification produces a marked decrease in ϕ as well as large increases in $\bar{r}$ and $\bar{\delta}$.

A large intensity increase and some frequency shift of the 600 cm^{-1} "defect" peak were also observed by Bates et al.

Prolonged Bombardment by Alpha Particles

Exarhos [15] has subjected thin fused silica plates to 5.5 MeV alpha irradiation for 1 week using a ^{244}Cm source. This amounted to about 1×10^{16} α/cm^2, and produced damage to a depth of 0.16 μm.

A Raman intensity decrease between 0-450 cm^{-1} was observed, in addition to a small (5-6 cm^{-1}) increase in the frequency of the 800 cm^{-1} peak, and a small (3-4 cm^{-1}) frequency decrease of the 1060 cm^{-1} peak. Apparently, alpha bombardment decreases ϕ and increases $\bar{r}$ and $\bar{\delta}$, which suggests that a densification process might be involved.

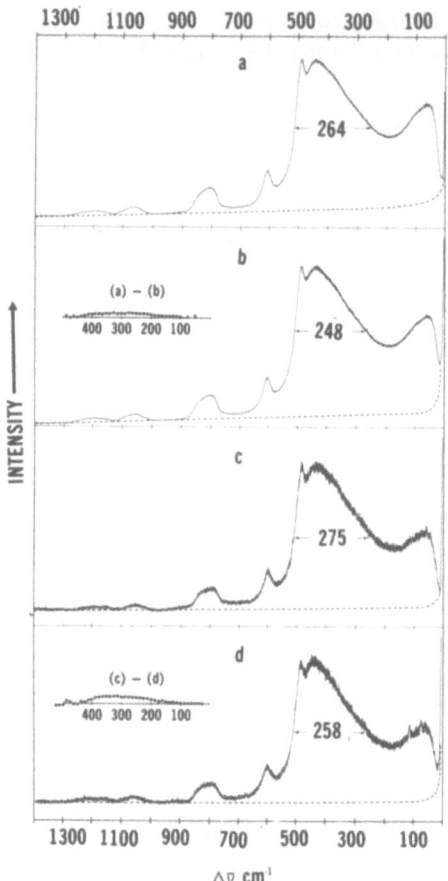

Fig. 8   Comparisons between Raman spectra from vitreous silica
         under large reversible torsion, (b), and irreversible
         uniaxial compaction, (d), where (a) and (b) are the respec-
         tive parent spectra from normal fused silica under com-
         parable sample and experimental conditions.

     Very small intensity increases were also suggested for the
490 and 600 cm^{-1} "defect" modes.

     All of the Raman effects described in this chapter for the
main network equilibrium structures of fused silica are listed in
Table III. Raman effects observed for the 490 and 600 cm^{-1} "defect"
modes are simply listed in Table IV, without making any attempt to
correlate them at this time.

Table III. Changes in the mean dihedral or tilt angle, the mean
            Si-O-Si bridging angle, and the mean Si-O bond distance
            as inferred from Raman data from vitreous silica under
            various conditions.

(a) -- Ref. 7   (e) -- Ref. 16
(b) -- Ref. 10   (f) -- Ref. 14
(c) -- Ref. 11   (g) -- Ref. 15
(d) -- Ref. 9

SYMBOL		$\bar{\delta}$	$\bar{\Phi}$	$\bar{r}$
MEANING		MEAN DIHEDRAL OR TILT ANGLE	MEAN Si-O-Si BRIDGING ANGLE	MEAN Si-O BOND DISTANCE
RAMAN EFFECT		$I_{0\text{-}450}$ DECREASES	$\Delta\bar{v}_{800}$ INCREASES	$\Delta\bar{v}_{1060,1200}$ DECREASES
ACTUAL EFFECT		(+), INCREASE OF $\bar{\delta}$	(−), DECREASE OF $\bar{\Phi}$	(+), INCREASE OF $\bar{r}$
		− ACTUAL EFFECTS −		
REVERSIBLE TENSILE STRESS	(a)	−	NOT AVAILABLE	NOT AVAILABLE
REVERSIBLE HYDROSTATIC	(b)	+	−	+
COMPRESSION	(c)	+	−	+
IRREVERSIBLE UNIAXIAL COMPACTION		+(d)	NOT AVAILABLE	+(e)
REVERSIBLE TORSION	(f)	+	NOT AVAILABLE	NOT AVAILABLE
IRREVERSIBLE NEUTRON COMPACTION	(g)	+	−	+
IRREVERSIBLE ALPHA BOMBARDMENT	(h)	+	−	+

Table IV. Changes in the intensities of the "defect" components near ~ 490 cm^{-1} and ~ 600 cm^{-1} for vitreous silica under various conditions.

(a) -- Ref. 2   (e) -- Ref. 3
(b) -- Ref. 7   (f) -- Ref. 14
(c) -- Ref. 10   (g) -- Ref. 15
(d) -- Ref. 9

CONDITION	INTENSITY	
	$\sim$490 cm^{-1}	$\sim$600 cm^{-1}
INCREASING (T$_F$), FICTIVE TEMPERATURE (a)	↑	↑
PROLONGED ANNEALING (a)	↓	↓
INCREASING OH CONTENT, CONSTANT T$_F$ (a)	↓	↓
REVERSIBLE TENSILE STRESS (b)	↑	NO SIGNIFICANT EFFECT
REVERSIBLE HYDROSTATIC COMPRESSION (c)	NO SIGNIFICANT EFFECT	↓
IRREVERSIBLE UNIAXIAL COMPACTION (d)	↓	NO SIGNIFICANT EFFECT
LARGE REVERSIBLE TORSION (e)	↓(?)	NO SIGNIFICANT EFFECT
IRREVERSIBLE NEUTRON COMPACTION (f)	NO SIGNIFICANT EFFECT	↑
IRREVERSIBLE ALPHA BOMBARDMENT (g)	↑(?)	↑(?)

CONCLUSIONS

     A hypothesis has been advanced for fused silica, namely, that a decrease of the mean equilibrium Si-O-Si bridging angle, ϕ, is accompanied by increases in both $\bar{r}$ and $\bar{\delta}$, the mean equilibrium

Si-O bond length, and the mean equilibrium dihedral or tilt angle, respectively (or vice versa). This hypothesis has been tested experimentally by means of Raman spectra. When a frequency increase of the 800 cm^{-1} peak is interpreted as a decrease in $\bar{\phi}$; when frequency decreases at 1060 and 1200 cm^{-1} are interpreted as an increase in $\bar{r}$, and when an intensity decrease below 450 cm^{-1} is interpreted as an increase in $\bar{\delta}$, (or vice versa in all cases) all present Raman data fit the new hypothesis. Hence, it would appear that a wide body of data for fused silica can be summarized by the simple statement that both $\bar{r}$ and $\bar{\delta}$ are negatively correlated to $\bar{\phi}$, cf., Ref. 17. It also appears that $\bar{\phi}$ decreases, and $\bar{r}$ and $\bar{\delta}$ increase, in densification mechanisms, whereas $\bar{\phi}$ may increase upon elongation, such as produced by tensile stress, as $\bar{r}$ presumably decreases, and $\bar{\delta}$ decreases.

REFERENCES

1. R. H. Stolen, J. T. Krause and C. R. Kurkjian, Disuss. Faraday Soc. 50:103 (1970).
2. R. H. Stolen and G. E. Walrafen, J. Chem. Phys. 64:2623 (1976).
3. M. S. Hokmabadi and G. E. Walrafen, J. Chem. Phys. 78:5273 (1983).
4. G. E. Walrafen, M. Abebe, F. A. Mauer, S. Block, G. J. Piermarini, and R. Munro, J. Chem. Phys. 77:2166 (1982).
5. G. E. Walrafen and P. N. Krishnan, Appl. Optics 21:359 (1982). See also, J. Stone and G. E. Walrafen, J. Chem. Phys. 76:1712 (1982), particularly pgs 1718-1719.
6. J. D. Jorgensen, J. Appl. Phys. 49:5473 (1978).
7. G. E. Walrafen, P. N. Krishnan, and S. W. Freiman, J. Appl. Phys. 52:2832 (1981).
8. P. W. Bridgman and I. Simon, J. Appl. Phys. 24:405 (1953).
9. G. E. Walrafen and P. N. Krishnan, J. Chem. Phys. 74:5328 (1981).
10. M. S. Hokmabadi, Doctoral Dissertation, Howard University (1981).
11. P. McMillan, B. Piriou, and R. Couty, J. Chem. Phys. 81:4234 (1984).
12. S. Mochizuki and N. Kawai, Solid State Commun. 11:763 (1972).
13. F. A. Seifert, B. O. Mysen and D. Virgo, Physics Chem. Glasses 24:141 (1983).
14. J. B. Bates, R. W. Hendricks, and L. B. Shaffer, J. Chem. Phys. 61:4163 (1974).
15. G. J. Exarhos, Nuclear Instruments and Methods in Physics Research 229:498 (1984).
16. I. Simon, Chapter 6 in Modern Aspects of the Vitreous State, J. D. MacKenzie Ed., Butterworths, London (1960), 120-151, see particularly p. 146.
17. A. G. Revesz and G. V. Gibbs, in Proc. Conf. Physics of MOS Insulators, eds. G. Locovsky et al., Pergamon Press, New York 92 (1980).
18. P. T. T. Wong, D. J. Moffatt and F. L. Baudais, Appl. Spectrosc. 39:734 (1985).
19. P. T. T. Wong, private communication.

NOTE ADDED IN PROOF

After this chapter was written, strong conformation for the proposed Raman structural correlation was obtained from the recent work of Wong et al. [18] and Wong [19]. An infrared band from crystalline alpha-quartz at 801 cm^{-1} was found to shift to about 830 cm^{-1} under a pressure of 50 kbar [18]. Also a strong shift to lower frequency with increasing pressure [19] was observed for another band at about 1144 cm^{-1}. The 1144 cm^{-1} is assigned to Si-O stretching [19], and the 801 cm^{-1} band to Si-O-Si bending, this work. These spectral changes in alpha quartz correspond to a decrease in the Si-O-Si bridging angle, and to an increase in the Si-O bond length. They confirm the correlation proposed here, and augment the Jorgensen observations [6].

VIBRATIONAL STUDIES OF GLASS STRUCTURE AND LOCALIZED INTERACTIONS

Gregory J. Exarhos

Battelle
Pacific Northwest Laboratory
Richland, WA 99352

Infrared absorption and Raman spectroscopic measurements of binary alkali oxide containing silicate, phosphate, and borate glasses yield features assignable to predominantly ionic cation site interactions in addition to network vibrations characteristic of the glass former oxide. Intense bands observed in Far IR spectra are assigned to vibrations of cations at localized sites in the glass. These features exhibit marked frequency shifts with cation mass, cation charge and glass stoichiometry but show little dependence on the nature of the glass former oxide. Bulk glass properties, including glass transition temperature and conductivity activation energy, scale with cation-site frequency which is an important parameter in models describing such phenomena.

In contrast to the localized mode observed in Far IR spectra of binary oxide glasses, Raman spectra of these materials reveal features arising from the more covalent interactions of the network former oxide. The presence of glass modifier cations serves to slightly perturb the network vibrational frequencies through cation size effects. Alkali content correlates with the fraction of nonbridging oxygen atoms present in the glass and is manifested by marked changes in measured Raman spectra. These observations form a basis for application of this technique to framework structural studies of simple alkali silicate glasses.

The capability of Raman spectroscopy to discern silica and silicic acid rich regions in $Na_2O.3SiO_2$ glass fibers subjected to hydrothermal dissolution conditions will also be discussed.

INTRODUCTION

The physical properties of binary metal oxide containing glasses
$[xM_mO_n \cdot yA_aO_b]$ are controlled to a large extent by the amount of
modifier oxide $[M_mO_n]$ present. Large radii cations act to disrupt
covalent bonding intrinsic to the highly cross-linked network former
oxide through creation of localized ionic sites. The extent of this
structural perturbation on bulk glass properties is manifested by
changes in glass transition temperature, cation mobility, leach-
ability, and radiation damage behavior. These localized charge
centers control many of the exchange phenomena observed in such
disordered phases.

Vibrational spectroscopy is a site specific probe that has been
used to study composition dependent structural changes in glass and
relate local glass structure to bulk physical properties. Bridging
and nonbridging oxygen sites in glass can be examined by vibrational
spectroscopy as a function of glass composition and following exposure of glass to conditions which can induce structural alterations.

Modifier cations are localized in ionic sites and act to bind
the network former oxide through ionic interactions with nonbridging
oxygen atoms. Features characteristic of localized cation vibrations in addition to modes assigned to vibrations of the covalent
network are observed. Far infrared active cation modes, a direct
measure of the cation ionic-site interaction energy, exhibit mass
and charge dependent frequencies that correlate with bulk glass
properties such as T_g, and cation mobility activation energies. Network former vibrations, infrared and Raman active, are characteristic
of both localized and extended vibrational modes of the glass former
oxide and exhibit frequency perturbations which depend on the nature
of the cation modifier present.

The application of vibrational spectroscopy to glass structural
studies is well documented. Raman studies of binary oxide glasses
have inferred correlations between glass structure and stoichiometry [1,2,3] and far infrared measurements have probed localized
cation site interactions [4,5,6,7]. A significant portion of this
article concerns fundamental vibrational studies of the metal-
metaphosphate glass forming system which consists of chain-like
anions $(PO_3^-)_n$ held together by predominantly ionic interactions
with modifier cations. Detailed vibrational studies of this system
have previously been reported. The earlier work will be reviewed
and augmented with recent studies which are directed toward developing a fundamental understanding of localized ionic sites in simple
glasses and their modification by leaching and irradiation. Raman
results are high-lighted by current work which proposes to understand
glass structure in the $Na_2O.3SiO_2$ system by following selective
leaching of the alkali phase from the parent glass.

EXPERIMENTAL

All glasses were prepared by conventional melt/quench techniques described previously [6,8]. Compositions were verified by elemental analysis and phase homogeneity was established from measured Raman spectra. Bulk glass samples as well as 25 micron diameter fibers and thin films (1 micron) were isolated for most compositions. Far infrared absorption spectra were obtained from thin films, or powdered specimens dispersed in low density polyethylene. Far infrared spectra were measured using a Nicolet 7199 FTIR spectrometer consisting of a high pressure Hg arc source, mylar beamsplitter (6μ or 12μ), and pyroelectric DTGS detector. Raman spectra were acquired from either 90° forward or 180° backscattering geometries using 200 mW of 514.5 or 488.0 nm Ar+ excitation. Red excitation from a tunable dye laser was used to excite Raman spectra from highly colored glasses or in some cases to minimize sample fluorescence. A conventional double monochromator equipped with photon counting electronics was used. Slitwidths were maintained at 400μ.

FUNDAMENTAL INFRARED STUDIES

The infrared absorption spectrum of a binary metal oxide glass may be separated into two regions as shown in Figure 1 for vitreous silver metaphosphate. For a given glass stoichiometry, the high frequency region characteristic of network former vibration modes is relatively insensitive to the kind of cation modifier present. Band

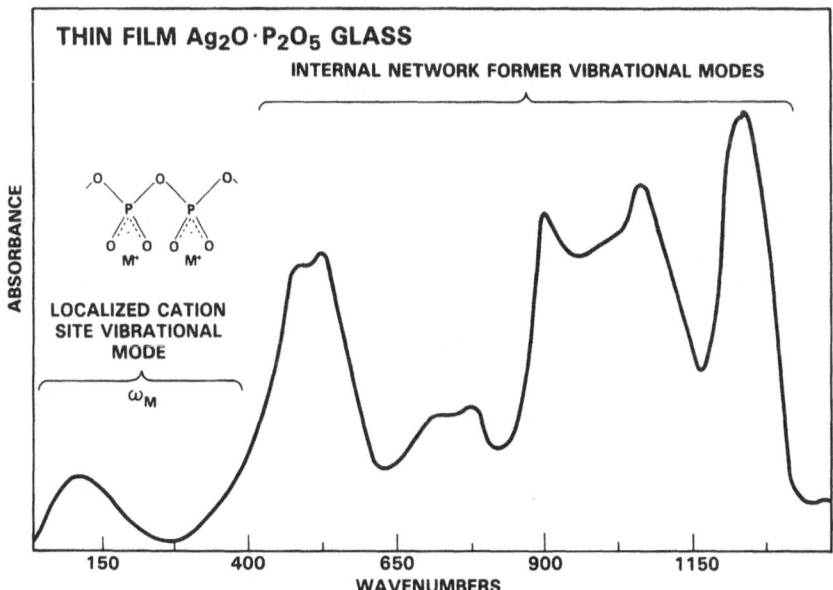

Fig. 1   Infrared absorption spectrum of silver metaphosphate glass.

intensities and frequency maxima show slight perturbations with different cations. However, vibrational bands below 500 cm^{-1} are effectively uncoupled from the network vibrations and exhibit marked dependence on cation mass and charge. These features have been assigned to localized cation-site vibrational modes and are a direct measure of the cation site interaction energy [6,8].

Cation mode frequencies have been measured for a large number of metal metaphosphate glasses and appear in a "periodic table" representation in Figure 2.

Observed bands exhibit large bandwidths which vary linearly with cation vibrational frequencies. The assignment of these bands to primarily cation motion is based upon far infrared measurements of crystalline isomorphs which demonstrate that the cation band constitutes an envelope of IR-active optical phonon modes [6]. Structural studies of single crystal metal metaphosphates have shown the cations to be localized in octahedral oxygen sites characterized by a mean cation-oxygen separation (R_{MO}) [9,10]. Similar localized centers are present in the amorphous state, however, such centers are spatially disordered and their interaction with other cation oscillators results in the observed far IR absorption feature which represents this orientationally disordered vibrational density of states.

Empirical correlations between cation frequencies, mass and charge may be inferred from data presented in Figure 2, but an additional parameter is necessary to adequately model the cation-site interaction energy. Cation size determines an energetically favorable packing arrangement as well as charge separation distance between a cation and its anionic site. One model based upon a simplified version of the Born-Mayer potential [11] was developed for understanding cation-site interactions in ionic oxide glasses [8].

LOCALIZED CATION VIBRATION FREQUENCIES (cm^{-1}) IN THE BINARY METAL METAPHOSPHATE GlASS FORMING SYSTEM, $M(PO_3)_x$ x = 1, 2, 3, 4

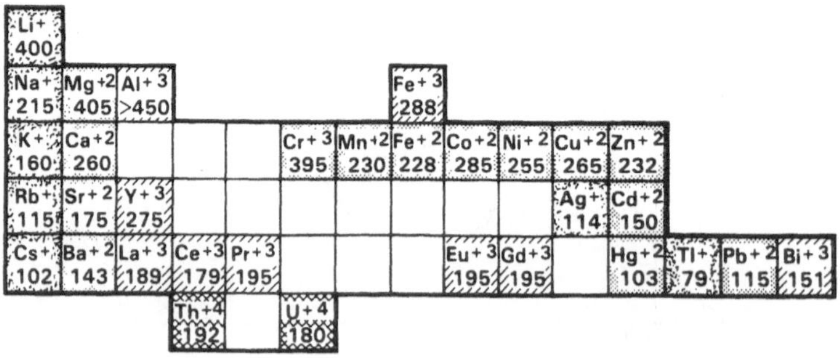

Li+ 400																	
Na+ 215	Mg+2 405	Al+3 >450						Fe+3 288									
K+ 160	Ca+2 260			Cr+3 395	Mn+2 230	Fe+2 228	Co+2 285	Ni+2 255	Cu+2 265	Zn+2 232							
Rb+ 115	Sr+2 175	Y+3 275							Ag+ 114	Cd+2 150							
Cs+ 102	Ba+2 143	La+3 189	Ce+3 179	Pr+3 195				Eu+3 195	Gd+3 195		Hg+2 103	Tl+ 79	Pb+2 115	Bi+3 151			
			Th+4 192	U+4 180													

Fig. 2 Periodic representation of cation frequencies in metaphos-
        phate glasses.

For a predominantly ionic interaction, the attractive potential may be written in terms of the magnitude of the ionic charges (q_M, q_o), the separation of the ions (R_{MO}), the repulsion constant ρ, and a constant (λ) intrinsic to the system under investigation.

$$U = -q_M q_o / R_{MO} + \lambda EXP(-R_{MO}/\rho) \tag{1}$$

From this relationship, the functional dependence of cation vibrational frequencies to parameters intrinsic to the cation is obtained.

$$\omega_0 = A \left[\frac{q_M}{MR_{MO}^3} \right]^{1/2} \tag{2}$$

Here, ω_0 is the cation frequency, A is a constant and M is the cation mass. · Equation (2) is valid when the repulsion constant (ρ) scales with the internuclear separation (R_{MO}). Cation vibrational frequencies vary linearly with

$$\left[\frac{q_M}{MR_{MO}^3} \right]^{1/2}$$

for the majority of metal metaphosphate glasses. However, small, highly charged cations (Li^+, Mg^{+2}, Al^{+3}) exhibit larger vibrational frequencies than predicted by Equation (2), indicating that they may exist in a different localized environment in the glass (tetrahedral) where the R_{MO} distance is expected to be smaller.

Far infrared absorption due to cation vibrations have been observed in other binary glass systems containing SiO_2, B_2O_3, GeO_2, or V_2O_5 [5]. The effect of the glass former oxide was observed to perturb the cation frequencies as seen in Table 1. In a given glass forming system, however, glass composition can significantly alter cation motion frequencies. Sodium phosphate glasses, $(Na_2O)_x \cdot P_2O_5$, can be prepared over a wide composition range ($0<X\leq2$) and exhibit marked changes in cation frequency as a function of composition. Figure 3 exhibits a linear frequency dependence for sodium and strontium phosphate glasses as a function of $1/X$. The parameter $1/X$ is proportional to the average volume available to a single cation oscillator for a particular composition. For small values of $1/X$, the glass network becomes "depolymerized" in order to satisfy the bonding constraints of the cation. Far infrared absorption bands sharpen with increasing cation content implying that cation-cation interactions become more significant than cation-network coupling interactions. For low alkali containing glasses, the network becomes more cross-linked and couples more strongly with the cation mode. The infrared band broadens and shifts to lower frequency. More configurational entropy is available at low

alkali contents giving rise to a higher degree of orientational
disorder for the cation oscillators.

Table 1.   Cation Vibrational
Frequencies in Oxide Glasses

Glass Composition	$\omega(cm^{-1})$
$Na_2O \cdot V_2O_5$	195
$Na_2O \cdot P_2O_5$	215
$Na_2O \cdot B_2O_3$	220
$Na_2O \cdot 2B_2O_3$	220
$Na_2O \cdot SiO_2$	230
$K_2O \cdot P_2O_5$	160
$K_2O \cdot 2B_2O_3$	175
$K_2O \cdot GeO_2$	165
$Cs_2O \cdot P_2O_5$	102
$Cs_2O \cdot B_2O_3$	108
$Cs_2O \cdot 2B_2O_3$	103
$Cs_2O \cdot Ge_2$	112

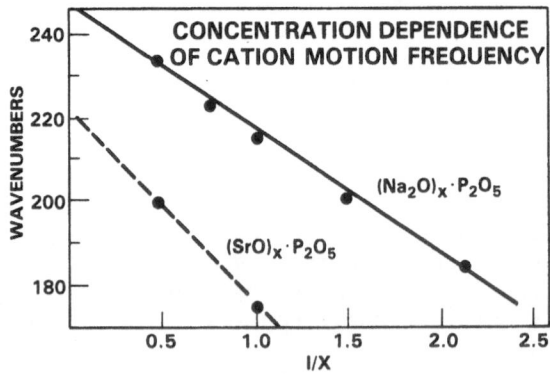

Fig. 3   Variation of band frequency with stoichiometry.

Invert glasses are characterized by compositions where the modifier oxide content exceeds the glass former oxide content. Trap and
Stevels [12] suggested that an inversion in the structural role of
the modifier cations occurs at high modifier contents where the continuity of the network is disrupted. Strong ionic cation site
interactions act to stabilize the disordered state, however, the
local cation symmetry may be different than in lower modifier oxide
containing glasses. Changes in cation mode frequency for a number
of polyphosphate glasses are shown in Figure 4. A shift to higher
frequency with increasing modifier content is always observed which
indicates increased cation-site interaction. A change in cation
local symmetry (lowering of coordination number) would also account
for cation band shifts to higher frequencies.

APPLIED INFRARED STUDIES

The magnitude of the cation-site interaction influences bulk
glass properties such as glass transition temperature and ion
mobility. Far infrared measurements have been used to correlate
molecular vibrational properties with measurable glass physical
properties and develop models for exchange phenomena in glass.

The glass transition phenomenon in alkali metaphosphates has
been explained in terms of measured vibrational parameters associated
with the cation site interaction [6]. Molecular rearrangement becomes possible at a temperature, T_g, where the available thermal
energy becomes comparable to the interaction energy between a cation
and its localized site. The bond stretching force constant for this
interaction is evaluated from vibrational data and increases linearly with glass transition temperature through the series of alkali
metaphosphate glasses [6].

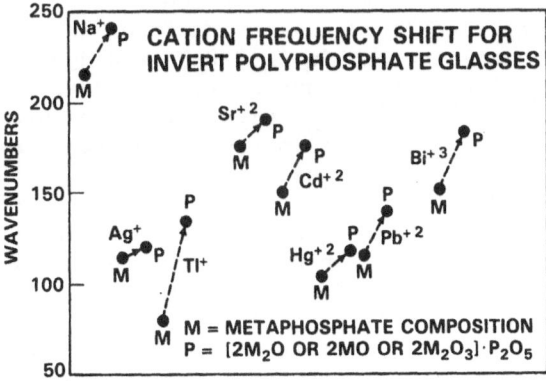

Fig. 4   Frequency correlation chart for invert glasses.

Ion mobilities in metal oxide containing glasses depend upon the strength of the cation-site interaction, the distance between available sites, and the number of attempts an ion executes in changing sites. The Rice and Roth model for ionic conductivity in isotropic media [13] is based upon excitation of a localized ion to a free state where it propagates over a mean free path until encountering an unoccupied site. If the mean free path (ℓ_o) is associated with the average cation site separation in glass, then the activation energy (E_a) for conduction may be written in terms of vibrational parameters of the localized cation site [14].

$$E_a = \frac{1}{2} M\ell_o^2 \nu_o^2 , \tag{3}$$

where M is the cation mass, ℓ_o is the site separation, and ν_o is the localized cation site vibrational frequency. Calculated activation energies for nine binary oxide glasses using measured vibrational data agree with measured E_a's to within 5% [7,14] and show the utility of vibrational parameters for the determination of bulk glass properties.

Measured vibrational data for localized cation vibrations have also been used in models which explain enhanced conductivity in mixed alkali glasses [15]. The "mixed alkali" effect may also explain the increased Ag^+ conductivity in the cation substituted borate glass $(.28-x)Tl_2O \cdot Ag_2O \cdot (.72)B_2O_3$. Sakka, et al. [16] suggested the existence of Ag_2O clusters as an explanation for enhanced Ag^+ mobility in this system. Far IR spectra shown in Figure 5 do not support this contention as the mixed cation glass spectra are precisely shown to be the summation of the appropriately weighted spectra for the single component metal oxide glasses. No vibrational mode assignable to Ag_2O was detected. The enhanced conductivity probably can be explained in terms of the vibrational "mixed-alkali" effect model developed previously [15].

Vibrational spectroscopy finds particular application to studies of exchange reactions at glass surfaces in contact with reactive gases. Metal cations in $Ag_2O \cdot P_2O_5$ and $HgO \cdot P_2O_5$ glasses are reduced to metal when the glasses are subjected to NH_3 at temperatures below 100°C [17]. _In situ_ far IR measurements indicate a marked intensity decrease in the cation bands and growth of a broad feature at _ca_ 200 cm^{-1} which has been assigned to the ν_σ hydrogen bonding mode of an amine group bound to the site originally occupied by the cation [18]. This observation supports a dissociative adsorption model for NH_3 on oxidic surfaces. Far IR measurements have also identified Ag_2S, a reaction product formed when silver metaphosphate glass is subjected to H_2S vapor [19]. The analytical and mechanistic applications of far IR measurements to reactive gas interactions with localized cation sites in oxide glasses have been demonstrated by these and similar studies.

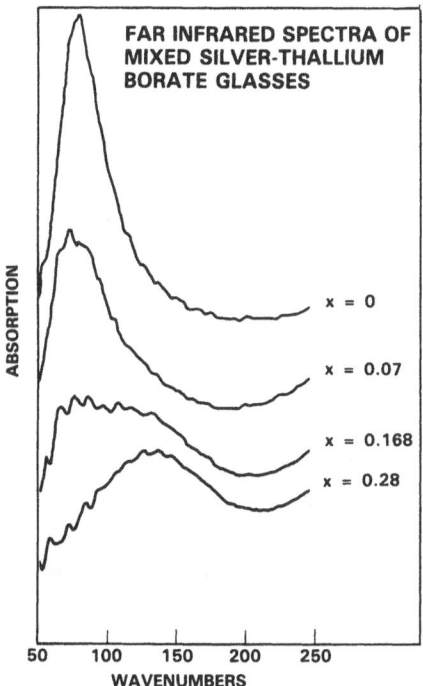

Fig. 5   Cation motion band dependence on concentration for
$(.28-x)Tl_2O \cdot XAg_2O \cdot B_2O_3$, where $0 < X \leq .28$.

FUNDAMENTAL RAMAN STUDIES

Structural aspects of the network former oxide are manifested
in measured Raman spectra which can also indicate glass stoichiometry, homogeneity, and phase purity [3,20]. While the cation
vibrational band is not Raman active, the network vibrational modes
are perturbed by cation mass, size, and charge as depicted in
Figures 6 and 7 for the metal metaphosphate system. In this glass
forming system, the strong feature between 1100 and 1200 cm^{-1} is
assigned to a symmetric $>PO_2^-$ terminal mode, while the band at
ca 700 cm^{-1} is assigned to in-chain $\sim P \sim O \sim P \sim$ symmetric stretching.
Previous work has shown that cation size effects are responsible
for frequency shifts of the $>PO_2^-$ vibrational mode for cation
substituted glasses [21], while the $\omega_{PO_2}/\omega_{POP}$ intensity ratio is
indicative of the electron withdrawing power of localized cations
from the phosphorus nonbridging oxygen bond [8]. For small, highly
charged cations (Al^{+3}), band broadening and shifts to higher frequency indicate a more covalent interaction with the network former
oxide. Raman studies of binary silicate and more complex glasses
have been reported [3,22] and localized structural elements have
been categorized on the basis of measured vibrational bands.

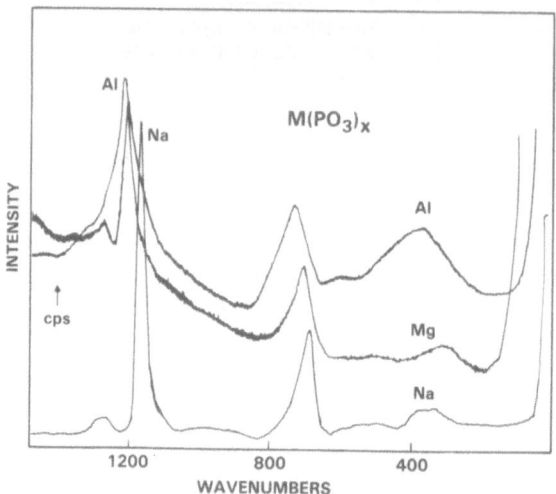

Fig. 6   Effect of cation charge on network vibrational modes in
         Raman spectra of metaphosphate glasses.

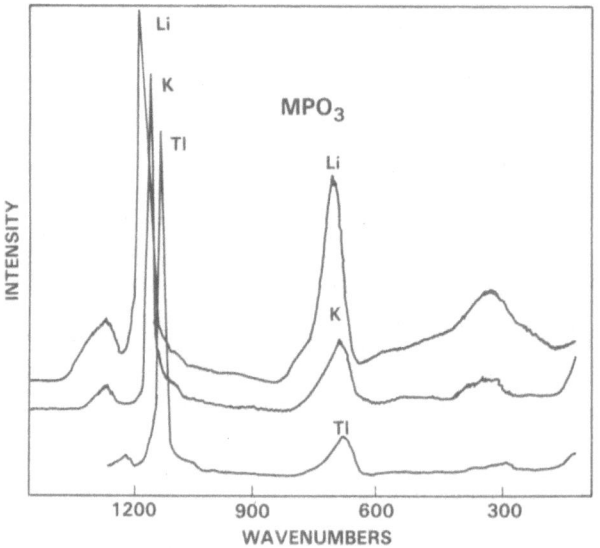

Fig. 7   Effect of cation mass on network vibrational modes in
         Raman spectra of metaphosphate glasses.

Raman spectroscopy, then, is a site specific probe for investigating
chemical effects on bonding at both bridging and nonbridging oxygen
sites in binary metal oxide containing glasses.

The effect of metal oxide additions to a glass former results in a depolymerization of the network owing to the formation of ionic nonbridging oxygen sites. Marked changes in measured Raman spectra are observed and can be associated with localized structural elements for a particular stoichiometry. Figure 8 exhibits Raman spectra for two silver polyphosphate glass compositions. The metaphosphate composition yields the two major Raman lines associated with vibrations of bridging oxygen atoms (network chain) and nonbridging oxygen ionic sites where the silver cations are localized. The pyrophosphate composition still exhibits a vibrational mode near 700 cm^{-1} associated with movement of bridging oxygen atoms but a strengthening of the in-chain bridging P-O-P interaction is inferred since the band is shifted to higher frequencies. The phosphorus nonbridging oxygen bond is apparently weaker in the pyrophosphate glass than in the metaphosphate glass since the 1142 cm^{-1} band shifts to much lower frequencies. This is consistent with far infrared spectral changes which suggest that the cation-site interaction is enhanced for the pyrophosphate glass. Understanding trends exhibited by the Raman spectra for variable metal oxide containing glasses forms a basis for interpreting vibrational spectra in other glass forming systems.

APPLIED RAMAN STUDIES

Two experiments are reported which cause alteration of glass structure and demonstrate the ability of Raman spectroscopy to probe localized bond changes resulting from external treatment of the glass. Subtle structural changes have been caused in metaphosphate glass by exposure to high doses of ionizing radiation while exchange phenomena in an alkali silicate glass resulting from

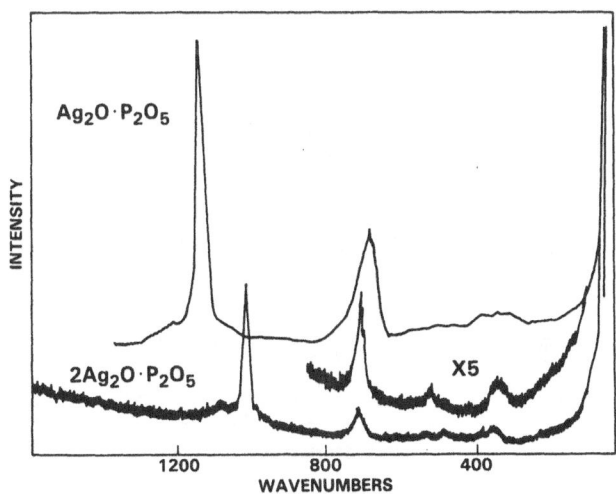

Fig. 8 Raman spectra of high silver oxide content phosphate glasses.

hydrothermal leaching lead to gross structural changes in the glass.
Measured vibrational changes are used to infer changes in localized
chemical bonding for these systems.

An yttrium metaphosphate glass, $Y(PO_3)_3$, was subjected to gamma
irradiation (1,333, 1.172 MeV) from a ^{60}Co source to a cumulative
dose of 1.5×10^{10} R inducing ionization damage in the glass and
coloring it pink. The Raman spectrum of the unirradiated glass is
shown in Figure 9 with band assignments for the two major features
indicated. Following irradiation, band shifts and intensity changes
were observed in measured Raman spectra, and are detailed in Table
2. The changes are consistent with a model requiring radiation
damage to be localized at ionic sites in the glass. Band shifts
result from radiation induced $>PO_2$ bond angle opening which causes
the glass to respond by closing the in-chain P⌐O⌐P bond angle [23].
Therefore, the chain contracts while the terminal ionic sites expand.
Interpretations are based on results of previous work where ion
size was shown to affect bond angles at localized ionic sites in the
glass [21].

Following annealing at 300°C, radiation induced color centers
are bleached and the glass again becomes colorless. However, the
$>PO_2$ vibrational band does not fully recover. Following annealing,
the bandwidth is ca 20% larger and peak frequency ca 2 cm^{-1} less
than the undamaged glass. The glass structure has been irreversibly
altered as a result of irradiation. Similar results have been
observed for glasses subjected to high energy alpha particle irradiation [23].

One approach to understanding glass structure in multicomponent
systems involves selective removal of a water soluble component by

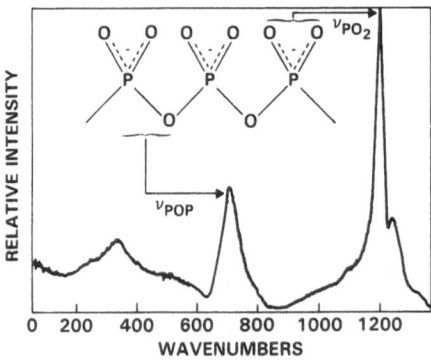

Fig. 9   Raman spectrum of yttrium metaphosphate glass with band
         assignments.

Table 2.   Raman Band Changes for Gamma
Irradiated $Y(PO_3)_3$ Glass

	Unirradiated	Irradiated
$\omega(PO_2)$	1202 cm^{-1}	1198 cm^{-1}
$\omega(POP)$	696 cm^{-1}	702 cm^{-1}
$I(POP)/I(PO_2)$	0.397	0.435

hydrothermal leaching followed by analysis of the leached material.
Raman spectroscopy has been used to follow selective removal of
alkali from $Na_2O \cdot 3SiO_2$ glass during aqueous leaching of 25μ diameter
glass fibers at 25° and 90°C [24].   Spectra are reproduced in Figure
10 and compared with Raman spectra of vitreous silica and silicic
acid, $(0.8)H_2O \cdot SiO_2$.   The leached glass spectrum is seen to be a
linear combination of the silica and silicic acid spectra indicating
that all Na^+ in the original glass has exchanged with H^+.   However,
features at 495 and 606 cm^{-1} assigned to "defect modes" or ring
vibrations [25] are not apparent in the leached fiber spectrum.
Apparently, structural elements producing such features were not
present in the starting material or bonding defects (unsatisfied
coordination) are susceptible to attack by aqueous species.   As
expected, the OH^- stretching region exhibits marked changes during
leaching eventually forming a broad diffuse band centered at 3450
cm^{-1} which is comparable to the feature exhibited by silicic acid.

CONCLUSION

     Vibrational spectroscopy is a sensitive probe of localized
structural elements in binary oxide glasses.   Far infrared active
vibrations assigned to modifier cation-site vibrational modes have
been determined for many systems and measured frequencies correlate
with bulk glass properties such as glass transition temperature
and conduction activation energy.   Raman spectra exhibit features
particular to the network former oxide and can provide information
on both bridging and non-bridging oxygen sites in the glass.   The
capability for in situ measurements makes vibrational spectroscopy
an attractive tool for glass structural studies and investigation
of surface exchange phenomena.

ACKNOWLEDGMENTS

     This work was partially supported by the U.S. Department of
Energy, Office of Basic Energy Sciences under contract DE-AC06-76RLO
1830.   Drs. B. N. Nelson, F. Y. Minabe, and I. G. Plotzker performed
many of the far infrared measurements which formed a basis for their
doctoral theses from Harvard University.

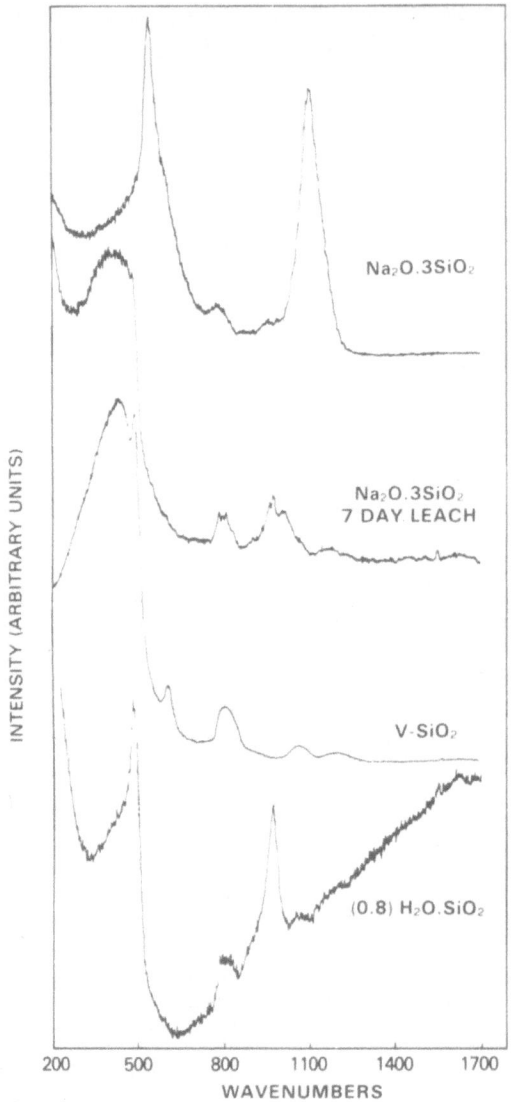

Fig. 10   Comparison of leached glass Raman spectrum with the pristine glass, silica and metasilicic acid.

REFERENCES

1.   S. A. Brawer and W. B. White, J. Chem. Phys., 63:2421 (1975).
2.   T. Furukawa, K. E. Fox, and W. B. White, J. Chem. Phys.,
       75:3226 (1981).
3.   D. W. Matson, S. K. Sharma, and J. A. Philpotts, J. Non-Cryst.
       Sol., 58:323 (1983).

4. G. J. Exarhos and W. M. Risen, Jr., Chem. Phys. Lett., 10(4):484 (1971).
5. G. J. Exarhos and W. M. Risen, Jr., Sol. State Comm., 11:755 (1972).
6. G. J. Exarhos, P. J. Miller, and W. M. Risen, Jr, J. Chem. Phys., 60(11):4145 (1974).
7. L. W. Panek, G. J. Exarhos, P. J. Bray, and W. M. Risen, Jr., J. Non-Cryst. Solids, 24(1):51 (1977).
8. B. N. Nelson and G. J. Exarhos, J. Chem. Phys. 71(7):2739 (1979).
9. K. H. Jost, Acta Cryst., 14:844 (1961).
10. K. H. Jost, Acta Cryst., 14:779 (1961).
11. C. Kittel, Introduction to Solid State Physics, Fourth Ed., Wiley, New York, 111, (1971).
12. H. L. Trap and J. M. Stevels, Phys. Chem. Glasses, 1:107, 181 (1960).
13. M. J. Rice and W. L. Roth, Sol. State Chem., 4:294 (1972).
14. G. J. Exarhos, P. J. Miller, and W. M. Risen, Jr., Sol. State Commun., 17:29 (1975).
15. G. B. Rouse, J. M. Gordon, and W. M. Risen, Jr., J. Non-Cryst. Solids, 33:83 (1979).
16. S. Sakka, K. Matusita, and K. Kamiya, Phys. Chem. Glasses, 20:25 (1979).
17. I. G. Plotzker and G. J. Exarhos, J. Phys. Chem., 83:2496 (1979).
18. G. L. Carlson, R. E. Witkowski, and W. G. Fateley, Spectrochim. Acta, 22:1117 (1966).
19. B. N. Nelson and G. J. Exarhos, J. Phys. Chem., 84:2867 (1980).
20. G. J. Exarhos and W. M. Risen, Jr., J. Am. Ceram. Soc., 57(9):401 (1974).
21. G. B. Rouse, P. J. Miller and W. M. Risen, Jr., J. Non-Cryst. Solids, 28:193 (1978).
22. S. A. Brawer and W. B. White, J. Non-Cryst. Solids, 23:261 (1977).
23. G. J. Exarhos, Nucl. Inst. and Methods in Phys. Res., 229(B1):498 (1984).
24. G. J. Exarhos and W. E. Conaway, J. Non-Cryst. Solids, 55(3):445 (1983).
25. F. L. Galeener, J. Non-Cryst. Solids, 49:53 (1982).

STUDY OF AMORPHOUS TO MICROCRYSTALLINE STRUCTURE CHANGE OF

HYDROGENATED Si -- WHAT STABILIZES THE AMORPHOUS PHASE?

Akio Hiraki

Department of Electrical Engineering
Osaka University
Suita, 565 Osaka, Japan

I. INTRODUCTION

Recently, hydrogenated amorphous Silicon (a-Si:H) films have been investigated in many laboratories [1] especially from the prospect of solar cells. Role of H in these films has been widely recognized to terminate the Si dangling bonds through forming Si-H bonds reducing greatly gap states. With this reduction of the gap states, the addition of reasonable concentration of dopants such as P or B can move the Fermi level to give either n-or p-type a-Si:H, respectively. On the other hand, another important role of H for possible stabilization of amorphous phase of the a-Si:H has been discussed in less extent.

In this chapter, the present author tries to stress on the positive role of certain kinds of impurities including H to stabilize the amorphous phase. This is based upon the structural study on transformation of microcrystalline state of hydrogenated silicon to amorphous one due to the presence of more electro-negative impurities.

Structure of tetrahedrally-bonded elemental amorphous semi-conductor such as present Si is assumed to be simply understood by the concept of continuous random network (CRN). But this assumption is far from the reality in the case of a-Si:H, as clearly seen in Fig. 1. In the figure is shown Si-density together with H-content of a-Si:H films prepared under various fabrication conditions of glow-discharge (GD) and r.f.-sputtering [diode (2SP) and tetrode (4SP)]. Obviously the densities are lower in all films than that of Si-crystal whose density is as dense as calculated CRN-structured amorphous Si with no H-content.

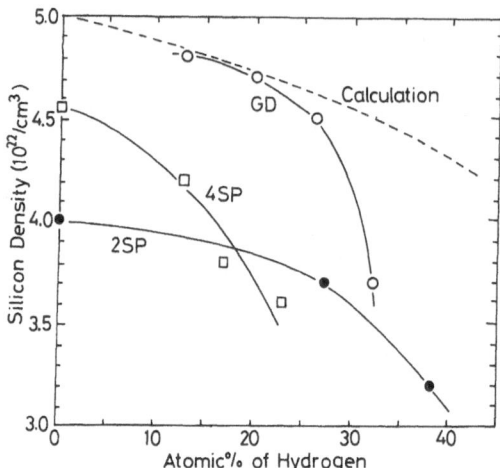

Fig. 1   Atomic density of silicon and hydrogen concentration for
         GD, 4SP and 2SP films. The calculated curve (dotted line)
         is described in the text.

     Density of the film can well be correlated with its film
morphology observed by transmission electron microscopy (TEM).
Namely, the films with far less Si-density (for example, 2SP-
films in Fig. 1) exhibit columnar morphology and only films of
relatively high density (like GD films of ∿10 atomic % of H in
Fig. 1) indicate uniform morphology.

     The above facts that a-Si:H films are generally less dense than that
of conceptionally assumed CRN-structured amorphous Si is understood
as follows. As has been well established, the basic unit of both
the crystalline and amorphous Si is the same diamond-like tetrahedron
with only small variation in bond angle [$\pm$ 10° (rms) off the normal
tetrahedral angle of 109°28'] and bond length (less than 1%). In
other words, the tetrahedral unit is very rigid and the construction
of the CRN-structured amorphous Si from the units is very likely
to induce strain. Consequently, without the presence of positive
causes in the film to stabilize or relax this CRN-type amorphous Si
of uniform film morphology, the film tends to take non-uniform
structure. Indeed, as mentioned with regard to Fig. 1, low Si-
density films of non-uniform or columnar morphology are frequently
fabricated. This non-uniform structure is obviously in less
strained or lower energy state with Si dangling bonds at the columnar surfaces passivated by H.

     In the course of study on fabrication of a-Si:H films by r.f.-
sputtering, we found that the sputtering in pure H_2-gas atmosphere
produces films composed of microcrystalline hydrogenated Si (μc-Si:H)
with non-uniform morphology [2]. The infrared (IR) spectra of Si-H
vibrations in these films, due to the crystalline nature, are

composed of well separated peaks which enable us to investigate
clearly the Si-H bonding configurations and also structure of the
films. However, when either the film thickness is very thin or a
slight fraction of nitrogen (N_2) is added into H_2 during the
sputtering, the film exhibits quite uniform morphology and at the
same time IR spectra become broad and structureless to indicate that
the film is amorphous instead of microcrystalline.

In this Chapter, at first (in Section II) the fabrication
of the μc-Si:H films and their interesting features including IR
property are introduced. Then in Sections III and IV, the description and discussion are presented about how the film thickness and
addition of small amount of N_2 into H_2, without changing other
fabrication parameters for μc-Si:H films, control the amorphous
and microcrystalline states. And in Section IV, the concluding
remark on the stabilization mechanism of the amorphous films of
Si is made.

II.   MICROCRYSTALLINE HYDROGENATED SILICON (μc-Si:H) PRODUCED BY
      r.f.-SPUTTERING IN HYDROGEN ATMOSPHERE FROM Si-TARGET AND ITS
      STRUCTURAL PROPERTIES

We have found by chance in the study of amorphous hydrogenated
silicon (a-Si:H) for possible use to photovoltaic devices fabricated by sputter-deposition technique that r.f.-sputtering in hydrogen
(H_2) atmosphere produces film of crystallized hydrogenated silicon
(μc-Si:H) as explained briefly below [2]. Usually sputter-deposition
of a-Si:H from Si-target is carried out in argon (Ar) atomsphere
containing hydrogen. The deposited material includes several
atomic % argon [3] which is supposed to induce microvoids (or
defects) in the films due to its large atomic size compared with
the open space in the tetrahedral network of the solid state silicon. The preparation of defectless or voidless amorphous silicon,
however, is desired for good photovoltaic devices. Since the
inner surface of a microvoid relates with the dangling bond, formation of the microvoid will increase the gap state density and
consequently degrade electric properties of the film.

Owing to the small atomic size, helium mixed with hydrogen is
expected to settle in the open space of the silicon network without
the crucial deformation that will induce the microvoid. Therefore,
the reactive sputtering in the atmosphere of helium (He) and hydrogen (H_2) molecules is expected to be favorable for fabrication of
hydrogenated silicon with smaller density of microvoids. With this
hope of obtaining the defectless film, we performed sputtering in
ambient gas of He+H_2. During the course of this study, crystalline
hydrogenated silicon (μc-Si:H or C-Si:H) has been fabricated at low
substrate temperatures below 250°C.

An r.f.-sputtering apparatus, NEVA type EP-21, was used to prepare the films. Sputtering conditions expect ambient gas contents were the same as those described elsewhere [4]. Reactive sputtering was carried out in the mixed gas of He+H_2 where total pressure was 1-3x10^{-1} Torr with partial mole fraction of hydrogen ranging from 0 to 100 mole%. Input r.f.-power was 3.6W/cm^2, and substrate temperatures (Ts) were 200-250°C, and sometimes about 80°C when the substrate holder was water-cooled.

Figure 2 is a typical X-ray diffraction pattern of a crystallized film deposited on a slide glass under the sputtering conditions described in the figure. The appearance of the film surface was somewhat milky or cloudy. The thickness was about 3 μm. The background of the diffraction pattern due to the substrate was subtracted. Positions of 2θ angles for (111), (200) and (311) were exactly the same as those in powdered crystal, but line widths were wider than those of the crystal shown by dotted curves. Further study using high resolution electron diffraction [5,6] has indicated that the films fabricated under He+H_2 atmosphere with H_2 larger than several mole% are composed of microcrystalline hydrogenated Si(μc-Si:H) whose sizes range from ∿ 50 nm to ∿ 5 nm in diameter. One example is clearly shown by the image photograph in Fig. 3 with exactly the same {111} plane spacing as that of pure crystalline Si.

Before going into the discussion why such μc-Si:H is produced, it is worth noting about the positive role of H_2 for the sputtering or film deposition process. In Fig. 4, deposition rate of the film

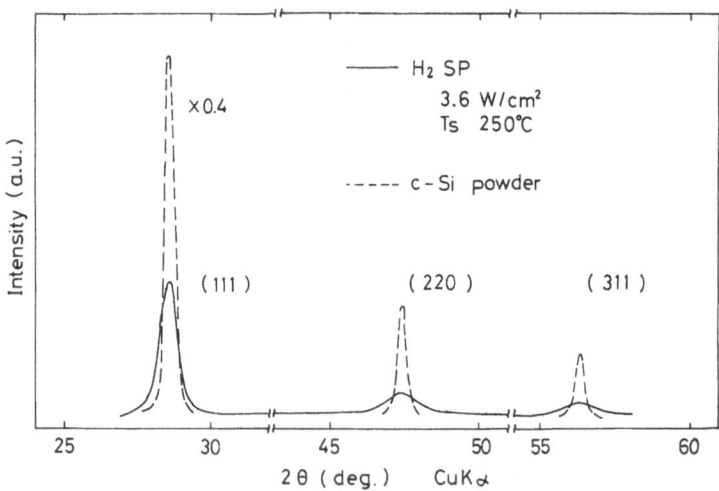

Fig. 2   X-ray diffraction pattern of crystallized film fabricated in pure H_2 gas at power of 3.6 W cm^{-2} and substrate temperature of 250°C. Dotted curves show that of powdered crystal.

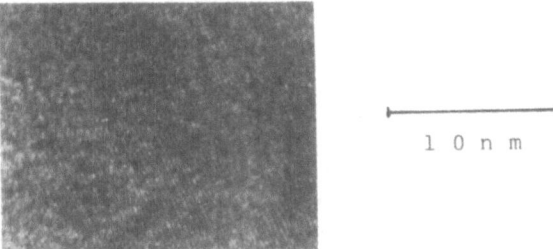

Fig. 3   High resolution lattice image photograph of µc-Si:H.

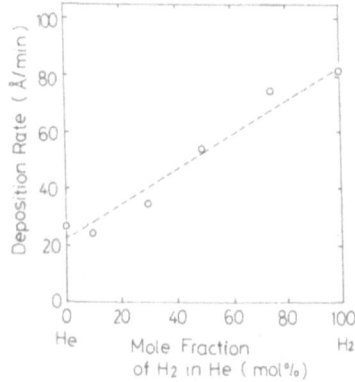

Fig. 4   Deposition rate of the films fabricated with various mole
fractions of H_2 in He.

is shown for various mole fractions of H_2 in sputtering atmosphere.
Deposition rate increased linearly with the concentration of H_2
and in highly concentrated gas (80 mole% or more of H_2 or pure H_2)
the sputtering proceeded with fairly large rate (8 nm/min) which
is almost as large as the deposition rate by pure Ar atmosphere.
This fact is interesting because H is smaller in mass than He, of
course than Ar, and consequently suggests that not only physical
but also chemical situations control the sputtering process.   In
other words, chemical reaction of hydrogen with solid silicon at
target surface play an important role in the course of the film
deposition process.

Due to the crystalline nature of the films, the infrared (IR)
spectra of Si-H vibrations [7] are composed of well separated peaks
which enable us to better understand the Si-H vibration in amorphous
hydrogenated silicon (a-Si:H).

The c-Si:H or μ_c-Si:H films described hereafter in this chapter were
prepared, unless specified, by r.f.-sputtering in H_2 gas under H_2

pressure $(1 \sim 5) \times 10^{-1}$ Torr with r.f.-power ranging from 1.2 to 3.8 W/cm^2.

Depending upon fabrication conditions, c-Si:H films with variety of IR properties can be fabricated. As an example, IR spectrum of a c-Si:H film deposited on a water cooled substrate is shown in Fig. 5 together with that of an a-Si:H film for comparison. The a-Si:H film was fabricated by conventional sputtering in Ar+H$_2$ atmosphere with Ar fraction larger than $\sim$ 70 mole%.

In addition to sharp and fine structured spectrum of c-Si:H compared with broader and structureless one of a-Si:H, complete absence of the absorption at 2000 cm^{-1} is found in the c-Si:H. The reason why this 2000 cm^{-1} absorption is not present will be discussed later.

In the stretching vibration region, at least six peaks are clearly seen as shown in Fig. 6(a). For the assignment of these peaks, the above situation is of great help since, if one assumes the absorption at 2000 cm^{-1} is due to an isolated Si-H bond stretching, we only have to consider =Si=H$_2$, -Si≡H$_3$ and their combinations. Namely, they are seven configurations (five =Si=H$_2$ related and two -Si≡H$_3$ related ones) as illustrated in Fig. 7.

An IR study [8] on Si-H stretching modes in substituted silane molecules has indicated that the stretching frequency ν_{Si-H} can be expressed by the following empirical equation:

$$\nu_{Si-H} = a + b \sum_{i=1}^{3} E(Ri)$$

where a and b are constants, E(Ri) is a measure of the electro-negativity of the i-th substituted atom or group R.

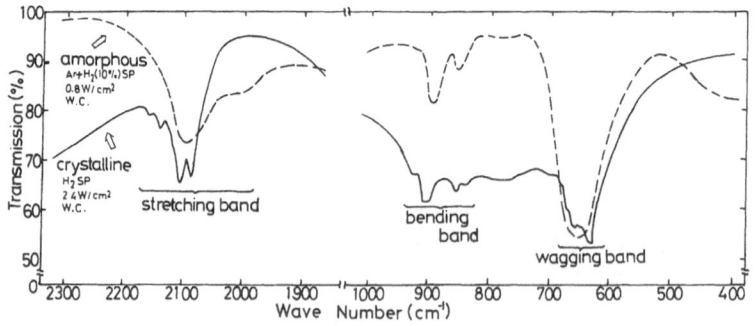

Fig. 5   IR spectra of c-Si:H and a-Si:H.

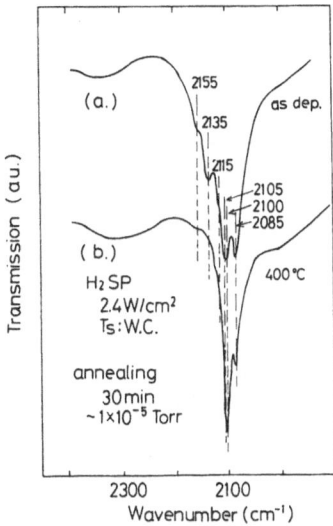

Fig. 6   Stretching mode region of c-Si:H (a) as deposited, (b)
         annealed at 400°C for 30 min.

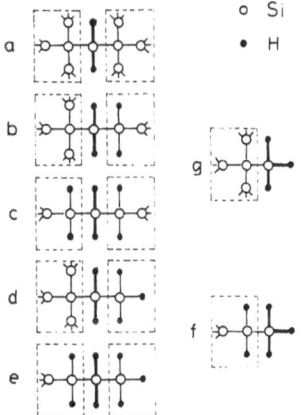

Fig. 7   Schematic of Si-H bond configurations in c-Si:H.

The applicability of the empirical relationship to the Si-H
stretching frequency in a-Si:H has been claimed by Lucovsky [9]
who has estimated ν_{Si-H} (in cm^{-1}) through underwritten expressions
in the following three types of substituted silane molecules:
SiHXYZ, SiH_2XY and SiH_3X, where X, Y, and Z represent substituted
atoms or groups.

$$\nu_{Si-H}(SiHXYZ) = 1740.7 + 34.7 \sum_{R=X,Y,Z} SR(R) \tag{1}$$

$$\nu_{Si-H}(SiH_2XY) = 1953.6 + 25.4 \sum_{R=X,Y} SR(R) \tag{2}$$

$$\nu_{Si-H}(SiH_3X) = 2086.1 + 22.5 \times SR(X) \tag{3}$$

where SR(R) is stability-ratio electro-negativity of the substituted atom or group R.

Defining R's indicated by the dotted rectangles (Fig. 7), we calculated the SR(Ri)'s in the seven configurations and from equations (2) and (3) estimated ν_{Si-H}'s which are listed in Table I with assigned experimental values taken from the peak positions in Fig. 6.

The appropriateness of the assignment and also effectiveness of (2) and (3) are recognized from the apparent linear relation between electro-negativity sum $\sum SR(R)$ and ν_{Si-H}(SiH,XY and SiH$_3$Y) shown in Fig. 8 -- open circles represent ν_{Si-H}'s in substituted silane molecules reported in reference 9.

TABLE I: ν_{Si-H} calculated and experimental.

Si-H configuration		sum of SR(R)	calculated	$\nu_{Si-H}(cm^{-1})$ experimental
=Si=H$_2$	a.	5.24	2089	2085
	b.	5.67	2100	2100
	c.	6.10	2111	2115
	d.	5.91	2106	2105
	e.	6.34	2117	--
-Si≡H$_3$	f.	3.05	2155	2155
	g.	2.62	2145	2135

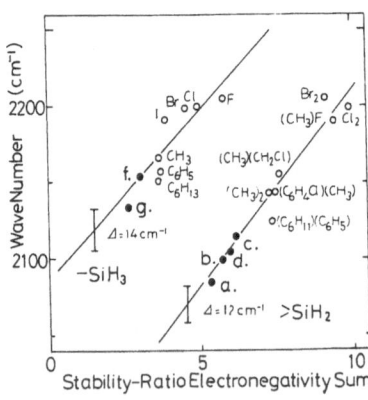

Fig. 8 ν_{Si-H} versus SR sum.   The error bars indicate standard deviations.

From Eq. (1) and taking X, Y and Z as sole Si atoms, ν_{Si-H} for an isolated $\equiv$Si-H configuration is calculated to be 2013 cm^{-1}, which supports the above made assumption of the absorption at 2000 cm^{-1} due to an isolated $\equiv$Si-H within experimental error.

Further, an annealing experiment on the c-Si:H film at 400°C for 30 min. confirmed the assignment of the -Si$\equiv$H$_3$ related stretching ν_{Si-H}(SiH X). Namely, as shown in Fig. 6(b) three peaks at 2105, 2135 and 2155 cm^{-1} disappeared indicating that -SiH$_3$ configuration was destroyed by the annealing.

One of the interesting properties of the present C- or μc-Si:H films is, as mentioned already, the absence of 2000 cm^{-1} IR absorption or $\equiv$Si-H configuration. So, we have pursued this cause through structural study of the film from the initial deposition stage [5] by transmission electron microscopy (TEM) correlated with sensitive IR absorption spectroscopy of Fourier transform (FT) type.

III. STRUCTURAL CHANGE (AMORPHOUS TO CRYSTALLINE) OF μc-Si:H
      FILM AS A FUNCTION OF FILM THICKNESS

Films for the TEM observation were self-supported on a grid after deposition on cleaved surfaces of KBr. Morphology of films thinner than 0.1 μm was studied with high resolution TEMs (200 kV, Hitachi and 1 MV at the University of Tokyo, JEOL), and films thicker than 0.1 μm with the ultra-high voltage TEM (3 MV, operating at 2 MV) at Osaka University. An a-Si:H film with a thickness of 0.5 $\sim$ 1 μm usually used for the a-Si device purposes is easily observed by this 3 MV TEM. High-resolution electron micrographs (see Fig. 3) were taken with a TEM, JEOL200CX (resolution for the lattice image: 0.14 nm). The image was taken with the direct beam and the next two outer diffracted rings. Optical densities of the negative photographic films of reflection high energy electron diffraction (RHEED) patterns were measured with a micro-photometer. In the RHEED experiment mirror-polished stainless steel was used as the substrate.

An FT/IR absorption spectrophotometer (Digilab FTS-20E) was used for the characterization of the thin film because of its fast scanning rate and high sensitivity by multiple integrations of 200 $\sim$ 5000 times. Samples for the IR absorption measurement were deposited on substrates of crystal silicon with high electrical resistivity.

Figure 9 shows the morphological change of the μc-Si:H with the film thickness observed by TEM. Interesting finding to be discussed later is that when the film is about 10 nm thick it is not crystalline but amorphous; no microcrystals are observed there. Moreover, the transmission electron diffraction (ED) pattern is

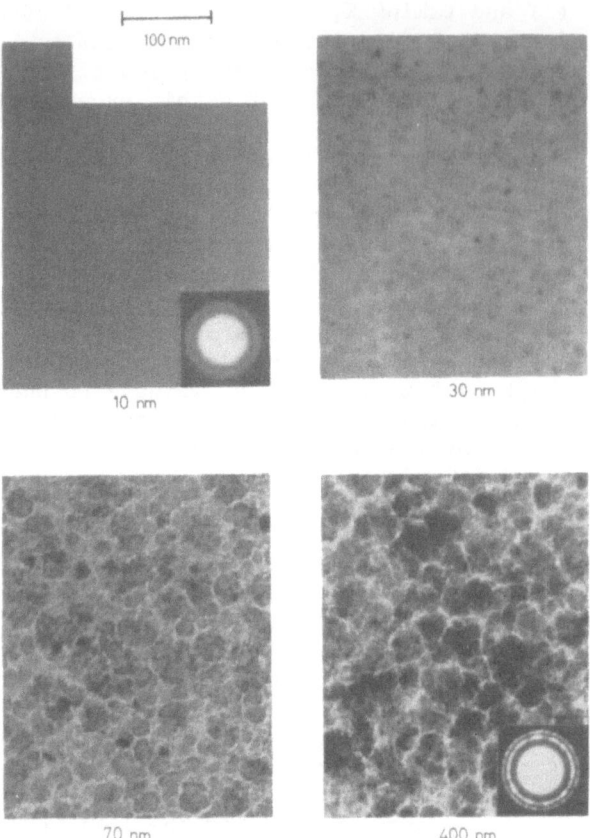

Fig. 9.   Morphological change of SP μc-Si with film thickness.

composed of diffuse haloes. Mycrocrystals 3-10 nm in diameter
begin to appear, making a mosaic pattern in the film 30 nm thick,
but an amorphous regions still remain over the film. The mosaic
pattern develops rapidly with the increase in film thickness as
shown in those films 70 nm and 400 nm thick. In the 400 nm
thick film, the size of the μc-Si grain is observed to be in a
range very roughly from 5 nm to several tens of nm. The feature
observed in the thick film of μc-Si suggests the occurrence of
a columnar aggregation composed of longitudinally associated μc-Si
grains with rather irregular shapes.

High resolution electron micrographs were taken for a μc-Si
film of roughly 0.1 μm thick. Lattice images arising from the
{111}plane are observed, as already shown by Fig. 3, throughout
the film with spacings of about 0.3 nm, showing that {111} is normal
to the film surface. A lattice image from the {220} plane with
spacings of about 0.2 nm is rarely observed. The shape of the
μc-Si grain is judged to be irregular from the area covered with

a uniform spread of the lattice image, and the size ranges
roughly from 5 to 50 nm. The appearance of the regular lattice
image implies that hydrogen does not exist inside the μc-Si grain
but it may be on the surface of the grain or among the columns as
schematically illustrated in Fig. 10. This is a quite different
situation from that in a-Si:H where hydrogen exists inside the
amorphous network as will be discussed in Section IV.

Figure 11 shows the change of the RHEED pattern with the
increase in film thickness. As the film thickness increases,

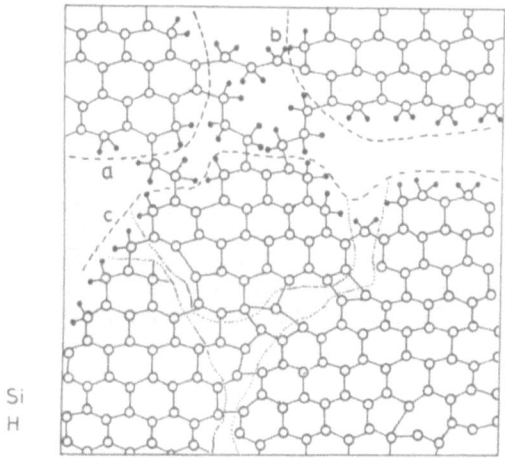

Fig. 10   Structure model of μc-Si:H film: Areas represented by
broken lines, a, b, c are columns. Each column (for
example c) is composed of crystalline regions represented
by dotted lines.

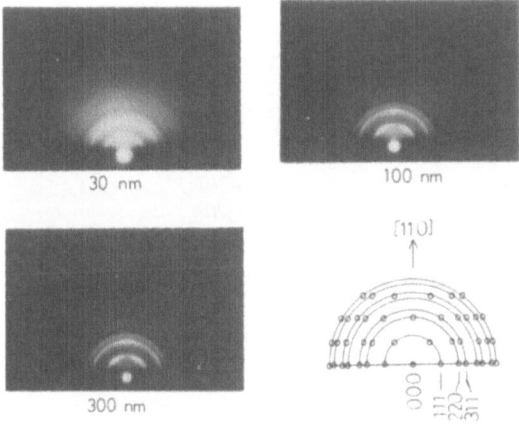

Fig. 11   RHEED patterns of SP μc-Si with film thicknesses.

diffraction rings become sharper, showing the increase in the
region of μc-Si grains on the film surface. In the films of 100
nm and 300 nm thick, the diffraction rings show the {110}
preferred orientation. The spot positions to appear for the
ideal {110} orientation are inserted in Figure 11. Negative
photographic films of these RHEED patterns have been analyzed
with a micro-photometer to estimate the degree of the orientation.
The spot intensity on the {220} diffraction ring relative to that
on {111}, at the 90° direction, increases steeply in the thickness
range from 0.05 to 0.1 μm, showing the enhancement of the {110}
orientation.

    Figure 12 shows IR absorption spectra for the Si-H stretching
bands of μc-Si films with various thickness. In the thin film
where the amorphous region is mixed, broad structured absorption
is observed with a peak near 2100 cm^{-1} and a shoulder near 2000
cm^{-1}. The presence of the shoulder at 2000 cm^{-1}, which can be
assigned to the ≡Si-H or SiH monhydride structure, is a characteristic feature of a-Si:H. As the film grows, the absorption intensity
near 2000 cm^{-1} (due to SiH) decreases and finally disappears, while
the intensity near 2100 cm^{-1}, assigned to =Si=H$_2$ or SiH$_2$ dihydride
structures, increases. Finally, in the sufficiently thick film
the absorption near 2100 cm^{-1} is narrowed and resolves into two
peaks at 2089 and 2104 cm^{-1}, which can be assigned to -SiH$_2$- and
-(SiH$_2$)$_2$-, respectively. In other words, these sharp and fine
structured IR absorption spectra discussed in Section II concerning
Fig. 5 are only expected from the thick deposited films. The
reason is discussed in the next section (IV).

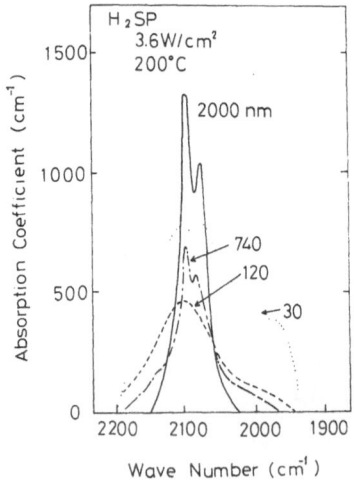

Fig. 12   Change of IR spectra of   μc-Si:H with film thickness.

IV. STRUCTURAL CHANGE (CRYSTALLINE TO AMORPHOUS) OF μc–Si:H
FILM DUE TO ADDITION OF MORE ELECTRO–NEGATIVE NITROGEN (N)
IMPURITIES

In the preceding sections, we have learned interesting features
of the μc–Si:H films fabricated by r.f.–sputtering of Si target
in H_2 atmosphere. They are: (a) IR spectra arising from Si–H
vibrations in the μc–Si:H films are composed of well–separated
peaks. From the peak positions clear assignment of the corresponding Si–H configurations can be made. They are seven configurations
(five $=Si=H_2-$ related and two $-Si\equiv H_3-$ related ones), but $\equiv Si-H$
configuration is not at all present.

(b) However, when the film is sufficiently thin ($\lesssim$ 30 nm), IR
spectra do show the presence of $\equiv Si$–H or SiH configuration and at
the same time the TEM indicates that the film is amorphous with
uniform morphology rather than microcrystalline of non–uniform
structure.

(c) With the increment of the film thickness, the SiH configuration gradually disappears in the IR spectra and no more SiH
configuration can be seen at 1 μm thickness.

(d) The morphology of the film also changes with the thickness
from uniform amorphous to non–uniform columnar structure and
finally microcrystalline structure with {110} orientation.

From these features a simple conclusion can be drawn that the SiH
configuration is at least necessary to stabilize uniform amorphous
structure against non–uniform columnar or microcrystalline structure.

This is understood as follows. Since the tetrahedral unit of
Si is very rigid, construction of amorphous structure by random
connection of these units of CRN–structured amorphous Si induces
high strain. So, to reduce this strain some connecting bonds
must be broken and hydrogen atoms attach to these broken bonds
giving rise to the SiH configurations as schematically shown in
Fig. 13. But this reduction of strain by the SiH becomes

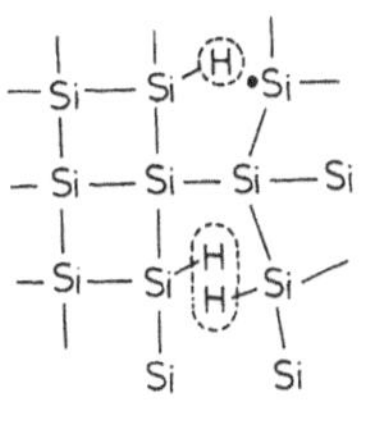

Fig. 13 Stabilization role of H in uniform structured a–Si:H.

insufficient when a uniform amorphous region exceeds 10 nm in dimension. Consequently, with the increment of film thickness the structure gradually changes to a less strained one like columnar and finally microcrystalline.

This statement seems to be evidenced by the observation of microcrystalline to amorphous phase transition [5] of the film through putting a small fraction of N_2 into H_2 gas for the sputtering atmosphere. Because in this case nitrogen (N) can play as a strain-relieving element, as explained later, due to its lower coordination number and higher electron negativity than Si (fabrication conditions: see Table II). This microcrystalline to amorphous phase transformation by addition of small amount of N_2 into H_2 gas is clearly seen from TEM (transmission electron micrograph) in Fig. 14. The characteristic columnar morphology of

TABLE II

Back pressure	$3 \sim 4 \times 10^{-7}$ Torr
Sputtering pressure	0.1 Torr
N_2 fraction in $H_2 + N_2$	$0 \sim 5$ mol%
rf power	3.8 W/cm²
Substrate temperature	250 °C

(N_2 fraction: 0 mol%)

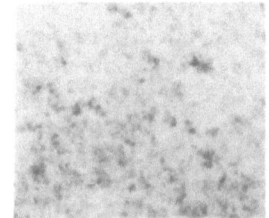

(N_2: 0.5 mol%)

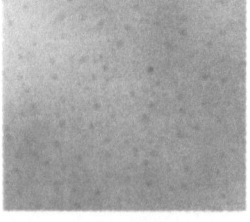

(N_2: 1 mol%)   0.2 μm

(N_2: 2 mol%)

Fig. 14 Morphological change from microcrystal to amorphous due to addition of nitrogen by TEM.

the microcrystalline phase changes to uniform amorphous morphology at an N_2 fraction of only $\sim$ 2 mole%. This is well contrasted by the fact as seen in Fig. 15, that amorphous phase is only introduced in various inert atom + H_2 gas, like Ar + H_2 gas, with a far higher inert atom fraction (in the vicinity of $\sim$ 80 mole%) under similar sputtering conditions (Table II). In addition, corresponding to the transition, in IR spectra (Fig. 16) SiH configuration appears

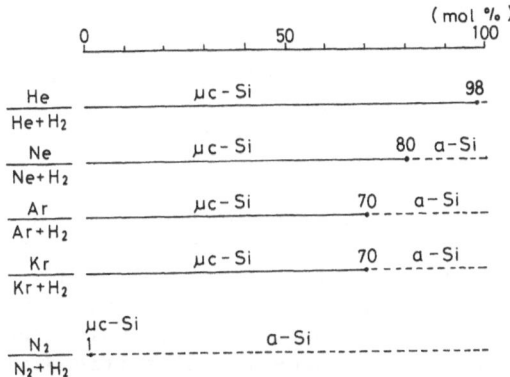

Fig. 15   Remarkable difference between inert (He, Ne, Ar, Kr) and highly electro-negative nitrogen (N) atoms for the fabrication of a- or µc-Si:H films.

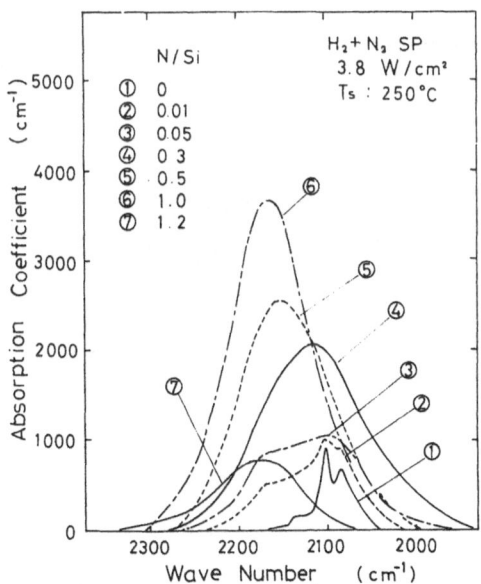

Fig. 16   Change of IR spectra of µc-Si:H due to addition of nitrogen.

although the vibration frequency shifts due to the presence of
nitrogen.

In this section, it has been shown that the strain in the
r.f.-sputter deposited Si film due to the rigidness of the Si
tetrahedron greatly influences the morphology. Except for
strain-relieving atoms or impurities (like N)/ or atom groups
(like SiH), the film morphology is a columnar one (including
microcrystal) which is less dense than the uniform amorphous film
as mentioned in Section I. The strain-relieving impurity must be
more electro-negative than Si and the coordination number should
be 2 or 3. Therefore, in addition to nitrogen (N), oxygen (O) is
also expected to play the same role. The requirement for a
higher electro-negative atom with a lower coordination number is
understood as discussed below.

The rigidness or inflexibility of the Si-tetrahedral is due
to the presence of the high density of the so-called bond charge [10]
in the middle of an Si-Si covalent bond. So when more electro-
negative impurity atoms like N attach to the Si-atoms, the
electrons tend to go to the impurity atoms, which results in a
reduction of bond charge and makes the Si tetrahedral more flexi-
ble. Lower coordination gives more freedom for the impurity atom
to interconnect two or three amorphous regions whose size must
be ~ 10 nm in dimension as already discussed. Therefore, these
impurities only in the concentration of $\sim 10^{18}$ atoms/cm^3 are
effective to interconnect continuously the amorphous regions to
stabilize morphologically uniform amorphous films from the (micro-)
crystallization.

In other words, a strain-relieving impurity is a kind of
connector as shown in Fig. 17. Of course the bond angle (θ)

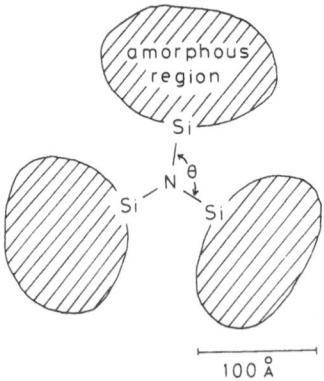

Fig. 17   Role of nitrogen (N) as a connector between amorphous
          regions of Si to result in uniform a-Si:H.

between the connector atom and connected Si atoms (in the figure N⟩θ with Si and Si) can vary flexibly due to its higher electro-negativity. In this respect H is an imperfect connector.

The morphological structure of the a-Si:H films is closely related to the density of silicon. This Si-density of the films is also related to their electrical and photoelectric properties. Namely, films with good properties for solar devices should be almost as dense as crystal in Si-density which is ~ 5×10^{22} Si atoms/cm^3.

Si-density of the thick films with non-uniform columnar structure and also microcrystalline structure shown by the photographs (Fig. 9(d)) are always more than 30% less dense than that of crystalline Si. However, a-Si:H films fabricated by glow-discharge deposition in silane (SiH_2) gas to be used for solar cells exhibit very uniform amorphous structure and consequently the density is only about 5% less than that of the crystal.

In this respect, for the fabrication of such uniform amorphous films, addition of small amount of O_2 or N_2 gas must be quite effective. In fact, several attempts in this direction have been reported to be successful.

V. CONCLUSIONS

It has been shown in this chapter that to fabricate CRN-structured a-Si:H with uniform morphology is generally very difficult due to rigidness of Si tetrahedron which is a basic unit of both crystalline and amorphous Si phases with only small variation in bond angle and bond length in the two phases. Therefore, to produce uniform amorphous Si films the presence in the film of proper impurities such as N or O to act as strain relaxing agents is required. Otherwise, the film tends to be composed of micro-crystalline Si (5 ~ 50 nm in diameter) with non-uniform columnar structure and far less dense in Si-density. Especially when r.f.-sputtering deposition in pure H_2-gas from Si target is employed for film fabrication, the microcrystalline structure of the films can be extensively studied from IR, TEM and other techniques.

The role of H in a-Si:H films is also important, besides the well recognized Si dangling bond termination to reduce gap states. H stabilizes the amorphous phase through breaking the strained Si-Si bond and then terminates the broken bonds in form of Si-H or monohydride configuration. However, its effect is only valid until the amorphous film is as thick as ~ 10 nm in film thickness. So, H is an imperfect strain releaser. To be a perfect strain releaser or connector in the sense explained in connection with Fig. 17 in Section IV, the impurity must satisfy the following requirements:

(1) It should be more electro-negative than Si.

(2) Its coordination number should be 2 or 3.

(3) Its concentration in the film must be larger than $\sim 10^{18}/cm^3$.

Drastic effect of putting N_2, perfect strain releaser, into H_2 atmosphere is described in Section IV. And in addition, more pronounced microcrystalline to amorphous transformation at substrate temperature (Ts) as low as $\sim 100°$ K was recently reported from the author's laboratory [11]. Namely, at $\sim 100°$ K in pure H_2 atmosphere alone a powder-like substance can be fabricated. IR spectra from this material indicates main presence of $-Si\equiv H_3$ configuration [12]. But, putting only ~ 0.05 mole% of N_2 into the H_2 atmosphere, without changing other parameters, converts the powder-like film into dense and uniform amorphous film.

This dramatic observation clearly suggests that the determining factor of film morphology (either amorphous or microcrystalline) or atomic arrangement in the film is not the substrate temperature. In other words, the fabricated film takes the most stable morphology, since in the plasma where the r.f.-sputtering takes place a lot of energy for the atomic arrangement is available.

REFERENCES

1. See for example. W. Paul and D. A. Anderson: Solar Energy
        Materials 5:229 (1981), and also H. Fritzsche, Solar Energy
        Materials 3:447 (1981).
2. T. Imura, K. Mogi, A. Hiraki, S. Nakashima and A. Mitsuishi,
        Solid State Comm. 40:161 (1981).
3. T. Imura, K. Kubota, K. Ushita and A. Hiraki, Japan J. Appl.
        Phys. 19:99 (1980).
4. T. Imura, K. Ushita and A. Hiraki, Japan J. Appl. Phys. 19:L65
        (1980).
5. A. Hiraki, Y. Fukushima, T. Sato, H. Kiyono, H. Terauchi and
        T. Imura, J. Non-Cryst. Sol. 59-60:791 (1983).
6. T. Imura, H. Kaya, H. Terauchi, H. Kiyono, A. Hiraki and
        M. Ichihara, Japan J. Appl.Phys. 23:179 (1984).
7. A. Hiraki, T. Imura, K. Mogi and M. Tashiro, J. de Physique
        42:c4-277 (1981).
8. L. Smith and C. Angelotti, Spectrochimica Acta 15:412 (1959).
9. G. Lucovsky, Solid State Comm. 29:571 (1979).
10. J. C. Phillips, Bonds and Band in Semiconductors, Academic
        Press, New York, 91-92 (1973).
11. T. Miyasato, F. Tokumura, Y. Kawakami and A. Hiraki, Solid
        State Comm., in press.
12. S. Hashimoto, T. Miyasato and A. Hiraki, Japan J. Appl. Phys.
        22:L748 (1983).

STRUCTURAL IMPLICATIONS OF GAS TRANSPORT IN AMORPHOUS SOLIDS

James F. Shackelford

Division of Materials Science and Engineering
University of California
Davis, CA 95616

INTRODUCTION

The interstitial structure of vitreous silica and related materials is in a difficult size range for evaluation. Diffraction and spectroscopic techniques are generally limited to more local structural descriptions (bond lengths, coordination numbers, defect structures, etc.). Electron microscopy surveys somewhat larger geometry (e.g., phase separation). A careful analysis of the diffusion and solubility of inert gas atoms and molecules provides an experimental probe for interstitial structure.[1] Such information is of substantial interest because of the basis for structural classification (in the manner of Bernal's canonical hole model[2] of liquid structure). There is also a close connection to some commercially important processes, such as the thermal oxidation of silicon.[3]

A number of studies reported at this symposium have touched upon problems related to gas transport. Griscom[4] presented evidence for the molecular diffusion of radiolytic H_2O in γ-irradiated high water silicas. Brown[5] has related these concepts to the annealing of radiation damage in MOS devices. Rabinzohn and Rochet[6] have characterized the step-by-step mechanism of O_2^{18} transport in silica films. For some time, the author has investigated the correlation between the structure of amorphous solids and gas transport. This current paper will be divided into three areas corresponding to the major segments of those studies. "Gas Probe Studies" will cover the experimental tools available for this work. "Modeling Studies" will cover the complementary investigation of interstitial structural models. "Applications" will relate the fundamental studies to some current problems of technological importance.

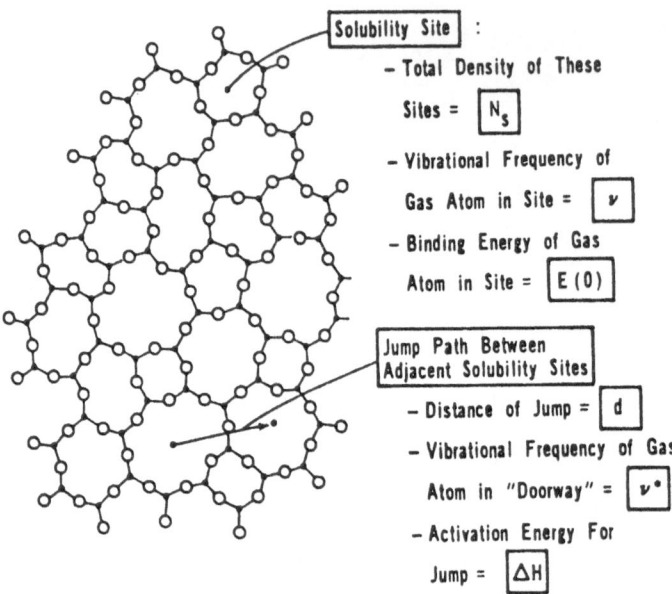

Fig. 1   Definition of gas probe parameters[1] against the Zachariasen
         schematic[7] of the random network structure.

GAS PROBE STUDIES

Figure 1 is a schematic summary of the "gas probe" analysis of
structure in amorphous solids.[1] The illustration is made relative
to the classical Zachariasen schematic[7] for oxide glasses. There
are six terms that can be derived from a statistical mechanical
analysis of the gas transport process. Shackelford et al.[8] developed
a solubility equation for describing the equilibrium solution of
non-reactive gases:

$$\frac{n_s}{p} = (h^2/2\pi mkT)^{3/2}(kT)^{-1} N_s \, x[e^{-h\nu/2kT}/(1-e^{-h\nu/kT})]^3 e^{-E(o)/RT} \quad (1)$$

where n_s is the number of gas atoms (or molecules) dissolved per unit
volume of solid under a gas atmosphere pressure, p. Also, h is
Planck's constant, m the mass of one gas atom (or molecule), k the
Boltzmann's constant, T the absolute temperature, N_s the number of
solubility sites per unit volume of solid, ν the vibrational frequency of the dissolved atom (or of the center of mass of a dissolved
molecule), R the universal gas constant, and E(o) the binding energy
of the dissolved gas atom (or molecule) to an interstitial solubility
site. The terms n_s, p, and T are experimental variables. By best-
fitting experimental (n_s/p) versus T data to Equation (1), it is
possible to determine values for N_s, ν, and E(o). This is summarized

relative to a typical solubility site in Figure 1. Masaryk and Fulrath[9] extended the approach of Shackelford et al. in developing a statistical mechanical expression for gas diffusivity, D:

$$D = \frac{1}{6}\frac{kT}{h}d^2 \frac{(e^{h\nu/2kT} - e^{-h\nu/2kT})^3}{(e^{h\nu*/2kT} - e^{-h\nu*/2kT})^2} e^{-\Delta H/RT} \tag{2}$$

where d is the jump distance between adjacent sites, $\nu*$ the vibrational frequency of the gas atom (or molecule) in the "doorway" between adjacent sites, and ΔH the activation energy for diffusion. All other terms are the same as in Equation (1). By best-fitting experimental D versus T data to Equation (2), it is possible to determine values for d, $\nu*$, and ΔH. It is important to note that a distribution of interstitial solubility site sizes is inherent in the Zachariasen schematic[7] of Figure 1. (These interstices correspond to the "canonical holes" of the Bernal model[2] of liquids.) Although the structural information from Equations (1) and (2) is restricted to an "average" interstice, an indication of the nature of the size distribution can be obtained by comparing results for various gas probe atoms and molecules of various sizes.

The experimental solubility parameter, S, is defined by:

$$n_s = S p \tag{3}$$

where n_s and p were defined relative to Equation (1). The diffusivity, D, is defined by Fick's first law:

$$J_x = -D\left(\frac{\partial c}{\partial x}\right) \tag{4}$$

where J_x is the net flux of gas atoms (or molecules) in the x-direction. D is the proportionality coefficient between this flux and the concentration gradient ($\partial c/\partial x$) of the gas species. An alternative gas transport parameter is the permeability, K, defined by:

$$R = K\left(\frac{\Delta p}{t}\right) \tag{5}$$

where R is the steady-state rate of gas flow across a membrane of thickness, t, due to a pressure drop Δp. It is straightforward to show that, for unreactive gases,

$$K = DS \tag{6}$$

The equality in Equation (6) is representative of the fact that the overall mechanism of steady-state permeation involves first the equilibrium solubility of the gas with the surface on the high-pressure side of the membrane, followed by the diffusion of the gas through the membrane.

Gas transport measurements generally fall into one of two categories. <u>Permeability-based</u> measurements are centered on the determination of K during the steady-state permeation of a membrane.[10] The diffusivity, D, is determined by analyzing the approach to or departure from steady-state permeation. Solubility can be obtained indirectly by rearrangement of Equation (6) (i.e., S = K/D). <u>Solubility-based</u> measurements involve direct determination of the solubility from the total amount of gas absorption or desorption from a sample. Diffusivity is determined by the rate of absorption or desorption. As much of the author's work has concentrated on equilibrium behavior and the related solubility parameters, our laboratory has used solubility-based measurements. Specifically, we measure a small pressure drop in a closed chamber as gas dissolves into a known volume of glass. Figure 2 illustrates typical experimental data and their relationship to solubility, S, and diffusivity, D.

An early hypothesis about the nature of the distribution of interstitial site sizes as suggested by gas probe data was made by Shackelford and Masaryk.[11] Table I summarizes a few key data that are consistent with a wide distribution of site sizes. In this case,

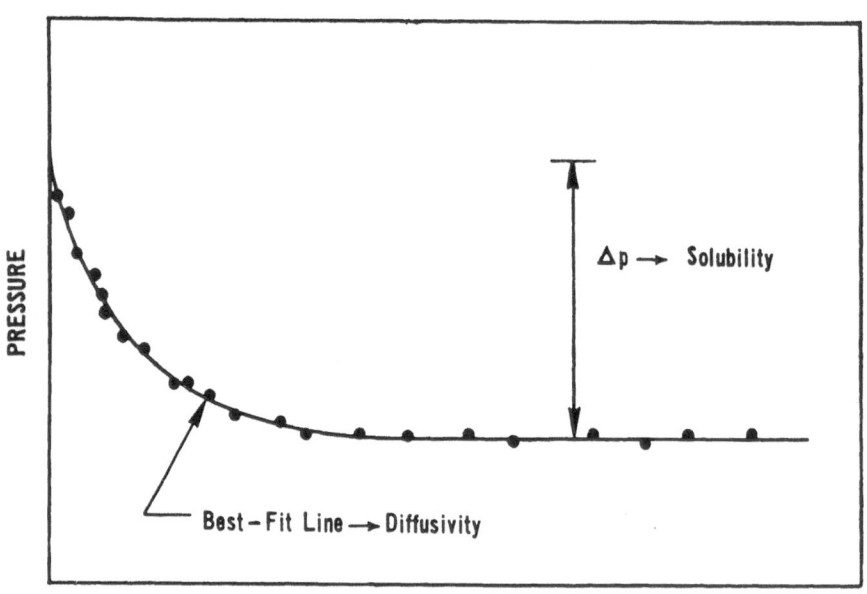

Fig. 2 Relationship of experimental data to solubility and diffusivity for a "solubility-based" measurement. The pressure in a closed chamber drops as some of the gas from the atmosphere slowly dissolves into the glass specimen.

the experimental data are N_S values first reported by Shelby[12] using a solubility-based experiment at high (up to 1400 atm) pressures. By having p as a significant variable, the N_S data could be cited with a high degree of confidence. (Data in the range of one atmosphere produce similar N_S data, but with relatively large uncertainty bars.) One should note that the N_S value is a function of the gas probe species. The density of sites accessible to He is 2.3×10^{21} cm^{-3}, whereas the density accessible to Ne is 1.3×10^{21} cm^{-3}. Both numbers are substantially less than 2.22×10^{22} cm^{-3}, the number calculated for cristobalite,[8] which is commonly taken as the crystalline analog of vitreous silica. A useful first-approximation is to take this N_S as the total number of sites per unit volume in vitreous silica. In this case, He atoms are accessible to roughly 10 percent of the total interstitial sites in vitreous silica, and Ne atoms (slightly larger in diameter than He) are accessible to roughly 6 percent. Shackelford and Masaryk[11] reviewed various, common probability distribution functions (PDF) that could be fitted to the data in Table 1. The most realistic was found to be the log-normal, as shown in Figure 3. In addition, the log-normal PDF frequently appears in physical systems. A strong analogy to the interstitial size distribution is the common occurrence of a log-normal PDF for particle-size distributions in crushing and grinding of powders.[13] Both systems describe space filling by randomly divided volume elements. Recently, Kurtz and Carpay[14,15] have demonstrated theoretically and experimentally the stability of the log-normal distribution for grain size in metallic and ceramic microstructures, another related and "random" space-filling system. The repeated occurrence of the log-normal PDF leads us to consider it (as illustrated by Figure 3) as a useful definition of the "random network" first proposed by Zachariasen.[7] Further support for this concept will be raised in the next section on modeling.*

---

* The concept of the Zachariasen model being reinforced by recent research is generally counter to the tone of this Symposium as a whole. A consistent theme has been the growing body of evidence for nonrandomness in amorphous solids. However, these observations are not inconsistent. It is the author's firm belief that the best hope for ordering the increasingly diffuse debate over structure in amorphous solids is to first establish the ideally random structure. The log-normal PDF appears to be a prime candidate for this purpose. Then, non-random structure can be cataloged systematically as defects relative to the ideally random structure. This situation is completely analogous to that long established in crystallography in which non-crystalline defects are defined relative to the ideally crystalline structures based upon the Bravais lattices. For practical purposes, defect-free crystals do not exist in nature. Similarly, we should not be surprised that ideally random structures do not exist. This does not diminish the importance of the ideal structure relative to which real, defective structure can be defined.

TABLE I

Solubility Site Densities in Vitreous Silica

Accessible to He[12]:	2.3×10^{21} cm^{-3}
Accessible to Ne[12]:	1.3×10^{21} cm^{-3}
Total[8]:	2.22×10^{22} cm^{-3}

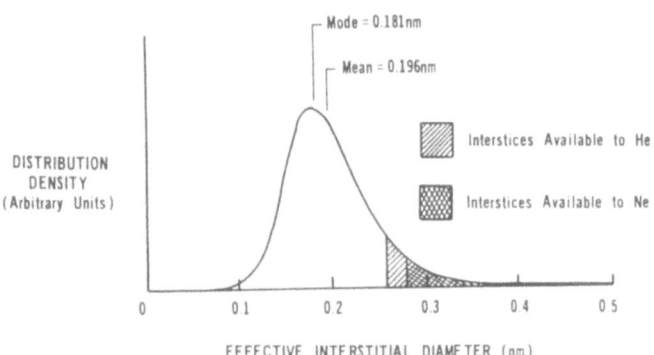

Fig. 3   Distribution of interstitial site size in vitreous silica
shown as a log-normal PDF for the data of Table 1.[11]

        Concentrating on the large diameter "tail" of the log-normal
PDF (Figure 4) allowed Shackelford[3] to determine that Ar and O_2
should be near the threshold of accessibility to the vitreous silica
network.   Comparative experimental results from Flores and Shackelford[16] bear this out (see Figure 5).   The current experimental goal
of our laboratory is to provide a sufficiently high sensitivity
pressure measurement to allow a precise determination of the likely
variation of solubility and diffusion of Ar and O_2 as a function of
thermal history.   The experimental precision has recently been
established, as illustrated by Figure 6.   In this model experiment,
a helium solubility measurement was carried out at a small total
atmospheric pressure (p_{final} = 180 torr) in order to test the
experimental system's ability to monitor a small pressure drop, in
the range expected for Ar and O_2 experiments at atmospheric pressure.
As seen in Figure 6, the experimental system can now monitor such
pressure variations quite adequately (to a precision of $\approx 10^{-3}$ torr).
In fact, it is now necessary to simultaneously monitor experimental
temperature variations in different parts of the system (Figure 6).
Fluctuations of such temperature on the order of a few tenths of a
degree centigrade produce noticeable fluctuations in the system
pressure.   This state-of-the-art experimental system is composed
of a Validyne DP-15 variable reluctance, differential pressure

transducer coupled to an IBM personal computer via a Keithley 195 multimeter and a Keithley 705 scanner (using a National Instrument GPIB-PC Instrumentation Interface Board, also known as an IEEE 488 bus), allowing on-line data acquisition in the form illustrated by Figure 6. High precision measurements of Ar and O_2 solubility and diffusion in vitreous silicas of various thermal histories are now under way.

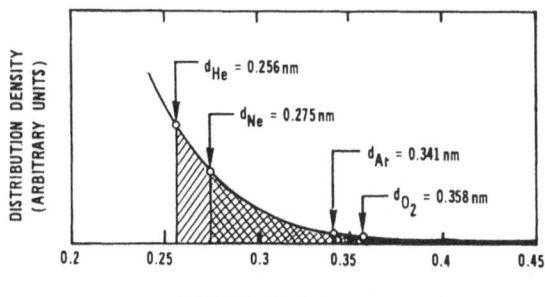

Fig. 4. A closer view of the "tail" of the distribution curve of Figure 3 indicates that a small fraction of total sites would be large enough to accommodate the relatively large species Ar and O_2.[3] These gases are near the threshold of accessibility to the vitreous silica network.

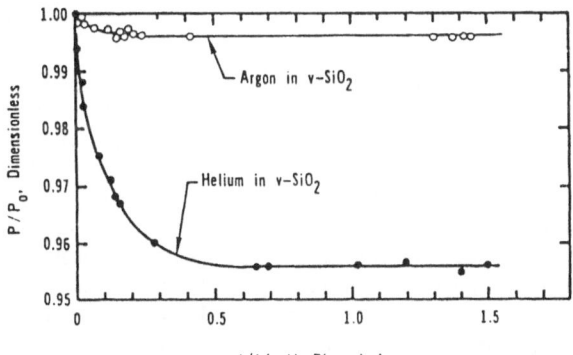

Fig. 5 The difficulty of measuring argon solubility is illustrated by comparative plots of relative pressure drops in typical experiments for argon and helium.[16] The small pressure drop for argon is due to (i) an inherently low solubility in vitreous silica and (ii) less efficient packing of vitreous silica fibers (for argon measurements) compared to the packing of 1 mm diameter rods (for helium).

MODELING STUDIES

The use of gas atoms and molecules as structural probes of
interstitial geometry can be quite effective, as demonstrated above.
However, these probes are somewhat limited in number and size. Only
a handful of gas species are highly permeable even to relatively
open silicate networks. As Figure 3 indicates, a significant number
of interstitial sites are too small to be probed by even the smallest
of the inert gases, helium. As a result, theoretical modeling is
helpful to characterize some of the more subtle details of interstitial geometry.

As a preliminary step, we have found that two-dimensional
schematics of the type developed by Zachariasen[7] are highly useful
for illustrating the nature of this interstitial geometry. Figure
7 defines a "triangle raft." The triangularly-coordinated network
model of Zachariasen is, of course, a two-dimensional schematic of
the vitreous network structure in which silica tetrahedra are
"randomly" connected. Cooper[17] has extensively analyzed the geometry of such networks of connected triangular building blocks.
Following his method, we replace the AO_3^{3-} triangular building block
with a simple triangle (Figure 7). Figure 8 shows the result of
systematically constructing an extended Zachariasen schematic with
300 rings (or two-dimensional interstices).[18] This structure was
produced by a systematic, step-wise addition of individual rings
following "random" construction rules. An important feature of the
construction process is that connectivity constraints eliminate many
of the randomly-chosen n-membered rings. The distribution of
n-membered rings demonstrated by Figure 8 is, then, not uniform but,
instead, skewed as illustrated by Figure 9(a). By plotting these
"ring statistics" (log n) on probability paper (Figure 9(b)), a
straight line plot indicates the distribution is log-normal. This
result has proven to be reproducible and reinforces the statement
in the previous section that the log-normal PDF is a useful working
definition of the "random network." Further reinforcement for this
was found in analyzing the ring statistics of the Bell and Dean[19]
model of vitreous silica, a three-dimensional structure of randomly-
connected tetrahedra. Shackelford and Brown[18] found that the ring
statistics for this 3-dimensional random network were similarly
close to a log-normal distribution. Phillips,[20] among others, has
pointed out that some features of the Bell and Dean model, such as
the radial distribution function, are not entirely consistent with
experimental data for vitreous silica. I would again emphasize
that such disagreement is not the basis for abandonment of the
random network model. Rather, it is an additional basis for
refining our understanding of the ideally random network against
which real material structures can be compared. Again, the log-
normal distribution appears to be a useful first step in quantifying
such an ideal random network structure.

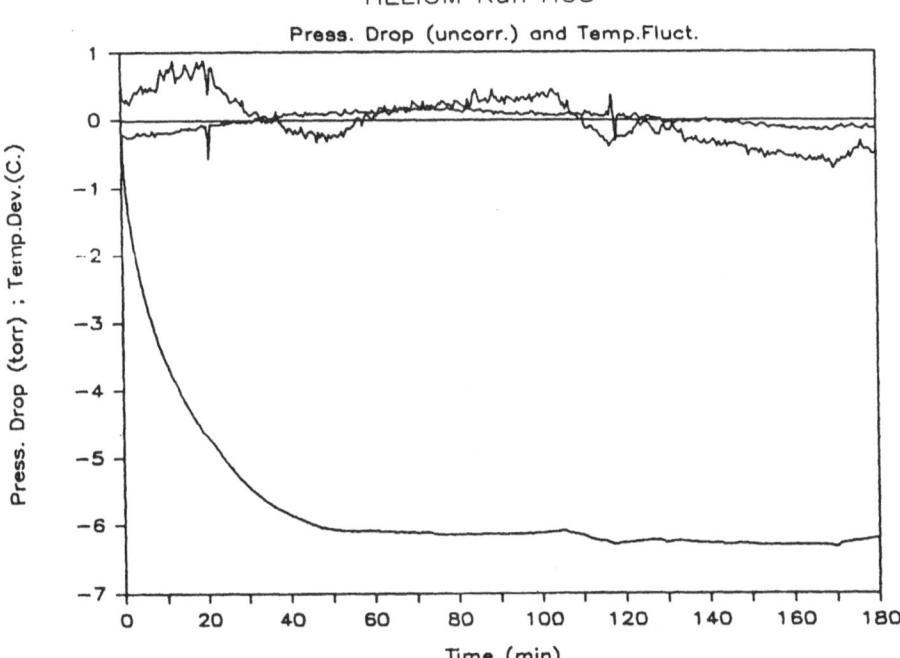

Fig. 6 A typical measurement of a small pressure drop solubility
experiment (see Figure 2) using a state-of-the-art experimental system. The general solubility curve indicates that
this approximately 6 torr pressure drop can be measured to
a precision of about 10^{-3} torr. The two irregular curves
at the top of the plot are traces of small temperature
fluctuations during the experiment in both the experimental
chamber and the surrounding environment. The result is a
noticeable fluctuation in the steady-state experimental
pressure.

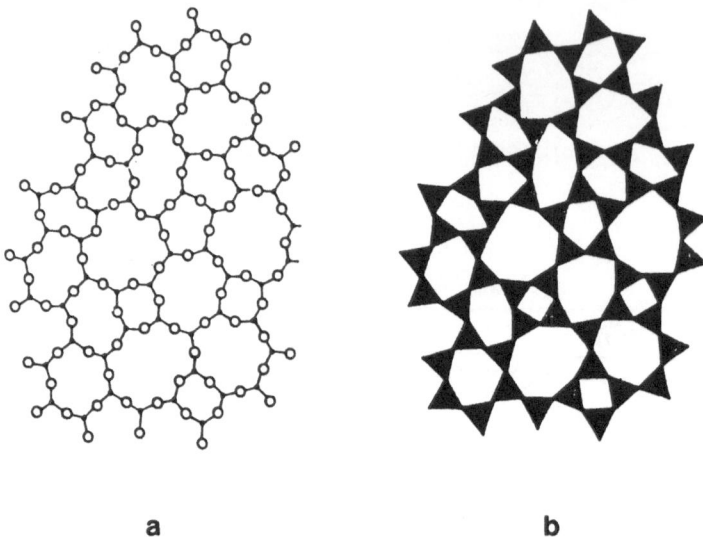

a   **b**

Fig. 7   Representation of (a) the Zachariasen schematic[7] by (b) an
array of triangles, i.e., a "triangle raft."[18]

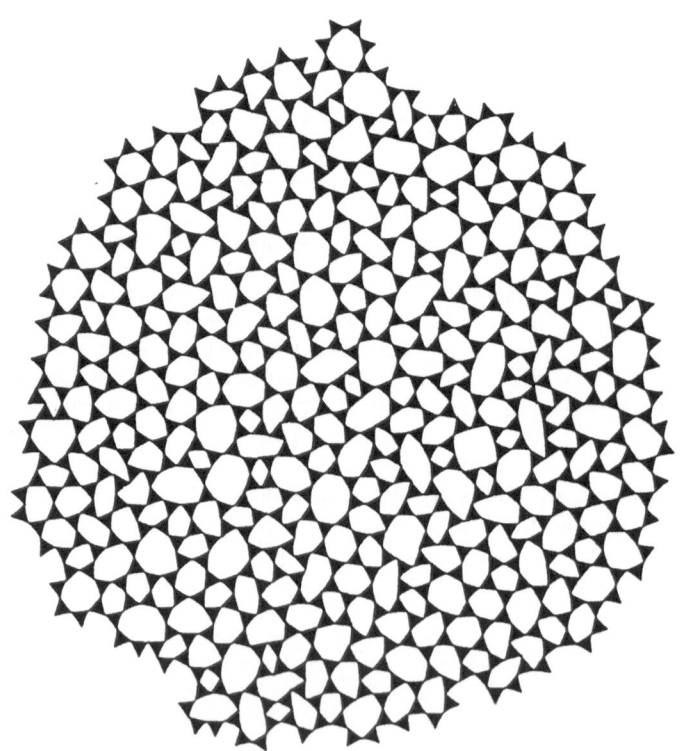

Fig. 8   A 300-ring triangle raft.[18]

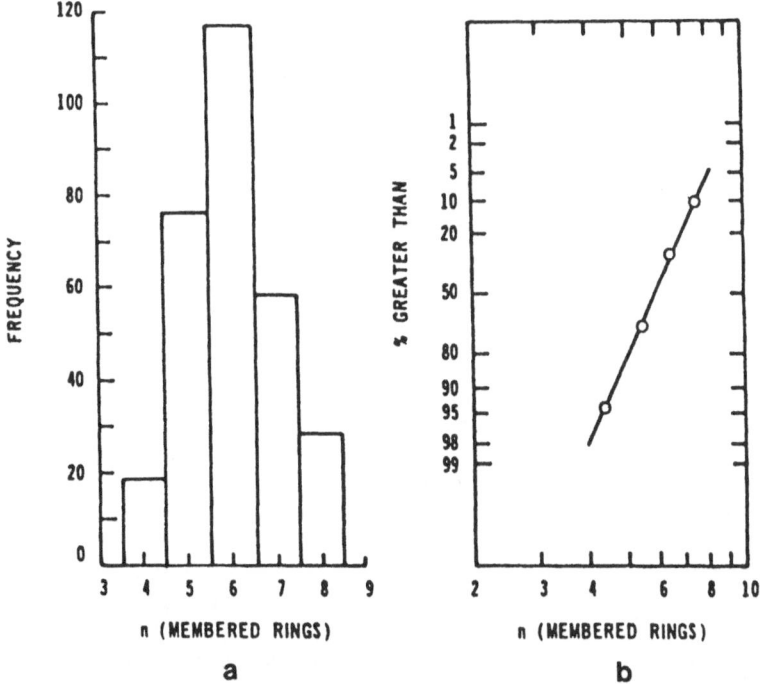

Fig. 9 (a) Histogram of the distribution of n-membered rings in the 300-ring triangle raft of Figure 8.[18] (b) Data from (a) replotted on log-normal probability paper.

APPLICATIONS

The characterization of interstitial structure in amorphous solids is useful, in itself, as a means of defining the nature of those materials.[21] However, there are also numerous problems of technological importance in which gas transport is a central mechanism. Perhaps the most obvious is the general field of vacuum technology.[22,23] Gas accessibility to structural components can serve as a practical limit to the level of vacuum attainable. This is an increasing problem with the increasingly wide use of ultra-high vacuum systems. A critical step in the thermal oxidation of silicon is frequently the diffusion of molecular oxygen through vitreous silica thin films.[3,24,25] Substantial research is currently directed toward a better understanding of the nature of this oxidation process. In addition, atomic gas transport can serve as a useful analogy for certain kinds of ionic diffusion in amorphous silicates. Rothman et al.[26] have used such an interstitial percolation model to explain their data for the diffusion of alkali ions in vitreous silica. We are extending these concepts in finding a successful model for describing the effect of trace impurity levels on the viscosity of vitreous silica.[27]

ACKNOWLEDGMENTS

This work has been made possible by National Science Foundation Grant DMR 82-04394.

REFERENCES

1. J. F. Shackelford, J. Non-Cryst. Solids, 42:165-174 (1980).
2. J. D. Bernal, Proc. Roy. Soc. Lond. A, 280:299-322 (1964).
3. J. F. Shackelford, J. Non-Cryst. Solids, 49:299-307 (1982).
4. D. L. Griscom, "Annealing of Radiation-Induced Defect Centers in High-Purity Bulk Fused Silicas by Molecular Diffusion Mechanisms," to be published this Proceedings.
5. D. B. Brown, "Annealing Radiation-Induced Damage in MOS Devices," to be published this Proceedings.
6. P. Rabinzohn and F. Rochet, "New Ways of Silicon Oxidation by Low Energy Oxygen Ions and Under Electron Bombardment. Study of Growth Mechanisms and Structure," to be published this Proceedings.
7. W. H. Zachariasen, J. Am. Chem. Soc., 54:3841-3851 (1932).
8. J. F. Shackelford, P. L. Studt, and R. M. Fulrath, J. Appl. Phys., 41 [7]:2777-2780 (1970).
9. J. S. Masaryk and R. M. Fulrath, J. Chem. Phys. 59 [3]:1198-1202 (1973).
10. J. E. Shelby, in Treatise on Materials Science and Technology, Vol. 17, Glass II, Academic Press, New York, (1979).
11. J. F. Shackelford and J. S. Masaryk, J. Non-Cryst. Solids, 130 [2]:137-134 (1978).
12. J. E. Shelby, J. Appl. Phys. 47 [1]:135-139 (1976).
13. G. Herdon, M. L. Smith, W. H. Hardwick, and P. Conner, in Small Particle Statistics, 2nd Ed., Academic Press, New York, (1960).
14. S. K. Kurtz and F. M. A. Carpay, J. Appl. Phys. 51 [11]:5725-5744 (1980).
15. S. K. Kurtz and F. M. A. Carpay, J. Appl. Phys. 51 [11]:5745-5754 (1980).
16. J. S. Flores and J. F. Shackelford, "The Solubility of Argon in Vitreous Silica," to be published in J. Non-Cryst. Solids.
17. A. R. Cooper, in Conf. on Boron in Glass and Glass Ceramics, L. D. Pye, V. D. Frechette, and N. J. Kreidl, eds., Plenum Press, New York, 167-181 (1978).
18. J. F. Shackelford and B. D. Brown, J. Non-Cryst. Solids, 44 [2/3]:379-382 (1981).
19. R. J. Bell and P. Dean, Phil. Mag., 25:1381-1398 (1972).
20. J. C. Phillips, "Topological Structure of Network Glasses," to be published this Proceedings.
21. J. L. Finney and J. Wallace, J. Non-Cryst. Solids, 43:165-187 (1981).
22. F. J. Norton, J. Appl. Phys. 28 [1]:34-39 (1957).
23. J. F. Shackelford, "Gas Permeation and the Outgassing and Leak Testing of Vacuum Systems," to be published in the Handbook of Nondestructive Testing.

24. N. F. Mott, Phil. Mag. A, 45 [2]:323-330 (1982).
25. A. G. Revesz, Phys. Stat. Sol. (a), 57:235-243 (1980); 57:657-666
    (1980); 58:107-113 (1980).
26. S. J. Rothman, T. L. M. Marcuso, L. J. Nowicki, P. M. Baldo,
    and A. W. McCormick, J. Amer. Ceram. Soc., 65 [11]:578-582 (1982).
27. P. P. Bihuniak and J. F. Shackelford, Amer. Ceram. Soc. Bull.
    62 [3]:416 (1983).

A. C. CONDUCTION IN CHALCOGENIDE GLASSES

S. R. Elliott

Department of Chemistry
University of Cambridge
Cambridge, U. K.

ABSTRACT

All noncrystalline semiconductors exhibit a frequency dependent conductivity which varies almost linearly with frequency. An extensive review is made of experimental data for one class of amorphous semiconductor, namely those formed partly or wholly from chalcogen (Gp. VI) atoms, for which it is generally believed that many of their opto-electronic properties are determined by the presence of charged dangling bond defects. Various theories for the phenomenon are reviewed, including the correlated barrier hopping (CBH) of electrons between charged defects, quantum-mechanical tunnelling (QMT) of electrons between defects and atomic hopping. It is concluded that on the basis of the present experimental evidence, the CBH model offers the most general satisfactory explanation of the behavior.

I. INTRODUCTION

A feature common to all amorphous semiconductors (and some other disordered systems) is a frequency dependent conductivity which increases approximately _linearly_ with increasing frequency. The origin of this behavior has been, and remains, a matter of some dispute. The phenomenon has at various times been ascribed to the motion of electrons or atoms, either hopping or tunnelling between sites.

The purpose of this article is to review the available experimental evidence for one class of amorphous semiconductor, namely glasses (produced by quenching from the melt) or amorphous thin films composed solely or mainly of chalcogen (Gp.

IV) elements, and to decide thereby on a probable mechanism. We
choose to single out chalcogenide materials for study for several
reasons. They have been much studied in the past and a wealth of
relevant experimental data exists from which it can be discerned
that whilst the A.C. behavior is similar to that exhibited by
other amorphous semiconductors, nevertheless certain distinct
differences exist. Furthermore, it has been argued that certain
structural defects control many of the opto-electronic properties
of these materials, and it is natural to discuss A.C. conductivity
in the light of these models. An additional feature which makes
the interpretation of experimental data less ambiguous is the
simple activated nature of the D.C. conductivity, indicative of
conduction by carriers in extended states beyond the mobility
edge, rather than the T 1/4 temperature dependence expected for
variable-range hopping (tunnelling) of electrons between states
at the Fermi level [1]. This is of great importance since the total
measured conductivity σ_T is then composed of both frequency
independent (D.C.) and frequency dependent (A.C.) components:

$$\sigma_T = \sigma(o) + \sigma(\omega) \tag{1}$$

Hence, the A.C. conductivity ($\sigma(\omega)$) can be obtained simply by
subtracting the D.C. value ($\sigma(o)$) from the measured conductivity
(see Fig.]), since it is universally believed that, for amorphous
chalcogenide materials at least, A.C. and D.C. conductivities arise
from completely different mechanisms. This is not so necessarily
for other systems (e.g., tetrahedrally bonded amorphous semiconductors such as evaporated Si or Ge) where the D.C. conductivity

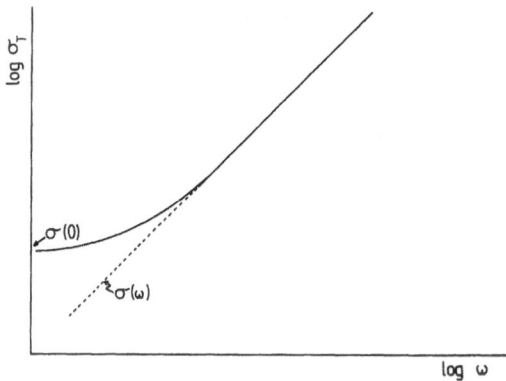

Fig. 1   Schematic plot of the total measured conductivity, σ_T, versus
         frequency for an amorphous semiconductor (solid line). The
         dotted line indicates the A.C. conductivity, $\sigma(\omega)$, obtained
         from σ_T by subtraction of the D.C. conductivity, $\sigma(0)$, when
         this is a valid procedure. The slope of the dotted line is
         s = 1.

follows a T 1/4 temperature dependence indicative of hopping, and hence the D.C. conductivity is the zero frequency limit of the A.C. conductivity; subtraction of the D.C. value to obtain the A.C. conductivity is <u>not</u> a valid procedure in this case. However, since $\sigma(\omega)$ is a monotonically increasing function of frequency and is generally only weakly temperature dependent, $\sigma(\omega)$ dominates at high frequencies or low temperatures. The frequency range of interest is typically 10- 10^6 Hz since this is the range covered conveniently by conventional bridge techniques. However, the range of frequencies over which the behavior

$$\sigma(\omega) = A\omega^s \tag{2}$$

is observed is much wider, ranging from > 10^{-4} Hz (e.g., for evaporated amorphous (a-) SiO [2] to $\sim$ < 10^9 Hz (e.g., for a-As$_2$Se$_3$ [3]), the first achieved by Fourier transforming the dielectric response measured in the time domain, and the latter obtained using microwave techniques.

In what follows, we discuss first in general terms how a linear frequency dependent conductivity might arise, and then present experimental data for a variety of chalcogenide glasses. Finally, specific models for A.C. conduction are discussed in detail in the light of experimental results, and in particular the "correlated barrier hopping" (CBH) model [4] is compared and contrasted wit the "quantum-mechanical tunnelling" (QMT) model [5].

II. ORIGIN OF A LINEAR FREQUENCY DEPENDENT CONDUCTIVITY

Consider the application of a harmonically varying electric field E(t) - E sin ωt to a sample producing a polarization P(t). In the frequency domain these are related via Fourier transforms to the complex quantities E(ω) and P(ω), where:

$$P(\omega) = \epsilon_o \chi(\omega) E(\omega) \tag{3}$$

The dielectric susceptibility $\chi(\omega)$ is in general a complex function of frequency

$$\chi(\omega) = \chi'(\omega) - i\chi''(\omega) \tag{4}$$

The imaginary component $\chi''(\omega)$ is the "dielectric loss" (since the current due to this is in phase with the driving field). $\chi(\omega)$ is related to the dielectric permittivity (or constant) by:

$$\epsilon_r'(\omega) = 1 + \chi'(\omega) \text{ and } \epsilon_r'' = \chi''(\omega) \tag{5}$$

where the factor of unity arises from the free space contribution. Furthermore, the A.C. conductivity is related to the dielectric loss by:

$$\sigma(\omega) = \varepsilon_o \omega \chi''(\omega) \tag{6}$$

to which should be added any D.C. component if applicable. The
real and imaginary parts of the susceptibility (or equivalently
the permittivity) are related via Kramers-Kronig (or Hilbert)
transforms [6].

We consider now the Debye model [7] in which is calculated the
dielectric response to an alternating electric field of either
an isolated charge oscillating thermally between two preferred
localized sites (represented by a double potential well) or the
analogous case of an inertialess dipole constrained to assume one
of two discrete spatial orientations (Fig. 2). A Debye response
is obtained when the polarization decay on removal of the exciting
field obeys a first-order rate equation with time constant τ (equal
to the oscillation period)

$$dP/dt = - P/\tau \tag{7}$$

which has the solution $P(t) = Po \exp(- t/\tau)$, where Po is the
initial value. Fourier transforming $f(t)-(Po/\tau) \exp(-t/\tau)$ yields:

$$\chi(\omega) = \frac{\chi(0)}{(1+i\omega\tau)} \quad \chi(0) \quad \frac{1}{1+\omega^2\tau^2} - \frac{i}{1+\omega^2\tau^2} \tag{8}$$

The dielectric loss $\chi''(\omega)$ thus exhibits a peak, since $\omega\tau/(1+\omega^2\tau^2)$
peaks at $\omega\tau = 1$ (Fig. 3). The A.C. conductivity $\sigma(\omega)$ is proportional
to $\omega^2\tau/(1+\omega^2\tau^2)$.

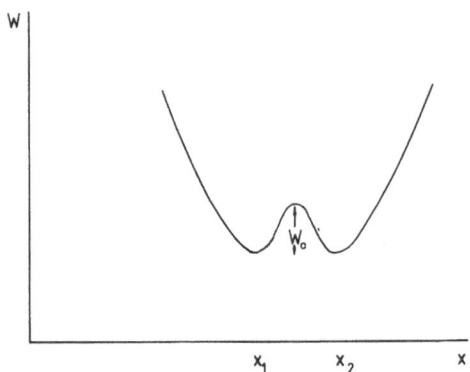

Fig. 2   Potential energy - configuration coordinate diagram for
         i)   dipole with preferred orientation (x≡0).
         ii) transitions of point charge between preferred positions
              (x≡r).
         In both cases, W_o is the activation energy of the process.

Note that $\sigma(\omega) \propto \omega^2$ for $\omega\tau<1$; this behavior must be carefully distinguished from the ω^2 frequency dependence that occurs [8,9,10] for a (contact) resistance in series with the equivalent parallel circuit of a lossy sample shown in Fig. 4. A dielectric loss peak is observed only for the case when the relaxation time τ has a <u>fixed</u> value; this has been assumed in the analysis so far. However, in random systems, and noncrystalline materials in particular, there must be a distribution of relaxation times rather than a fixed value. The A.C. conductivity must then be expressed as a sum over contributions from different discrete values of τ, or as in this case for a <u>continuous</u> distribution of τ, as an integral:

$$\sigma(\omega) = \int_{0}^{\infty} \alpha \; n(\tau) \; \frac{\omega^2 \tau}{(1+\omega^2 \tau^2)} \; d\tau \tag{9}$$

Here α is the polarizability of a pair of sites. To obtain the (approximately) linear frequency dependence that is observed in chalcogenide glasses (and other amorphous semiconductors), one requires that $n(\tau) \propto 1/\tau$, and hence $\sigma(\omega) \propto \int \frac{\omega d(\omega\tau)}{(1+\omega^2\tau^2)} \propto \omega$. For this

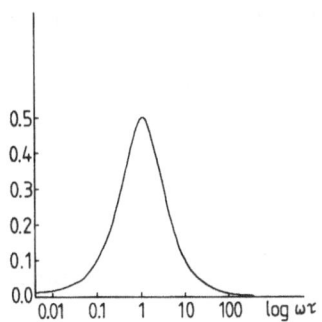

Fig. 3 Plot of the function $\omega\tau/(1+\omega^2\tau^2)$ versus $\omega\tau$, showing the sharply peaked nature of the function at $\omega\tau=1$.

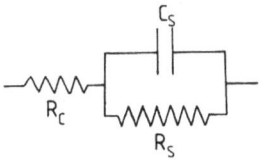

Fig. 4 Equivalent circuit of a lossy dielectric, represented as a capacitance C_S and resistance R_S, in series with a contact resistance R_C. This combination yields a ω^2 frequency dependence of the effective measured resistance [8].

condition to hold, the relaxation time must be an exponential
function of a random variable ζ

$$\tau = \tau_0 \exp(\zeta) \tag{10}$$

where τ_0 is the order of an inverse phonon frequency $_{ph}$ and
where ζ has a flat distribution, $n(\zeta)$ = constant. Thus,
$n(\tau) = n(\tau) \frac{d\zeta}{d\tau} \propto \tau^{-1}$, as required. In practice, $n(\zeta)$ and α may
be weakly frequency dependent, leading to a slightly sub-linear
frequency dependence. Two physical mechanisms that give rise to
the function form for τ are classical hopping of a carrier
over the potential barrier separating two sites (Fig. 2), in which
case ζ = W/kT, or alternatively quantum-mechanical tunnelling
through the barrier , in which case ζ = 2r/a, where it is assumed
that the wave function of the carrier localized at a site is
proportional to exp(-r/a). Both these cases will be considered in
interpreting the experimental data on chalcogenide glasses,
although we note here that information on the physical mechanism
responsible for the A.C. loss resides primarily in the
departure from unity of the frequency dependence and in the temperature dependence of the A.C. conductivity.

This treatment forms the basis of many theories of A.C.
conduction in amorphous semiconductors. It is assumed generally
that:

i) The "pair approximation" holds, in which motion of carriers
is contained within a pair of sites; the total A.C. response is
then obtained by summing over all pairs. It is important to note
that this approximation, which excludes multiple hopping or participation by more than two sites, cannot lead to a finite D.C.
conductivity, and thus D.C. and A.C. conductivities must arise
from different mechanisms, as is the case for chalcogenide glasses.

ii) The dielectric response of a pair of sites (or dipole) is
accurately Debye-like.

iii) The relaxation time is an exponential function of a random
variable, which may be the barrier height, intersite separation,
or a mixture of both [9]. We will examine later the consequences
of non-random distributions.

III. EXPERIMENTAL DATA FOR AMORPHOUS CHALCOGENIDE MATERIALS

A.C. conductivity data have been reported for the following
chalcogenides. Of the pure elemental materials, only amorphous
Se has been studied [8,12-18], although Se-Te [17,19] and Se-Te-Tl
alloys [20,21] have also been examined. Of the arsenic chalcogenide
alloys, stoichiometric As_2Se_3 [8,12-15,18,22-29], As-Se alloys [30-

32], stoichiometric As_2Se_3 [8,12,14,15,18,22,27,31,33,34], As-S alloys [12,31], stoichiometric As_2Te_3 [35-42], As-Se-Te [43,44] and As-S-Te alloys [45] have all been studied.   In addition in the As-Te system, As-Te-Se-Tl [26,27], As-Te-O [38,40], As-Se-Tl [39,40], As-Te-Si [36,39,40] and As-Te-Ge [15,18,42,46,47] alloys have been examined.   Quarternary alloys which exhibit switching behavior, As-Te-Si-Ge alloys, have also had A.C. data reported [35-37,48,49], along with a similar system As-Te-Ge-Ag [50].   Other chalcogenide alloys for which A.C. measurements are available are As-Sb-S alloys [51], Ge-Se alloys [14,15,18,52], and Ge-Sb-Se alloys [53]. Chalcogenide glasses doped with impurities have also received attention, As_2S_3: Ag [22,54], As_2S_3:In [55], As_2Se_3:Ag [22,56] and Ge-S:Ag,Cu [57].   In addition, the pressure dependence of A.C. conductivity has also been reported for As_2Se_3 and As-Se-Te alloys [58], and for As_2Te_3 and As-Te-Si and As-Te-Tl alloys [59]. If oxygen is regarded as a chalcogen (belonging to Gp. VI) then additional materials merit consideration, SiO [2,60,61] and SiO_2 [62].

Despite the apparent plethora of data evidenced above, there are unfortunately very few comprehensive reports on the A.C. properties of amorphous chalcogenide materials. Many accounts simply display a single curve showing the frequency dependence or temperature dependence (plotted versus inverse temperature to extract "activation energies") of the conductivity. Data of this kind is quite inadequate; what is required to gain a full under-standing of this complex phenomenon is a comprehensive joint frequency and temperature investigation of, preferably, a range of similar materials having different band gaps, or alternatively prepared in differing ways.   Nevertheless, we show in Figs. 5, 6, 7, certain data typical of chalcogenide materials and which illustrate

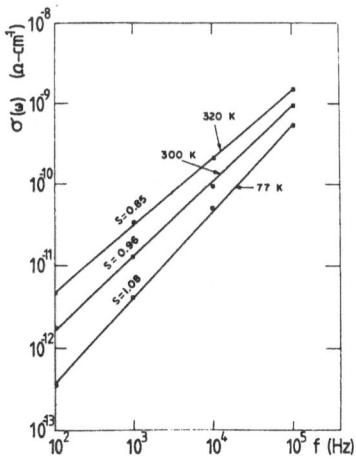

Fig. 5   Frequency dependence of the A.C. conductivity, $\sigma(\omega)$, of thin films of amorphous Se at different temperatures [8].

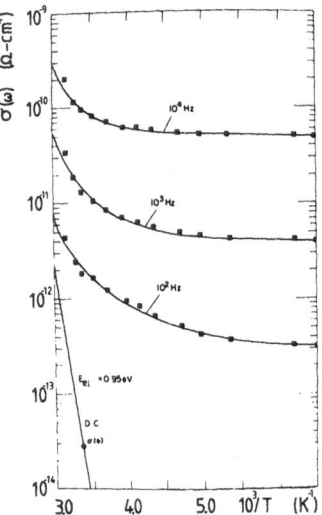

Fig. 6   Semilog plots of the total conductivity, σ_T, versus inverse
temperature for thin films of amorphous Se at different
frequencies [8].

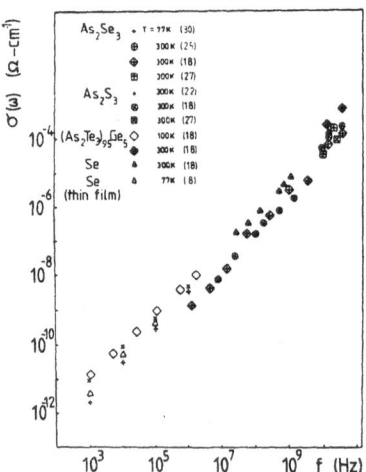

Fig. 7   A.C. conductivity, $\sigma(\omega)$, for a variety of chalcogenide
glasses.

many of the features we wish to discuss. Fig. 5 shows the frequency dependence of the conductivity of thin films of amorphous
selenium at three temperatures [8]. Points to note are the change
in frequency exponent s with temperature (this is a general
feature exhibited by chalcogenide materials) increasing to a
value in excess of unity at low temperatures, and the relatively
small overall temperature dependence of the conductivity. This
feature is emphasized in Fig. 6 where the temperature dependence
is seen to decrease at low temperatures [8]; at high temperatures
the temperature dependence of the total conductivity σ_T tends to
that of the D.C. conductivity $\sigma(o)$ in accordance with Eq. (1).
Note that the temperature dependence at low temperatures in this
case appears to be frequency dependent; this is simply a reflection
of the increase in the frequency exponent s with decreasing
temperature manifested in Fig. 5.

The A.C. conductivities of a variety of chalcogenide glasses [18]
are shown in Fig. 7. It can readily be seen that the magnitude of
the A.C. conductivity at a given frequency for these glasses varies
only by an order of magnitude or so, although we will see later
that this is not so necessarily for amorphous thin films.

A summary of relevant parameters obtained from the references
cited above is given in Table I for several important chalcogenide
materials. The parameters chosen are the frequency exponent s,
measured at room temperature and another (lower temperature) if
available, and the value of the A.C. conductivity at room temperature and a fixed frequency (usually 10^4 Hz). Such values are
entered only if given in the original reference. We also show
in Table II similar parameters for two chalcogenide alloys systems,
As-Se and As-S, as a function of composition.

The conclusions we draw from this comprehensive survey of the
experimental literature are the following:

TABLE I

Material	Glass/Film	s(T=300K)	s(T)	$\sigma(f=10^4 Hz; T=300K)$ $(\Omega^{-1}cm^{-1})$	Ref.
As_2S_3	f	0.9	-	1.1×10^{-10}	8
As_2Se_3	f	1	-	6×10^{-10}	8
"	g	0.86	0.75(423)		23
As_2Te_3	f	-	~1 (<100)	$2 \times 10^{-6}(10^5; 100)$	42
Se	f	0.96	1.08(77)	1.5×10^{-10}	8
'	g	0.63	0.72(200)	7.4×10^{-11}	17

TABLE II

System	x	s(T)	E_{e1}(eV)	$\sigma(f=10^3 Hz; T=300K)$ $(\Omega^{-1} cm^{-1})$	Ref.
$As_x Se_{100-x}$	10	0.77 (300)	0.74	7.8×10^{-10}	32
	20	0.70 "	0.67	2.2×10^{-9}	
	30	0.92 "	0.98	1.4×10^{-9}	
	40	0.78 "	0.93	2.5×10^{-9}	
	50	0.73 "	0.97	1.7×10^{-9}	
	60	0.90 "	1.03	6.3×10^{-9}	
$As_x S_{100-x}$	37.2	0.97 (315)	-	-	31
	38.5	1.00 "	-	-	
	40.0	1.09 "	-	-	
	42.2	1.12 "	-	-	
	42.6	1.11 "	-	-	

i) The conductivity increases monotonically with frequency in the frequency range typically from 10 Hz (although this may extend down to $\sim 10^{-4}$ Hz) to 10^8 Hz.

ii) Over this entire region, the conductivity can be accurately fitted by a power law of the form given by Eq. (2). At a fixed temperature, the frequency exponent s shows no change in value with frequency.

iii) The value of the exponent s is approximately unity, and generally slightly less than this. However, values for s slightly in excess of unity have also been reported [8,3],33,45].

iv) Generally, the exponent s is a function of temperature, increasing (usually towards unity) with decreasing temperature. No case of the opposite temperature dependence has been reported for chalcogenide materials.

v) The A.C. conductivity $\sigma(\omega)$, as defined by Eq. (1), is generally only weakly temperature dependent, much weaker than the D.C. conductivity $\sigma(o)$ which is simply activated with an acti- vation energy equal to approximately half the band gap. However, for the case of narrow gap materials (e.g., As_2Te_3), the tempera- ture dependence of $\sigma(\omega)$ is strong (but less than that of $\sigma(o)$) in an intermediate temperature range, but at low temperatures $\sigma(\omega)$ becomes weakly temperature dependent again. (This behavior is believed to arise from a change in mechanism for A.C. conduction with temperature, and will be discussed later). In some cases, σ(ω) is completely temperature independent [18,34]. In these cases, it is almost universally found that the exponent s equals unity.

vi) The magnitude of the A.C. conductivity at fixed frequency and temperature lies in a relatively narrow band ($\sim 10^{-9} - 10^{-8}$ Ω^{-1} cm^{-1}) for __glassy__ materials [39], although the range spanned by the same materials in __thin film__ form is much wider ($\sim 10^{-9} - 10^{-6}$ Ω^{-1} cm^{-1}). By contrast, the D.C. conductivity for these materials ranges from $\sim 10^{-14} - 10^{-4}$ Ω^{-1} cm^{-1}.

vii) In addition to the weak dependence on band gap (or σ(o)) shown by σ(ω) described in (vi), the exponent s also generally exhibits a band gap dependence, increasing with increasing band gap. This can be seen particularly from Table II.

IV. MODELS FOR A.C. CONDUCTION IN AMORPHOUS CHALCOGENIDE MATERIALS

A. Quantum-mechanical tunnelling (QMT)

The first report of an approximately linear frequency dependent conductivity of a disordered system (n-type doped crystalline Si, compensated to provide randomly situated ionized donor centers) was by Pollak and Geballe [5]. The A.C. conductivity increased monotonically with frequency on a log-log plot, and in the frequency range measured ($10^2 - 10^5$ s^{-1}) showed __no__ deviation from linearity. The frequency exponent, s, __increased__ with __decreasing__ temperature, being 0.74 at 12K and 0.79 at 3K. Pollak and Geballe interpreted this data in terms of electrons moving between majority impurity sites by means of quantum-mechanical tunnelling - 'impurity conduction' - (compensation ensuring empty sites to which to hop); the transition rate was obtained from the theory of Miller and Abrahams [63] and was exponentially dependent on a random variable (see Eq. (10), the variable in this case being the separation of centers, r. The polarizability for a pair of sites (assuming the pair approximation) was taken as [5]:

$$\alpha = \frac{e^2 r^2}{12kT\cosh^2(\Delta/2kT)} \tag{11}$$

where Δ is the energy separation between the states of the pair of sites. Assuming a __random__ distribution of donor centers

$$p(r)dr = 4\pi N_D r^2 dr \tag{12}$$

(where N_D is the donor density), the A.C. conductivity was calculated approximately using Eq. (9). The sharply peaked nature of the integrand of Eq. (9) with respect to r (for τ exponentially dependent on r) ensures that the major contribution to σ(ω) at a frequency ω arises from those paris for which ωτ≈. Equation (9) then becomes $\sigma(\omega) \propto \omega \ r^4 \int \dfrac{\omega\tau}{(1+\omega^2\tau^2)} \ dr$

the r^4 term (two powers from the polarizability, Eq. (11) and
two powers from the spatial distribution of centers, Eq. (12))
can be taken out of the integral, having the value for the
hopping distance, r_ω , for which $\omega\tau=1$,

$$r_\omega = \frac{a}{2} \ \ell n \ (\frac{1}{\omega\tau_o}) \tag{13}$$

Thus the frequency dependence predicted by the QMT model is
$\sigma(\omega) \propto \omega \ \ell n^4 \ (\frac{1}{\omega\tau_o})$. Several features emerge from this analysis.

i)   The frequency exponent, (defined by $s = \dfrac{d(\ell n\sigma(\omega))}{d(\ell n\omega)}$ is
given by:

$$s = 1-4/\ell n \ (\frac{1}{\omega\tau_o}) \tag{14}$$

This predicts that the exponent s is itself frequency dependent.
This is illustrated in Fig. 8 in which the departures of the
predicted curve from an $\omega^{0.8}$ (or ω^1) dependence are clearly seen.

ii)   No temperature dependence of the exponent s is predicted
for the QMT mechanism, unless $\tau_o = \dfrac{(1)}{\nu_{ph}}$ is itself temperature

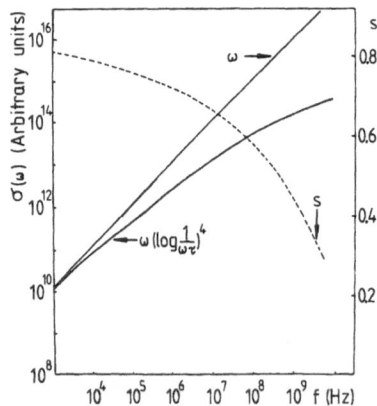

Fig. 8   Plot of the frequency dependent conductivity predicted
         by the QMT model as a function of frequency; a line of
         unit slope is shown for comparison.   The frequency depen-
         dence of the exponent s is also shown.

dependent. This could conceivably happen if polaron formation is favored [64], but it is found that the temperature dependence predicted for s is an _increasing_ function of temperature [4].

It thus appears that even for the system for which it was developed, namely doped, compensated crystalline silicon, the QMT theory fails on two counts. The frequency exponent s is experimentally found to be _independent_ of frequency (in contrast to (i)) and is a decreasing function of temperature (in contrast to (ii)). We shall see now how this theory has been adapted for the case of amorphous semiconductors, and how it fails again to account for experimental data, this time of chalcogenide materials.

Austin and Mott [64] were the first to apply the mechanism of QMT to the case of amorphous semiconductors, and assumed that tunnelling of electrons took place amongst a band of (defect) states at the Fermi level, E_f [65]. The full expression for the A.C. conductivity so derived is:

$$\sigma(\omega) = C \; N^2(E_f) kT \; e^2 \; a^5 \omega \; \ell n^4 \left(\frac{1}{\omega \; \tau_o}\right) \tag{15}$$

where C is a numerical constant, equalling $\pi/3$ for the treatment of Austin and Mott [64,65]. Subsequent workers have derived essentially the same equation with slightly differing values of C $\left(\frac{\pi^4}{96} \; [66], \; \frac{3.66 \; \pi^2}{6}\right) [67]$.

When attempting to explain experimental A.C. data for chalcogenides using Eq. (15), several of the difficulties encountered for the impurity conduction case recur.

i) The frequency exponent, s, is predicted to _decrease_ with increasing frequency, as in Fig. 8c (for a typical case of $\tau_o = 10^{-13}$). This is completely at variance with all the experimental evidence cited in Section III. If a change in s with frequency occurs at all, it is in the opposite direction, s tending to 2 at the very highest frequencies for some samples [27,29].

ii) Values of s for chalcogenide materials commonly lie in the range 0.9–1.0 (See Table I). For the case of s=0.95, for example, Eq. (14) indicates a value of $\nu_{ph} = \frac{1}{\tau_o} \simeq 10^{39}$ Hz, an unphysical value.

iii) Many chalcogenide materials exhibit a temperature dependent s, s increasing with decreasing temperature. This is not accounted for by Eq. (14).

iv) The temperature dependence of the A.C. conductivity is
predicted by the QMT model to be directly proportional to T
(Eq. (15)). In many cases, $\sigma(\omega)$ for chalcogenide materials is
temperature independent [18,34], and when temperature dependent,
in the few cases when it has been measured accurately, the
temperature dependence is sub-linear ($\sim T^{0.45}$) [42].

All these arguments lead to the conclusion that the mechanism
of QMT at the Fermi level is not applicable to the case of
amorphous chalcogenides. In fact, it is the author's opinion
that no system, amorphous or otherwise, has been reported which
exhibits all the features required by QMT theory. (A particularly
crucial test of the theory is the decrease in slope of a log-log
plot of conductivity versus frequency at high frequencies.) Faced
with this, we must turn to other mechanisms to explain the A.C.
loss in chalcogenides.

B. Atomic Hopping

It was suggested [68] some time ago that A.C. conductivity
and certain low temperature thermal anomalies [69] (such as
thermal conductivity or specific heat), which both appear to be
universal features of glasses, might have a common cause. The
model used to explain the thermal anomalies assumed that atoms
tunnelled between two nearly equivalent sites separated by an
energy difference Δ[70,71]; such a system constitutes a 'two-
level system', as in Fig. 2. Pollak and Pike [72] were the first
to adopt this model to explain the A.C. conductivity of amorphous
materials. They assumed that atoms, partially or fully ionized,
hopped over the barrier separating the two level systems: i.e.,
they considered the temperature regime higher than that in which
thermal anomalies are observed. The assumption of ionized atoms
is necessary to ensure coupling to the applied field and hence
a change in dipole moment, giving rise to a dielectric loss. The
relaxation time for the process is then

$$\tau = (W_{12} + W_{21})^{-1} = \tau_o \left[\frac{1}{(\exp(\Delta+W)/kT)^{-1}} + \frac{1}{(\exp(W/kt)^{-1})} \right]$$

$$\approx \tau_o \exp(W/kT) \tag{16}$$

where W_{12} and W_{21} are the transition rates between equilibrium
positions 1 and 2 (Fig. 2); the approximate expression holds
for $W \gg kT$, $\Delta \gg kT$. Using the expression for polarizability (Eq. (11))
and for the A.C. conductivity (Eq. (9)), one obtains [72]

$$\sigma(\omega) = \frac{\pi \omega e^2 r_o^2}{6 \Delta_o W_o} NkT \tanh(\Delta_o/2kT) \tag{17}$$

The quantities r, , W were considered to be random, independent variables, with a distribution

$$p(r,\Delta,W) = \delta(r-r_o)/r_o^2 \, W_o\Delta_o \text{ for } 0 < \Delta < \Delta_o, \ W' < W < W' + W_o$$

and $W_o >> kT$. Note that Eq. (17) predicts that the A.C. conductivity should be accurately linear in frequency, with a temperature dependence given by the function T tanh ($\Delta_o/2kT$); this is approximately proportional to T for $T<\Delta_o/2k$ and temperature independent for $T>\Delta_o/2k$. A similar treatment has been given by Le Cleac'h [18]. In both cases, values for N, the number of ionized atoms that hop, is $\sim 10^{19} - 10^{20}$ cm^{-3}; this is somewhat higher than the values deduced from the low temperature thermal experiments.

It can readily be seen that the atomic model explains observations such as those of Le Cleach'h [18] of a frequency-dependent conductivity whose frequency exponent s = 1 and which is temperature independent (as long as $\Delta_o/2k$ is less than the lowest temperature measured). The approximate constancy of $\sigma(\omega)$ for a variety of materials is also understandable if N is similar for different materials. What is not predicted by this model is a value for s differing from unity and which is temperature (and band gap) dependent. Deviations of s from unity occur when the hopping distance, r_ω, is a function of frequency, as it is for QMT (Eq. 13); for the atomic mechanism, r_ω is taken to be a constant ($=r_o$). It is also not clear why in some cases the same material should have $\sigma(\omega)$ temperature independent with s=1, whilst in other cases $\sigma(\omega)$ is temperature dependent with s less than unity. It was with such apparent contradictions in mind and with the need for a universal mechanism which considered in detail the special nature of the structural defects in chalcogenide materials that we developed the theory [4] outlined in the next section.

C. Correlated Barrier Hopping (CBH)

This model takes as its mechanism the hopping of electrons over the potential barriers separating defects, whose sites undergo lattice distortion upon change in electron occupation. An important difference from the previous theories is the assumption that the barrier height is coorelated with the intersite separation, so W and r are not independent variables. In addition, the effect of changes in preparation conditions are considered explicitly.

We begin with a brief discussion of defects in amorphous chalcogenide materials. It has long been recognized that chalcogenide materials are somewhat anomalous amorphous semiconductors. Whilst many different experiments (such as photoluminescence, A.C. conductivity, pinning of the Fermi level) point to the existence of states in the gap, most likely arising from dangling bonds, no

ESR or paramagnetic susceptibility is observed [68] (at least for samples kept in the dark). This suggests that spin-pairing is occurring in some manner. Anderson [73] was the first to suggest a general mechanism for spin-pairing employing the strong electron-phonon interactions likely to be present in the low-dimensional chalcogenide materials. Mott and coworkers applied this general theory to the case of specific defects, dangling bonds [74,75], and the nature of the resulting defect configurations was subsequently made clear [76]. Dangling bonds which contain a single unpaired spin (D^0 [74] or $C_3{}^0$ [76], where the superscript refers to the charge state, the subscript to the coordination number and C refers to a chalcogen atom) are unstable to the reaction [74] $2D^0 \rightarrow D^+ + D^-$ (or $C_3{}^+ + C_1{}^-$ [76]). The repulsive energy required to place two electrons on the same site (D^-) is supposed to be overcome by the lattice distortion (bonding) energy [74], or electronic energy [76, gained at the positively charged center by utilizing the lone-pair p orbitals on the chalcogen atom [74,76]. Both schools of thought are in agreement over the diamagnetic centers (D^+, D^- or $C_3{}^+$, $C_1{}^-$) but differ over the nature of the neutral center; Mott et al. believe it to be essentially a singly coordinated atom, but weakly bonded to a neighboring chain, whereas Kastner et al. believe the center to be symmetrically 3-fold coordinated, i.e., $C_3{}^0$. Recent self-consistent pseudo-potential calculations [77] indicate that in fact the lowest energy neutral defect (in a-Se) is simply singly coordinated (i.e., $C_1{}^0$). For our purpose, however, all we need to know is that in the ground state, the defects are equally and oppositely charged, spin-paired and diamagnetic.

Our mechanism for the process giving a dielectric loss is then the simultaneous hopping of two electrons from a D^- to a D^+ center [4], thereby giving a change in dipole moment (see Fig. 9). The electrons hop over a barrier separating the two sites, whose height is correlated with the intersite separation (Fig. 10). (A similar model has been proposed by Pike [78] to explain A.C. data of Sc_2O_3, but one-electron hopping was assumed there). This

Fig. 9   Schematic diagram illustrating the bipolaron transport assumed in the CBH model [4]. Two electrons hop simultaneously from a D^- to a D^+ center, thereby resulting in an interchange of position and consequent change of dipole moment.

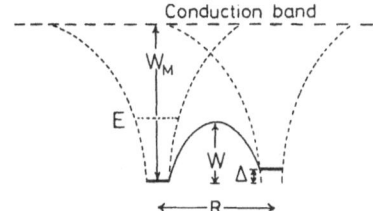

Fig. 10   Schematic diagram illustrating the overlap of Coulombic
          potential wells for two close, charge centers. The lower-
          ing of the height of the barrier for hopping in the CBH
          model is indicated.

correlation arises through the Coulomb term involved with such
charged centers and is given by (in S.I. units):

$$W = W_m - \frac{8e^2}{4\pi\varepsilon_o \varepsilon r} \tag{18}$$

where W is the height of the barrier over which the 2 electrons hop
between sites of separation r, W_m is the maximum barrier height
(for infinite separation) and ε is the dielectric constant. For
centers forming a two-level system, separated in energy by Δ, as in
Fig. 10, the relaxation time is given by Eq. (16), which reduces
to:

$$\tau = \tau_o \exp (W/kT)/(1+\exp (-\Delta/kT)) \tag{19}$$

for W>>kT and E>>Δ. The expression used previously for the
polarizability (Eq. 11) must be multiplied by a factor of 4 to
allow for 2-electron transport. The A.C. conductivity is calculated, using Eq. (9), as in Ref. 4, with the exception of the
$\text{sech}^2(\Delta/2kT)$ term (from Eq. 11) since previously it was assumed
that $\Delta = 0$ [4]. This term in Δ is integrated over the range
$0<\Delta<\Delta_o$, assuming a flat distribution [18,70,78] yielding an
additional factor $2kT/\Delta_o$ tanh $(\Delta_o/2kT)$. If it is further assumed
that the centers are randomly distributed in space, such that the
probability of finding two sites at a separation of r is given
by $4\pi Nr^2$ (where N is the concentration of centers), then the
conductivity can be written as

$$\sigma(\omega) \sim \frac{N^2}{2} \frac{2kT}{\Delta_o} \tanh(\Delta_o/2kT) \; \omega \; \frac{e^2 r^2}{3kT} \frac{\omega\tau}{(1+\omega^2\tau^2)} 4\pi r^2 dr$$

by equating the probability distribution in τ to that in r;
the additional factor of N/2 in this expression is the number
of hopping carriers (pairs of electrons in this case). The

integration variable, r, can be transformed to W (using Eq. 18) since $dW = \dfrac{2e^2}{\pi\varepsilon\varepsilon_o r^2}\, dr$, and using Eq. (19) (in which $\Delta=0$ is taken, since this produces the dominant contribution to the integral)

$$\frac{d\tau}{\tau} = \frac{dW}{kT} \text{ , yielding}$$

$$\sigma(\omega) \simeq N^2\varepsilon\varepsilon_o \frac{kT}{\Delta_o} \tanh (\Delta_o/2kT)\, \omega \; r_\omega^6 \int_{\tau_{min}}^{\tau_{max}} \frac{\omega\, d\tau}{(1+\omega^2\tau^2)}$$

The hopping distance r_ω can be taken out of the integral since the integrand is sharply peaked at $\omega\tau=1$. For $\tau_{min} \simeq \tau_o(\ll 1)$ and $\tau_{max} \simeq \tau_o \exp (W_m/kT)$ ($\gg 1$), the integral can be evaluated approximately to give $\pi/2$ [66]. Thus the final expression for the A.C. conductivity in the CBH model is

$$\sigma(\omega) \simeq \frac{\pi^3}{3} N^2\varepsilon\varepsilon_o \frac{kT}{\Delta_o} \tanh (\Delta_o/2kT)\omega \; r_\omega^6 \tag{20}$$

where the hopping distance is given (from Eq. 18) by

$$r_\omega = (\frac{2e^2}{\pi\varepsilon\varepsilon_o}) \frac{1}{(W_m - W_\omega)} \tag{21}$$

and $W_\omega = kt\ell n(1/\omega\tau)$. Equation (20) differs from that originally given in Ref. 4 in two ways: S.I. units have been used, and a distribution of site energy differences, Δ, has been assumed, rather than $\Delta_o=0$. The expression for the A.C. conductivity (Eq. 20) has several features and consequences. The frequency dependence is given by the factor r_ω^6, where $(W_m - W_\omega)^{-6}$ can be written to a first approximation [4,78] as $1/W_m^6 (1/\omega\tau_o)^\beta$, where

$$\beta = \frac{6kT}{W_m} \tag{22}$$

Note that the sixth power of r_ω has contributions from the polarizability (two powers), the <u>random</u> spatial distribution of centers (two powers) and (the differential of) the Coulombic nature of the barrier height-intersite separation correlation (Eq. 18 - two powers). If the potential is not Coulombic but varies as, say $1/r^m$, then the power of the hopping distance becomes $m+5$. Using the approximation leading to Eq. (22), the A.C. conductivity then can be written as

$$\sigma(\omega) \simeq \frac{\pi}{3} N^2 \varepsilon\varepsilon_o \frac{kT}{\Delta_o} \tanh (\Delta_o/2kT) \left(\frac{2e^2}{\tau\varepsilon\varepsilon_o}\right)^6 \frac{\omega^s}{\tau_o\beta} \tag{23}$$

where

$$s = 1 - \beta \tag{24}$$

It can be seen immediately that the CBH model for _random_ centers
(Eq. 23) predicts a sub-linear frequency dependence of $\sigma(\omega)$, for
which the exponent, s, is a _decreasing_ function of temperature.
This is in accord with many of the experimental data for chalcogenide materials. A much better approximation for r_ω (exact to
second order and a very good approximation to third order) is
given by [4]

$$(W_m - W_\omega)^{-6} \simeq \frac{1}{W_m^6} \left(\frac{1}{\omega\tau_o}\right)^\beta \exp\left[3\left(\frac{kT}{W_m}\right)^2 \ln^2\left(\frac{1}{\omega\tau_o}\right)\right] \tag{25}$$

The frequency exponent of $\sigma(\omega)$ $\left(S = \dfrac{d\ln\sigma(\omega)}{d\ln\omega}\right)$ is now given by

$$s = 1 - \beta_{-\gamma} \text{ where } \gamma = 6\left(\frac{kT}{W_m}\right)^2 \ln\left(\frac{1}{\omega\tau_o}\right)$$

In order to discuss further Eqs. (20) or (23), we need estimates for W_m, the maximum barrier height at infinite site separation
(Fig. 10), for chalcogenide materials. Mott et al. [75] show that
the (thermal) energy required to take _two_ electrons from a D^-
state into the conduction band is given by $B - W_1 + W_2$ where B is
the optical band gap, W_1 is the energy (distortion plus electronic)
to take an electron from the valence band to turn D^o into D^-, and
W_2 is the total energy to take an electron from D^o to the conduction band. This expression can also be understood as being
simply twice the energy difference between the Fermi level and the
conduction band. Amorphous chalcogenide materials are in general
p-type (the Fermi level lying nearer the valence band) and exhibit
simple activated D.C. conductivity [65], of activation energy ΔE.
Thus in this case we may write for W_m

$$W_m = 2(B - \Delta E) \tag{26}$$

Since the Fermi level often lies near mid-gap (so $\Delta E \approx B/2$), to a
first approximation we may take W_m equal to the band gap B [4];
actual values [80] for two chalcogenide materials are shown in
Table III, from which it can be seen that $B \approx W$ to within 10%.

TABLE III

	W_1 (eV)	W_2 (eV)	ΔE (eV)	B (eV)	$W_M = 2(B-\Delta E)$ (eV)
As_2Te_3	0.29	0.39	0.45	1.0	1.1
As_2Se_3	0.55	0.65	0.9	2.0	2.2

With this interpretation of W_m in mind, several surprising features emerge from an examination of Eq. (23) for $\sigma(\omega)$. The magnitude of the A.C. conductivity is predicted to be band gap dependent (through the W_m^{-6} term), and the frequency exponent is also predicted to be dependent on B (through the β term). This correlation emerges because of the nature of the defects present in chalcogenide materials (which undergo a distortion upon change in electron occupancy) and which pin the Fermi level near mid-gap; thus thermal excitation of these defects requires energies involving the band gap energy B. This correlation is not to be expected, however, for A.C. conduction arising from QMT at the Fermi level since activation of carriers does not occur. This unexpected correlation beteeen A.C. conductivity on the one hand, and D.C. electrical activation energy (or equivalently the D.C. conductivity) on the other, had been noted earlier by Davis & Mott [81] in an examination of experimental data for chalcogenide materials (see inset to Fig. 11). It has been demonstrated [4,82] that these

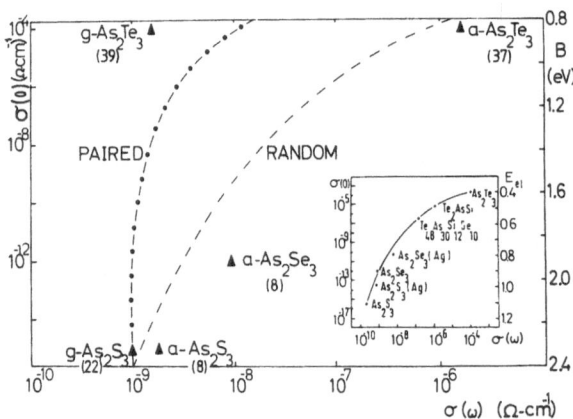

Fig. 11   Variation of A.C. conductivity, $\sigma(\omega)$, with electrical gap, ΔE (or equivalently D.C. conductivity, $\sigma(0)$) calculated on the CBH model for both random and paired centers; a constant value for the density of centers, N, was assumed in both cases. Experimental values for various amorphous chalcogenide materials, in both bulk glassy and thin film forms, are shown. Shown as an inset is the correlation between A.C. and D.C. conductivities of chalcogenide materials first demonstrated by Davis and Mott [81].

data are explained by the CBH model and cannot be accounted for
using a QMT mechanism.   In Eq. (23), it is seen that the only
parameters that might vary from one material to another are
N, ε, W_m (or B) (neglecting any variation in Δ_0, if non-zero).
The function log $(\sigma(\omega)\varepsilon^5 N^2)$ was plotted versus log B, yielding a
straight line of slope -7.5, in reasonable agreement with the
predicted value, -6 [4].   This plot has been criticized [41] since
the values for N used in it were obtained in the first place from
Eq. (20), and it is alleged that therefore this offers no proper
test of the CBH theory.   To circumvent this criticism, we showed in
Fig. 11 a revised plot [82] of Eq. (20) versus log B, assuming that
the concentration of defects giving rise to $\sigma(\omega)$ is the same in
every case.   This cannot be so, but we have no independent source
for obtaining values of N, and so to a zeroth approximation we
take N to be constant.   Values for β are calculated using Eq. (22)
and the appropriate band gap B; the error in using experimental
values for β (which in any case are very close to β_{calc}) in the
term $(\omega\tau_0)^\beta$ is very small.   A continuous variation in dielectric
constant with band gap B has been used, assuming the Moss rule [83]
to hold:

$$\varepsilon = (1/\kappa B)^{1/2} \tag{27}$$

where $\kappa = 3.4 \times 10^{-3}$ eV^{-1} for the $As_2 X_3$ series (X = S, Se, Te).   It
can be seen that a correlation between $\sigma(\omega)$ and B is indeed predicted for a random distribution of centers, and the experimental
points for amorphous thin films lie close to the theoretical
curve.   The small departures from the predicted curve are probably
due to variations in N between materials.   This interpretation has
assumed a random distribution of hopping centers; we believe this
situation to be most prevalent in thin films prepared by vacuum
evaporation or sputtering, in which the material produced is not
in thermodynamic equilibrium at the deposition temperature and
substantial atomic rearrangement after deposition is precluded.
This situation may not prevail in melt-quenched glasses (see
later).   Kocka [39-41] has argued that the high value of $\sigma(\omega)$ for
$As_2 Te_3$ thin films measured by Rockstad [35-37] is a consequence
of the "thin-film nature", and is not characteristic of the material
itself.   To the extent of believing that the preparation of thin
films introduces randomly situated defects, and that these give
rise to a large value of $\sigma(\omega)$ because of the small value of B,
we follow this point of view.

The case of a-$As_2 Te_3$ (and other small band gap materials
is complicated due to the existence of two mechanisms which
contribute to $\sigma(\omega)$ [37,38,41,84].   At temperatures above 200K, the
temperature dependence of the measured A.C. conductivity becomes
more rapid, as indicated in Section III.   If the extrapolated
contribution of $\sigma(\omega)$ due to CBH is subtracted from σ_T, it is found

that the resultant is simply activated with an activation energy
of 0.36 eV, some 0.03 eV lower than that observed for the D.C.
conductivity $\sigma(0)$ [84]. This extra contribution $\sigma'(\omega)$ is then
believed to arise from excitation of carriers to localized band-
tail states, where motion occurs by QMT [37,84]. The A.C.
conductivity in this case can then be written as [84,85]:

$$\alpha'(\omega) \simeq \frac{\pi^2}{12} \ln(2) \; N^2(E) \; \frac{a^5}{16} \; e^2 \; kT \; (n^4 (1/\omega\tau_0) \exp(-E/kt) \tag{28}$$

where QMT is assumed to occur amongst tail states at energy E from
the Fermi level (for As_2Te_3, E=0.36 eV) where the density of states
is N(E) and wavefunction dimension is a. (It must be pointed out
that $\sigma'(\omega)$ can only be obtained from σ_T by subtraction of the
contribution believed to arise from hopping of carriers amongst
defect states deep in the gap near the Fermi level, and this
process is very model dependent.) Thus, the dominant contribution
to $\sigma_T(\omega)$ at room temperature will be due to QMT in band-tail
states ($\sigma'(\omega)$) for which we expect s $\simeq 0.8$ (see Section IV.A.)
Experimental values for s in the range 0.75-0.87 (sometimes increasing to ~ 0.92 at very high frequencies) have been reported [36]
and support this assumption, although confusingly the values for
s predicted by the CBH model are also near 0.8 [4] for room
temperature. (This conclusion was not used in Ref. 4 where it was
assumed erroneously that $\sigma_T(\omega)$ at room temperature was due to
CBH.) The value of $\sigma(\omega)$ for a-As_2Te_3 was used in Fig. 11 is the
CBH value extrapolated to room temperature; such procedures are
not required for larger band gap materials since the energy of
activation to band-tail states is so large that $\sigma'(\omega)$ is much
smaller than $\sigma(\omega)$ due to CBH at all temperatures.

An alternative model for the strong exponential temperature
dependence of the A.C. conductivity observed particularly in
a-As_2Te_3 at high temperatures has been given by Shimakawa [86].
He has observed that the generation of D^0 centers produced by
thermal excitation of D^+ or D^- centers increases at high temperatures; the density of D^0 is given by

$$N_o = N_- \exp \left[\frac{-(E_F - W_1)}{kT} \right] + N_+ \exp \left[\frac{-(B-E_F - W_2)}{kT} \right]$$

$$= 2N_- \exp \left[\frac{-(E_F - W_1)}{kT} \right]$$

An additional mechanism leading to A.C. conduction now arises,
namely the hopping of a single electron between D^0 and D^- (or D^+).
This situation has been considered previously for the case of
amorphous arsenic [87,88], where both positive and negative
correlation energy defects (D^0 and D^+, D^-, respectively) are

detected by E.S.R.. The A.C. conductivity for single polaron hopping between D^o and D^- is approximately [87,88]:

$$\sigma_1(\omega) \simeq \frac{\pi^3}{3} N_1 N_2 \, \varepsilon\varepsilon_o \frac{kT}{\Delta_o} \tanh(\Delta_o/2kT) \, [\frac{e^2}{\Pi\varepsilon\varepsilon_o W_{M_1}}]^6 \, \frac{\omega^{s_1}}{\beta_1 \tau_o} \qquad (29)$$

where N_1 is the density of D^o centers ($\equiv N_o$) and N_2 is the density of D^- centers ($\equiv N_-$), W_{M_1} is the maximum thermal energy to place an electron from the valence band onto D^o ($\equiv W_1$). For materials in which the band gap and hence W_1 is small (e.g., As_2Te_3), Eq. (29) is not sufficiently accurate at high temperatures and the higher order term given by Eq. (25) must be used. It is then found that single polaron CBH should be dominant at high temperatures because of the activated dependence of the concentration of D^o (N_o) and the exponential term in Eq. (25). Larger band gap materials have larger values for W_1, and hence the density of thermally generated D^o centers is small, so bipolaron CBH is dominant.

The frequency exponent, s, is also predicted to be a function of B (Eqs. (22) and (24)). However, it is the departures from unity which yield useful information, and much of the experimental data in the literature is not sufficiently accurate for this purpose. In addition, it is always difficult to compare data on completely different materials preapred in different laboratories. What is of great use is a comparative study of different compositions in the same alloy system. This has only been performed on the following systems, Se_xTe_{1-x} [17], As_xSe_{1-x} [31,32] and As_xS_{1-x} [31], all of which were melt-quenched glasses. This is unfortunate for our present purpose since non-random defect-pairing effects (see later) may complicate the issue in these cases. Nevertheless, we show in Fig. 12 experimental data for the glassy

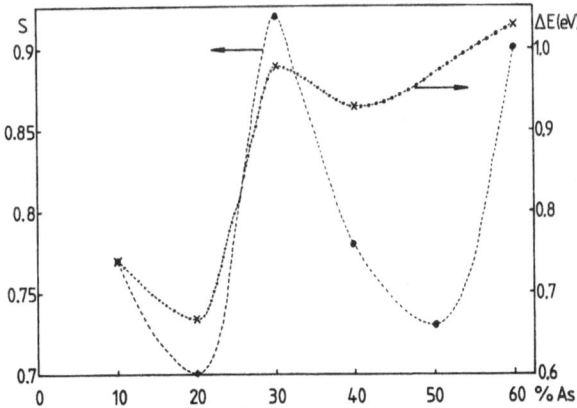

Fig. 12   Correlation between the frequency exponent s and the electrical gap for the As-Se alloy system [32].

As_xSe_{1-x} system [32] in which it is shown that there is a reasonably clear correlation of s with the electrical activation energy
E.   However, variations in the optical gap B are not reported and
so an unequivocal correspondence between s and W_M cannot be deduced;
however, if we suppose that B scales in the same manner as E with
composition, then the variation of s with W_M is qualitatively as
predicted by the CBH model.

The temperature dependence of $\sigma(\omega)$ on the CBH model has been
discussed previously [84] for the case of sites of equal energy
($\Delta_o=0$).   In this case the temperature dependence arises solely
from the factor $(1/\omega\tau_o)^\beta$ and obeys $\frac{d(\ln\sigma(\omega))}{dT} = \frac{6k}{W_M}(\ln(1/\omega\tau_o))$.
Many experimental data are plotted semi-logarithmically versus in-
verse temperature; the CBH model predicts that $\frac{d(\ln\sigma(\omega))}{d(1/T)} = \frac{-6kT^2}{W_M}$
$\ln(1/\omega\tau_o)$.

Experimentally, some slopes of $\ln\sigma(\omega)$ vs. $1/T$ plots do show
a decrease with decreasing temperature or increasing frequency [8,23].
However, these data in general only offer a qualitative comparison
with the CBH theory since the very weak temperature dependence
of $\sigma(\omega)$ is obscured by plotting it versus inverse temperature.   The
only temperature data which affords a detailed comparison are
by Hauser on $a-As_2Te_3$ films [42] (Fig. 13); the A.C. conductivity

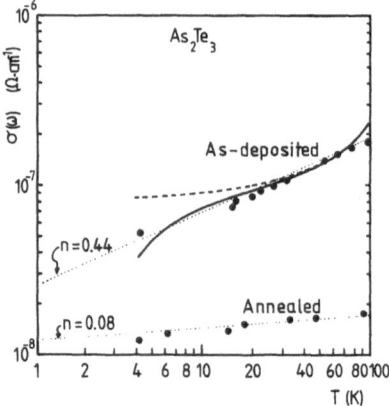

Fig. 13   Temperature dependence of the A.C. conductivity of $a-As_2Te_3$
          thin films sputtered at 77K measured at 10^4 Hz before and
          after annealing at 300K [42] (data points are filled circles).
          The theoretical curve calculated for the CBH model using the
          full expression (Eq. 20) including the site energy asymmetry
          factor ($\Delta_o = 1.5 \times 10^{-3}$ eV, $\tau = 1.6 \times 10^{-13}$s) is shown by the
          full curve; the dashed curve is for $\Delta_o = 0$. The dotted curves
          are the lines $\sigma(\omega) \propto T^{0.44}$ and $T^{0.08}$ proposed empirically by
          Hauser [42] to fit the data for the as-deposited and annealed
          material, respectively.

has been measured in the temperature range 4-100 K (i.e., well
below the region in which hopping in band-tails ($\sigma'(\omega)$) is expected).
He finds that samples sputtered onto substrates kept at 77K and
then cooled to 4.2 K exhibit an A.C. conductivity which has a
frequency exponent s = 1 and the temperature dependence on warming
is sub-linear, ($= T^{0.44}$); the as-deposited samples also exhibit
a $T^{1/4}$ temperature dependence of the D.C. conductivity, $\sigma(o)$,
characteristic of variable-range hopping and an ESR signal ($N_0 \sim$
2×10^{18} spins cm^{-3}).

The presence of D^o centers due to disorder complicates the
issue since one must now consider the effect of $\sigma(\omega)$ of the hopping
of one electron between D^o and D^+ or D^-. This situation has been
considered earlier, but the difference here is that we consider
the density of D^o centers to be constant, since they arise from a
non-equilibrium preparation rather than thermal generation. For
temperatures in the range of interest (4-100 K), the higher order
term in $\sigma(\omega)$ (Eq. (25)) is close to unity for both single and
bipolaron hopping, and also the ratio of A.C. conductivities for
the two mechanisms is given by the ratio of Eqs. (23) and (29):

$$\frac{\sigma_1(\omega)}{\sigma_2(\omega)} = \frac{N_1}{N_2} \left[\frac{W_{M_2}}{2W_{M_1}}\right]^6 \frac{\omega^{(s_1-s_2)}}{\tau_o^{(\beta_1-\beta_2)}} \tag{30}$$

W_{M_2} = 1.1 eV and $W_{M_1} \equiv W_1 \simeq 0.29$ eV for a-As_2Te_3 (Table III), and
$\beta_{1,2}$ = $6kT/W_{M_{1,2}}$ = 1-s, for single or bipolaron hopping. From
these data we find that $\sigma_1(\omega)/\sigma_2(\omega) \simeq 150 \times N_1/N_2$ at T = 50K, and
hence if $N_2 > 150 N_1$ (= 3.0×10^{20} cm^{-3}, $\sigma_2(\omega)$. Such a value of N_2
is probably not unreasonable considering the non-equilibrium
state of the 77K deposited films.

The temperature dependence is now given by Eq. (20) or (23)
and is approximately proportional to T $tanh\left(\frac{\Delta_o}{2kT}\right) \frac{1}{(\omega\tau_o)^\beta}$;
T $tanh (\Delta_o/2kT)$ is approximately proportional to T for $T < \Delta_o/2kT$
and roughly constant thereafter. We show in Fig. 13 the theoretical curve, calculated using Eq. (20) with Δ_o = 1.5×10^{-3} eV
which is seen to fit the data very well for temperatures below 80K;
the increase in the theoretical temperature dependence of $\sigma(\omega)$
above this temperature is caused by the exponential-like nature of
the factor $(1/\omega\tau_o)^\beta$, and it is likely that the experimental data
lies below this curve because on warming above 77K the sample
starts to anneal and the density of defects, and hence $\sigma(\omega)$,
decreases. This is shown rather dramatically by the lower curve
in Fig. 13 which is for the sample annealed to 300K. This has no

ESR signal nor a $T^{1/4}$ dependence of $\sigma(o)$, indicating that the metastable D^0 centers have been annealed away, and the dominant contribution to $\sigma(\omega)$ must now be from bipolarons on the CBH model. Thus
annealing may have three effects: the number of D^0 and D^+, D^-
centers is reduced leading to a decrease in the overall magnitude
of $\sigma(\omega)$, the distribution in intersite energy differences is
further sharpened (Δ_o is reduced) and atomic relaxations may lead
to effective defect pairing (see later), both of which act to
decrease the temperature dependence. Note that the QMT mechanism
cannot account for a sub-linear temperature dependence of $\sigma(\omega)$
nor for a frequency exponent $s \simeq 1$ (the bipolaron CBH estimate at
50K is 0.97).

From the preceding discussion we have seen that for randomly
distributed centers the CBH model accounts for many features of
the A.C. conductivity observed in amorphous chalcogenide thin
films: a correlation of $\sigma(\omega)$ and s with band gap (or D.C. conductivity), an increase in s with decreasing temperature, and a weak
(sub-linear) temperature dependence of $\sigma(\omega)$. What remains to be
explained are the seemingly conflicting data reported for chalcogenide glasses, where s=1 and a temperature independent $\sigma(\omega)$
are often reported. To understand these observations, together
with the superlinear frequency exponents sometimes observed, we
must consider non-random distributions of centers [86], and this
is the subject of the next section.

Amorphous materials differ from their crystalline counterparts
in that an isolated dangling bond (neutral center) is topologically
possible for a continuous random network describing the amorphous
structure, whereas defects in crystalline lattices cannot occur
single (e.g., a vacancy in a tetrahedrally coordinated lattice
produces four dangling bonds). We expect such neutral dangling
bonds in an amorphous network to be randomly distributed. This is
not to be expected for the defects believed to exist in chalcogenide glasses, namely equally and oppositely charged centers,
where the mutual Coulombic attraction will cause preferential
pairing of mobile centers in the melt [76]; on quenching to form a
glass this non-random distribution is assumed to be frozen-in.
The form of the resulting distribution of centers is difficult to
calculate precisely. In a discussion on recombination Street [89]
assumed that an earlier treatment by Reiss et al, [90,91] developed
for charged impurities in crystalline semiconductors was appropriate
for the case of chalcogenide glasses. For dilute systems the following approximation [90,91] has been used [82,92]:

$$G(r) = 4 \, N\Pi r^2 \, \exp(-4\Pi Nr^3/3) \, \exp(e^2/4\Pi\varepsilon_o kT_g r) \tag{31}$$

where N is the concentration of charged centers, the first
exponential term is an exclusion factor (ignored in the previous

treatment for <u>random</u> systems, since the hopping distance (Eq. (21)) is much smaller than the average intersite separation) and the second exponential term accounts for the attractive Coulombic interaction. It has been assumed that this Boltzmann distribution is frozen-in upon cooling through the glass-transition temperature T_g; rapid quenching may so inhibit atomic rearrangement that the value for the quench temperature may be taken instead of T_g. The form of Eq. (31) is shown in Fig. 14, where parameters appropriate to glassy As_2S_3 have been used, and a comparison is made with the random case. An arbitrary cut-off, $r_{min} = 4.5$ Å (corresponding to third nearest neighbors) has been used [89], and the area under the paired curve is normalized to unit area. It can be seen that $G(r)$ for the paired case is a sharply decreasing function of r for small r and then follows approximately the random distribution at larger r. The importance of the behavior of $G(r)$ at small r lies in the fact that values for the hopping distance, r_ω on the CBH model lie typically in the range 5–7 Å (from Eq. (21)) and as can be seen from Fig. 14 this is precisely the region where $G(r)$ for the paired case decreases rapidly. This has an important effect on $\sigma(\omega)$ since departures in the frequency exponent, s, from unity are caused by the hopping distance being a function of frequency and the size of such deviations from unity are determined by the functional dependence of r_ω in $\sigma(\omega)$. We can calculate the A.C. conductivity for non-random centers using Eq. (31) accepting, however, that this is only approximate; the analytic form for $G(r)$ on the other hand does allow $\sigma(\omega)$ to be calculated. We note first that since the hopping distance for bipolaron CBH is much smaller than the average separation of centers (given by the maximum of $G(r)$ in Fig. 14), the exclusion factor $\exp(-4\Pi N r^3/3)$ is almost unity and may be neglected, as was

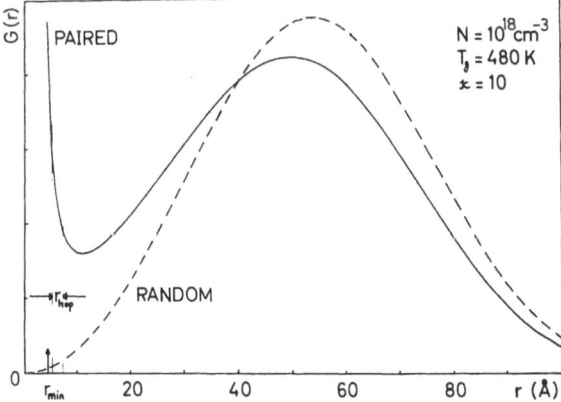

Fig. 14   Spatial distribution function for charged defects which are either randomly or pairwise distributed. Parameters pertaining to As_2Se_3 have been used.

done implicitly in the previous treatment for random centers.
(This is not so for single polaron CBH in which the hopping
distance is much bigger and the exclusion factor must be included
[78,93,94].) The A.C. conductivity is calculated in the same
way as before [92], yielding (in S.I. units):

$$\sigma_p(\omega) \simeq \frac{\pi^3}{3} N^2 \varepsilon \varepsilon_o \frac{kT}{\Delta_o} \ \tanh \ (\Delta_o/2kT) \left[\frac{2e^2}{\Pi \varepsilon \varepsilon W_M}\right]^6 \frac{\omega}{(\omega\tau_o)^\beta} \ (\omega\tau_o)^\delta \left(\frac{W_M}{8kT_g}\right) \quad (32)$$

where similar approximations to those used in obtaining Eq. (23)
have been used. The factor δ is given by:

$$\delta = \frac{T}{8T_g} \qquad\qquad\qquad\qquad\qquad\qquad\qquad\qquad (33)$$

and hence the frequency exponent is:

$$s = 1 - \beta + \delta \qquad\qquad\qquad\qquad\qquad\qquad\qquad\qquad (34)$$

Several new features emerge from Eq. (32).

    i) If $\beta \simeq \delta$, the frequency exponent s is very close to unity.

    ii) Since the temperature dependence of $\sigma(\omega)$ arises from the
factors involving β and δ (for temperatures $T > \Delta_o/2k$ such that
$T \tanh (\Delta_o/2kT)$ is constant which is true for all except the
lowest temperatures), the overall temperature dependence of $\sigma(\omega)$
is very small.

    It emerges that for many chalcogenide glasses, the factors
W_M and T_g are such that β is approximately equal to δ. As an
example, we show in Fig. 15 the A.C. conductivity, calculated for
both random and paired centers (without the $T \tanh (\Delta_o/2kT)$ term),
using realistic experimentally obtained parameters appropriate to
As_2S_3 $W_M=2.4$ eV, $\varepsilon = 10$, $T_g = 480$ K, $\tau_o = 10^{-13}$ s). It can be
seen readily that the A.C. conductivity for the paired case
exhibits no temperature dependence and has a frequency exponent
equal to unity, in marked contrast to the random case which leads
to a temperature dependent $\sigma(\omega)$ and s, as discussed previously.
This is, we believe the origin of the seemingly contradictory
experimental results obtained from thin films or glasses of the
same material; in the former case the D^+, D^- centers (ideally)
are randomly distributed since atomic relaxation is hindered,
whereas in the latter case pairing between D^+ and D^- centers can
readily take place in the fluid melt leading to a non-random
distribution.

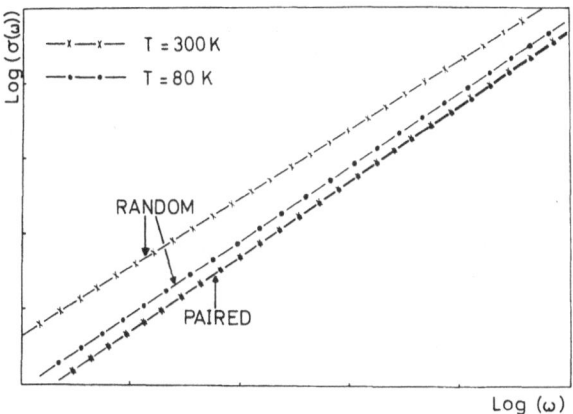

Fig. 15   Theoretical curves for A.C. conductivity, $\sigma(\omega)$, as a
function of frequency at two temperatures calculated for
the CBH model for both random and paired centers. Note
that $\sigma(\omega)$ is temperature independent and the frequency
exponent, s, equals unity for the paired case.

        A systematic variation of $\sigma(\omega)$ between materials (i.e., the
band gap dependence) can also be discerned for the paired case
from Eq. (32). The additional factor exp $[W_M/8kT_g]$ introduces
a dependence on W_M (and hence B) opposing the factor W_M^{-6}. Thus
we expect the band gap dependence of $\sigma(\omega)$ for paired centers to
be weakened compared with that for random sites. This is illustrated in Fig. 11, where the variation in $\sigma(\omega)$ with W_M B is
shown. The variation in T_g is accounted for by the empirical
relation between T_g and B discovered by DeNeufville and Rockstad [95]:

$$T_g = T_g^{\;o} + \;\; B \tag{35}$$

Here $T_g^{\;o}$ = 328 K and α = 68 K eV^{-1} for the As_2X_3 series (X = S,Se,Te).
The curves for paired and random centers have been computed for
the same value of N in Fig. 11. It can be seen that $\sigma(\omega)$ varies
only by about an order of magnitude for the paired case for a range
of band gaps 0.8 < B < 2.4 eV, whereas $\sigma(\omega)$ for random centers
varies over a much wider range, of the order of three decades.
Thus, on this model of paired centers, we expect the A.C. conductivity for chalcogenide materials to be roughly constant, although
the D.C. conductivity varies by $\sim 10^{12}$ for the same materials. This
is observed in many chalcogenide glasses [39,41], some data for
which are shown in Fig. 11, and has led to an atomic mechanism
being proposed [18,72], as we have seen in Section IV.B. This is
unnecessary in our view, since it appears that the <u>electronic</u>
CBH mechanism is capable of explaining the experimental results
when non-random distributions are taken into account, with the

result that <u>one</u> fundamental mechanism can be responsible for all
the A.C. behavior observed in chalcogenide materials.

The final point we wish to discuss concerns the <u>super-linear</u>
frequency dependence occasionally reported for chalcogenide
materials [8,31,33,45]. This behavior can be understood simply
from the discussion above, for if $\delta > \beta$ then Eq. (32) predicts a
super-linear frequency dependence. Thus, a non-random distribution of centers can give rise to a functional dependence of $\sigma(\omega)$
on r_ω such that $s > 1$; it has been argued that this behavior
should be more prevalent in materials having large band gaps [33],
and indeed this seems to be the case. If the simple analytic
form of the paired distribution function Eq. (31) is to be believed,
the occurrence of super-linear behavior ($\delta > < \beta$) is seen to depend
upon an interplay of the parameters W_M and T_g (or quench temperature). One well-documented case of this phenomenon is for amorphous
Se thin films [8], whereas the temperature is lowered the exponent,
s, increases from 0.85 at 320 K becoming equal to 1.08 at 77 K.
This continual increase of s to values in excess of unity as the
temperature is lowered is not accounted for by the simplified model
described previously. Super-linear behavior as a result of the
condition $\delta > \beta$ implies that $\sigma(\omega)$ is a <u>decreasing</u> function of
temperature, rather than the reverse. All we can assume is that
as T decreases the hopping distance decreases in accord with Eq. (21),
being on the right-hand 'random-like' side of the minimum of G(r)
in Fig. 14 at high temperatures, and on the left-hand 'paired' side
of the minimum at low temperatures, thereby allowing a super-linear
exponent to occur. The minimum in G(r) which separates paired from
random centers occurs at a distance $r_{min} = e^2/8\pi\varepsilon\varepsilon_o kT_g$ for the
approximate distribution, Eq. (31). For se, $\varepsilon = 6.3(8)$ and $T_g \simeq$
300 K giving a value for $r_{min} = 44$ Å; if this were the case for
the thin films concerned, then the hopping distance r_ω would lie
well to the left of the minimum in the decreasing 'paired' region
of G(r). In actual fact, the effective quench temperature for the
films must be considerably higher, which would decrease r_{min}, so
r may indeed lie near the minimum of G(r), leading to a super-
linear value for s at low temperature.

Interestingly, A.C. conduction with a frequency exponent near
unity has been observed in <u>single</u> <u>crystals</u> of chalcogenide
materials, $GeSe_2$ [18] (s=1) and As_2S_3 [96] (s=0.85). This rather
surprising result seems to rule out the mechanism of atomic hopping
in a two-level system [18,72] since the low-temperature thermal
anomalies, believed to result from the presence of two-level
systems, are not observed in crystalline chalcogenide materials.
However, there is evidence that the same type of structural
defects (D^+ and D^-) exist in both crystalline and amorphous form
of chalcogenide materials; photoluminescence spectra in particular
are very similar for the two different phases [97]. This suggests
that charged defects are involved in $\sigma(\omega)$ in both cases, and the CBH
mechanism may be responsible.

The CBH model (in both single and bipolaron forms) predicts a dielectric loss peak at low frequencies when the condition $\omega\tau_{max} \gg 1$ is no longer satisfied [4,78] (where $\tau_{max} = \tau_o \exp(W_M/kT)$). Such loss peaks have indeed been observed in amorphous As_2S_3 [12], As_4S_4 [12] and As_2Se_3 [12,86]. These have been ascribed to transitions between D^0 and D^- or D^+ when the D^0 centers are present either as a result of disorder in as-deposited films [4] (as in Ref. 12) or as a consequence of thermal excitation of D^- or D^+ centers [86].

The CBH model has been applied to systems other than the chalcogenides detailed in this review. It was originally applied to scandium oxide [78] in the single polaron form and then to other transition metal oxide glasses [93,94]. Also in single polaron form, it has been applied with success to amorphous SiO [60] (where Eqs. (22) and (24) for the temperature dependence of s are accurately obeyed), and amorphous SiO_2 [62] and amorphous Si [79]. The CBH model in its bipolaron form has also been applied to a-Si [98] and a-Ge [99], in the belief that spin-pairing of defects may also occur in these materials, perhaps at void surfaces.

The ultimate test of any theory is comparison with experiment. In this regard, A.C. conductivity fares badly at present, since many experimental accounts are not sufficiently detailed. For a proper understanding of this complex phenomenon, experiments over a wide frequency and temperature range are required and preferably on a series of materials with differing band gaps. Much of the physics of the process lies in the small departures of the frequency exponent, s, from linearity and in the weak temperature dependence, both of which need to be clearly described in any experimental account.

REFERENCES

1. N. F. Mott, Phil. Mag. 19:835 (1969).
2. M. S. Frost, A. K. Johscher, Thin Solid Films 29:7 (1975).
3. P. C. Taylor, S. G. Bishop and D. L. Mitchell, Sol. St. Comm. 8:1783 (1970).
4. S. R. Elliott, Phil. Mag. 36:1291 (1977).
5. M. Pollak, T. H. Geballe, Phys. Rev. 122:1742 (1961).
6. L. D. Landau, and E. M. Lifshitz, Electrodynamics of Continuous Media, Pergamon Press (1960).
7. P. Debye, Polar Molecules, Dover (1945).
8. A. I. Lakatos and M. Abkowitz, Phys. Rev. 3:1791 (1971).
9. R. A. Street, G. R. Davies, and A. D. Yoffe, J. of Non-Cryst. Sol. 5:276 (1971).
10. J. Kocka, Phys. Stat. Sol. a48:K59 (1978).
11. M. Pollak, U.S. Nat. Techn. Inf. Service AD 757097.
12. R. A. Street and A. D. Yoffe, J. of Non-Cryst. Sol. 8-10:745 (1972).

13. M. Abkowitz, A. Il Lakatos and H. Scher, Phys. Rev. B9:1813 (1974).
14. X. LeCleac'h and J. F. Palmier, Proc. 7th Int. Conf. on Amorphous and Liquid Semiconductors, CICL, Edinburgh, p. 580 (1977).
15. X. LeCleac'h, Sol. St. Comm. 21:309 (1977).
16. A. L. Dawar, J. C. Joshi and I. J. Makhija, J. Non-Cryst. Sol. 29:409 (1978).
17. R. M. Mehra, P. C. Mathur, A. K. Kathuria and R. Shyam, Phys. Rev. B18:5620 (1978).
18. X. LeCleac'h, J. Physique 40:417 (1979).
19. K. Shimakawa, S. Nitta, K. Naruse, T. Arizumi and A. Yoshida, Proc. 6th Int. Conf. on Amorphous and Liquid Semiconductors, Nauka, Leningrad, p. 226 (1976).
20. A. Yoshida, T. Arizumi, K. Shimakawa and S. Nitta, Proc. 6th Int. Conf. on Amorphous and Liquid Semiconductors, Nauka, Leningrad, p. 231 (1976).
21. S. Nitta, K. Shimakawa, K. Naruse, K. Sakaguchi, A. Yoshida and T. Arizumi, Proc. 6th Int. Conf. on Amorphous and Liquid Semiconductors, Nauka, Leningrad p. 221 (1976).
22. A. E. Owen, J. M. Robertson, J. Non-Cryst. Sol. 2:40 (1970).
23. E. B. Ivkin and B. T. Kolomiets, J. Non-Cryst. Sol. 3:41 (1970).
24. M. Kitao, F. Araki and S. Yamada, Phys. Stat. Sol. 37:K119 (1970).
25. C. Crevecoeur and H. J. de Wit, Sol. St. Comm. 9:445 (1971); ibid. 9:ix (1971).
26. D. L. Mitchell, S. G. Bishop and P. C. Taylor, J. Non-Cryst. Sol. 8-10:231 (1972).
27. U. Stron and P. C. Taylor, Proc. 5th Int. Conf. on Amorphous and Liquid Semiconductors, p. 375, Taylor and Francis (1974).
28. J. Kocka, S. R. Elliott and E. A. Davis, J. Phys. C12:2589 (1979).
29. K. G. Breitschwerdt and J. Hafner, J. Non-Cryst. Sol. 35-36:993 (1980).
30. M. Kitao, Jap. J. Appl. Phys. 11:1472 (1972).
31. R. B. South and A. E. Owen, Proc. 5th Int. Conf. on Amorphous and Liquid Semiconductors, p. 305, Taylor and Francis (1974).
32. R. Mohan, S. Mahadevan, and K. J. Rao, Mat. Res. Bull. 15:917 (1980).
33. S. R. Elliott, Sol. St. Comm. 28:939 (1978).
34. J. I. Polanco, G. G. Roberts and M. B. Myers, Phil. Mag. 25:117 (1972).
35. H. K. Rockstad, Sol. St. Comm. 7:1507 (1969).
36. H. K. Rockstad, J. Non-Cryst. Sol. 2:192 (1970).
37. H. K. Rockstad, J. Non-Cryst. Sol. 8-10:621 (1972).
38. L. Stourac, A. Abraham, J. Cocka, I. Kubelik, A. Triska and M. Zavetova, Proc. 5th Int. Conf. on Amorphous and Liquid Semiconductors, p. 297, Taylor and Francis (1974).
39. J. Kocka, Czech. J. Phys. B26:807 (1976).

40. J. Kocka, A. Triska, L. Stourac, Proc. 6th Int. Conf. on Amorphous and Liquid Semiconductors, Nauka, Leningrad, p. 249 (1976).
41. J. Kocka, Proc. Conf. on Amorphous Semiconductors, Pardubice, p. 252 (1978).
42. J. J. Hauser, Phys. Rev. Lett. 44:1534 (1980).
43. H. Segawa, J. Phys. Soc. Japan 36:1087 (1974).
44. S. Mahadevan, A. Giridhar and K. J. Rao, J. Phys. C10:4499 (1977).
45. J. Gazso, J. Kocka and F. Kosek, Proc. Conf. on Amorphous Semiconductors, Pardubice, p. 293 (1978).
46. X. LeCleac'h and J. F. Palmier, Non. Cryst. Sol. 18:265 (1975).
47. X. LeCleac'h and J. F. Palmier, Proc. 6th Int. Conf. on Amorphous and Liquid Semiconductors, Nauka, Leningrad, p. 393 (1976).
48. G. P. Carver, R. S. Allgaier and J. P. DeNeufville, Proc. 5th Int. Conf. on Amorphous and Liquid Semiconductors, Taylor and Francis, p. 313 (1974).
49. A. K. Johscher and M. S. Frost, Thin Solid Films, 37:267 (1976).
50. O. S. Panwar, M. Radhakrishna, K. K. Srivastava and K. N. Lakshminarayan, Phil. Mag. B41:253 (1980).
51. G. G. Roberts and J. I. Polanco, Sol. St. Comm. 10:709 (1972).
52. B. Fromm, W. Paul and W. Teubner, Thin Solid Films 59:337 (1979).
53. A. Giridhar, P. L. L. Narashimham and S. Mahadevan, J. Non-Cryst. Sol. 37:165 (1980).
54. P. L. Sherrell and J. C. Thompson, J. Non-Cryst. Sol. 24:69 (1977).
55. M. Burman, J. Hirsch and T. Ramdeen, J. Phys. C14:117 (1981).
56. M. Kitao, N. Asakura and S. Yamada, Jap. J. Appl. Phys. 19:L302 (1980).
57. S. Barta and M. Jergel, Proc. Conf. on Amorphous Semiconductors, Pardubice, 297 (1978).
58. E. B. Ivkin, B. T. Kolomiets, E. M. Raspopova and K. D. Tsendin, Sov. Phys. Semiconductors, 5:1558 (1972).
59. J. Kristofik, Proc. Conf. on Amorphous Semiconductors, Pardubice, 277 (1978).
60. M. Jourdain, A. dePolignac and J. Despujols, J. Phys. C12:4999 (1979).
61. M. Jourdain and J. Despujols, J. Phys. C13:1593 (1980).
62. M. Meaudre and R. Meaudre, Phil. Mag. B40:401 (1979).
63. A. Miller and E. Abrahams, Phys. Rev. 120:745 (1960).
64. I. G. Austin and N. F. Mott, Adv. Phys. 18:41 (1969).
65. N. F. Mott and E. A. Davis, Electronic Processes in Non-Crystalline Materials, O.U.P. (1979).
66. M. Pollak, Phil. Mag. 23:519 (1971).
67. P. N. Butcher and K. J. Hayden, Proc. 7th Int. Conf. on Amorphous and Liquid Semiconductors, CICL, Edinburgh, p. 234 (1976).
68. H. Fritzsche in Electronic and Structural Properties of Amorphous Semiconductors, Academic Press (1973).

69. R. C. Zeller and R. O. Pohl, Phys. Rev. B4:2029 (1971).
70. P. W. Anderson, B. I. Halperin and C. M. Varma, Phil. Mag. 25:1 (1972).
71. W. A. Phillips, J. Low Temp. Phys. 7:351 (1972).
72. M. Pollak and G. E. Pike, Phys. Rev. Lett. 28:1449 (1972).
73. P. W. Anderson, Phys. Rev. Lett. 34:953 (1975).
74. R. A. Street and N. F. Mott, Phys. Rev. Lett. 37:1293 (1975).
75. N. F. Mott, E. A. Davis and R. A. Street, Phil. Mag. 32:961 (1975).
76. M. Kastner, D. Adler and H. Fritzsche, Phys. Rev. Lett. 37:1504 (1976).
77. D. Vanderbilt and J. D. Joannopoulos, Sol. St. Comm. 35:535 (1980).
78. G. E. Pike, Phys. Rev. B6:1572 (1972).
79. G. Rieder, Phys. Rev. B20:607 (1979).
80. R. A. Street, Proc. 6th Inc. Conf. on Amorphous and Liquid Semiconductors, Nauka, Leningrad, p. 116 (1976).
81. E. A. Davis and N. F. Mott, Phil. Mag. 22:903 (1970).
82. S. R. Elliott, J. Non-Cryst. Sol. 35-36:855 (1980).
83. T. S. Moss, Photoconductivity in the Elements, Butterworth (1952).
84. S. R. Elliott, Phil. Mag. B37:553 (1978).
85. M. Pollak, private communication.
86. K. Shimakawa, Proc. 9th Int. Conf. on Amorphous and Liquid Semiconductors, to be published.
87. S. R. Elliott and E. A. Davis, Proc. 7th Int. Conf. on Amorphous and Liquid Semiconductors, CICL, Edinburgh, p. 637 (1977).
88. G. N. Greaves, S. R. Elliott and E. A. Davis, Adv. Phys. 28:49 (1979).
89. R. A. Street, Phys. Rev. B17:3984 (1978).
90. H. Reiss, J. Chem. Phys. 25:400 (1956).
91. H. Reiss, C. S. Fuller and F. J. Morin, Bell Syst. Tech. J. 35:535 (1956).
92. S. R. Elliott, Phil. Mag. B40:507 (1979).
93. B. E. Springett, J. Non-Cryst. Sol. 15:179 (1974).
94. A. Mansingh, J. K. Vaid, and R. P. Tandon, J. Phys. C8:1023 (1975).
95. J. P. DeNeufville and H. K. Rockstad, Proc. 5th Int. Conf. on Amorphous and Liquid Semiconductors, Taylor and Francis, p. 419 (1974).
96. M. Abkowitz, D. F. Blossey and A. I. Lakatos, Phys. Rev. B12:3400 (1975).
97. R. A. Street, Adv. Phys. 25:397 (1976).
98. S. R. Elliott, Phil. Mag. B38:325 (1978).
99. A. R. Long and N. Balkan, Phil. Mag. B41:287 (1980).

NMR STUDY OF STRUCTURE AND BONDING IN GLASSES*

P. J. Bray and M. L. Lui

Department of Physics
Brown University
Providence, RI 02912

INTRODUCTION

Nuclear magnetic resonance (NMR) techniques have been employed to study the bonding between atoms, and arrangements of atoms, in glasses. Oxygen bonding in SiO_2 and B_2O_3 has been investigated using the oxygen-17 isotope. The O^{17} NMR spectrum for vitreous SiO_2 (v-SiO_2) exhibits well-defined structure and indicates significant local order in the material. Matching of experimental and computer-simulated spectra permits extraction of the O^{17} quadrupole coupling parameters which are analyzed to obtain charge densities in the electronic orbitals and the Si-O-Si bond angles and distributions. Comparisons can be made with values obtained from the O^{17} NMR spectrum of α-cristobalite. Both boron and oxygen have been studied in v-B_2O_3. The B^{10}, B^{11}, and O^{17} NMR spectra yield quadrupole parameters, bond angles, and charge densities. Two distinct types of oxygen site are clearly revealed which are consistent with the boroxol ring oxygens and the ring-connecting oxygens. B^{10} studies of $Na_2O-B_2O_3$ and $Li_2O-B_2O_3$ glasses yield spectra which are beautifully simulated by adding the spectra from the crystalline compounds in the respective systems. The relative amount of the structural grouping (i.e., boroxol, diborate, metaborate) from each compound can then be determined in each glass and followed as a function of composition. B^{11} studies in the technologically important sodium borosilicate system have permitted construction of structural models for the glasses in this system. The models yield the amount of each type of borate grouping present in each glass.

Nuclear magnetic resonance (NMR) techniques have been employed extensively to study the microstructure of oxide glasses. Several reviews of this work have been published recently [1-4]. These

285

techniques make use of the sensitivity of quadrupolar perturbations of Zeeman energy levels to short-range order in the arrangements of atoms. Matching of computer-simulated and experimental spectra yields values for the quadrupole coupling constant Q_{cc}, and asymmetry parameter η. These values may then be interpreted in terms of the arrangements of atoms and the bonding between atoms. The theory and experimental methods for NMR studies have been presented elsewhere [3-6], and will not be repeated here.

O^{17} NMR IN SiO_2

Since the natural abundance of O^{17} is only 0.037%, the O^{17} NMR signal from samples unenriched in the O^{17} isotope cannot be easily detected by conventional wide-line NMR techniques. Procedures for preparing isotopically-enriched amorphous SiO_2 and B_2O_3 from water enriched to 40% or more in the O^{17} isotope have been developed [7] and the enriched materials have been studied by NMR.

The O^{17} derivative adsorption spectrum for amorphous SiO_2 is presented in Figure 1. The structure in the spectrum is caused by second-order quadrupolar effects and is typical of that due to a high quadrupole coupling constant Q_{cc} and low asymmetry parameter η.

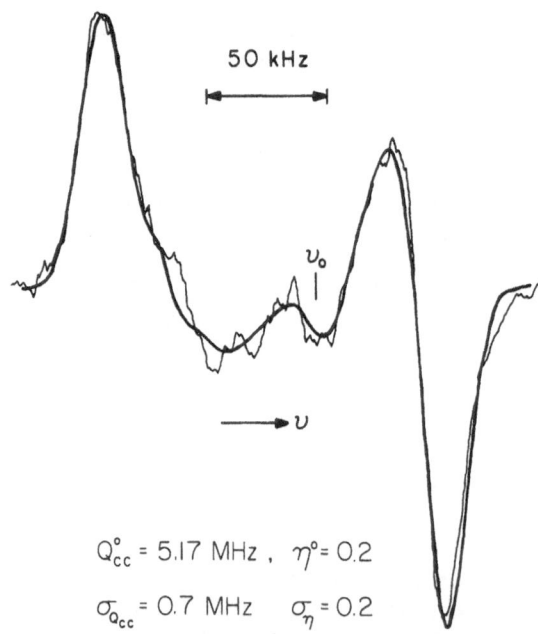

$$Q^{o}_{cc} = 5.17 \text{ MHz}, \quad \eta^{o} = 0.2$$

$$\sigma_{Q_{cc}} = 0.7 \text{ MHz} \quad \sigma_{\eta} = 0.2$$

Fig. 1 The O^{17} NMR experimental spectrum (narrow line) and its computer simulation (broad, smooth line) for amorphous SiO_2.

A good fit of the experimental spectrum to a computer simulation is obtained with a one-site fitting, i.e., all the oxygen atoms have similar atomic environments [8]. The computer simulation gives the following set of most probable values for the quadrupole parameters: Q_{cc}^o = 5.17 MHz with a distribution of $\sigma_{Q_{cc}}$ = 0.7 MHz, and η^o = 0.2 with a distribution of σ_η = 0.2. A Townes-Dailey calculation using the values of Q_{cc}^o and η^o yields a most-likely set of charge densities in the electronic orbitals which is consistent with silicon 3d orbital participation in bonding between the oxygen and silicon atoms. By using the measured standard deviations of the quadrupole parameters from their most probable values, an estimate of the variation of the Si-O-Si bond angle α is found to be within the range $130^o \lesssim \alpha \lesssim 180^o$. This is consistent with the results of x-ray investigation of vitreous SiO_2 [9].

Half of the amorphous SiO_2 sample has now been heat-treated at 1045^oC for 86 hours to produce α-cristobalite. The conversion was confirmed by x-ray studies. The heat treatment was continued until the amorphous background in the x-ray powder pattern had been reduced to a flat baseline. O^{17} NMR studies, however, found that saturation effects caused by the long spin-lattice relaxation time (T_1) in α-cristobalite did not allow the usual derivation absorption spectrum to be obtained. Instead, high power dispersion mode (HPDM) spectra were obtained in order to permit comparison of α-cristobalite with amorphous SiO_2. Under suitable conditions [10], the HPDM spectrum corresponds quite closely to the absorption curve. Figure 2 shows the HPDM spectra for amorphous SiO_2 and α-cristobalite recorded at an operating frequency of 8 MHz under identical experimental conditions. The rounding or broadening of the features in the spectrum of amorphous SiO_2 reflects the distribution in quadrupole parameters in the glass, while the sharply-defined spectrum for α-cristobalite reflects the well-ordered character of this material. The separation of the positions of the peaks in the spectrum of α-cristobalite is consistent with choosing as values for the quadrupole parameters the most probable values of those found for amorphous SiO_2.

NMR STUDIES OF VITREOUS B_2O_3

Three B_2O_3 glass samples with different isotopic enrichment were studied [11]. The B^{11} NMR spectrum of a B_2O_3 glass containing the natural abundance of B^{11} is given in Figure 3. The B^{10} NMR spectrum of a B_2O_3 glass enriched to 92% B^{10} is presented in Figure 4. The quadrupole parameters obtained from computer simulations of these spectra are summarized in Table 1. If the values of Q_{cc}^o and $\sigma_{Q_{cc}}$ for B^{10} are divided by 2.084 (the ratio of the quadrupole moments Q_{B10}/Q_{B11}), the values of Q_{cc}^o and $\sigma_{Q_{cc}}$ for B^{11} are obtained. Thus, only a single set of values for the quadrupole parameters is used to fit the B^{10} and B^{11} spectra separately. The large value of

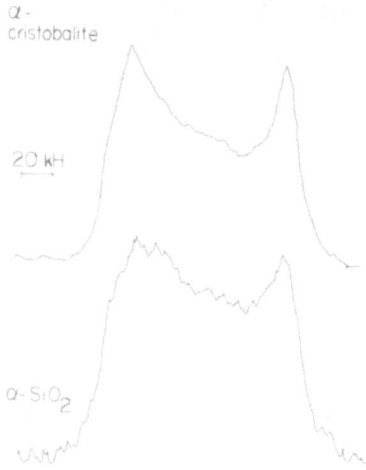

Fig. 2 O^{17} NMR spectra of SiO_2 (dispersion mode) at 8 MH_z.

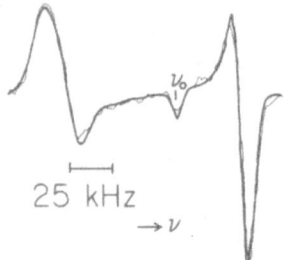

Fig. 3 Experimental B^{11} derivative and simulated spec-
trum of B_2O_3 glass. ν_0 =
8 MH_z.

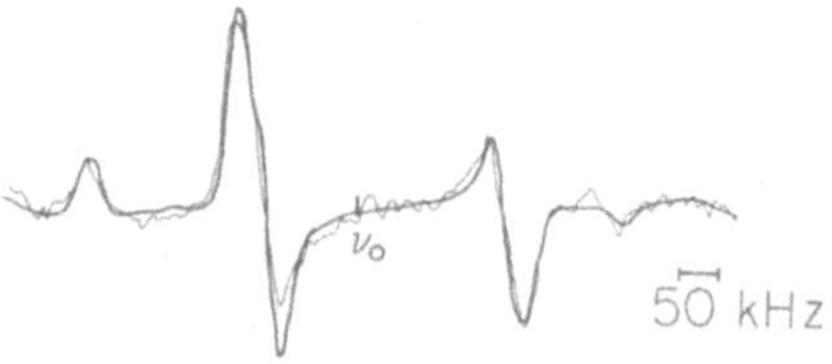

Fig. 4 B^{10} NMR derivative spectrum
and computer simulation of
B_2O_3 glass. ν_0 = 7 MH_z.

TABLE 1

Quadrupole parameters obtained for B^{10}, B^{11}, and O^{17} in B_2O_3 glass

	Q_{cc}^o (MH$_z$)	$\sigma_{Q_{cc}}$ (MH$_z$)	η^o	σ_η
B^{10}	5.51	0.21	0.12	0.043
B^{11}	2.64	0.10	0.12	0.043
O^{17} O(R)	4.69	0.10	0.58	0
O(C)	5.75	0	0.4	0.2

Q_{cc}^O indicates that the boron atoms in the glass are 3-coordinated. It should be noted here that the more detailed B^{10} spectrum permitted a better determination in the distribution of the quadrupole parameters.

The first derivative of the O^{17} NMR absorption spectrum for vitreous B_2O_3 prepared from water enriched to 37% O^{17} is presented in Figure 5. Second-order quadrupolar effects are responsible for the observed structure. A good fit to a computer simulation is obtained by invoking a two-site fitting, i.e., the simulation is a weighed sum of two spectra with two different sets of quadrupole parameters. This means that the oxygen atoms in the glass may be found in two different atomic environments. This is consistent with the boroxol model for B_2O_3 glass [12,13], in which oxygen atoms may be found in two distinct sites: a ring oxygen site $O(R)$, which is an oxygen atom contained in a boroxol ring, and a connecting oxygen site $O(C)$, which is an oxygen that is outside a boroxol ring. (See Figure 6). If the B_2O_3 glass is entirely made up of boroxol rings joined together by connecting oxygen atoms in a random network, then the ratio of $O(R)$ sites to $O(C)$ sites should be 2. Computer simulation yields the following values for the quadrupole parameters of the two sites. Site 1: Q_{cc}^O = 4.69 MHz, $\sigma_{Q_{cc}}$ = 0.10 MHz, η^O = 0.58, σ_η = 0. This is identified with the

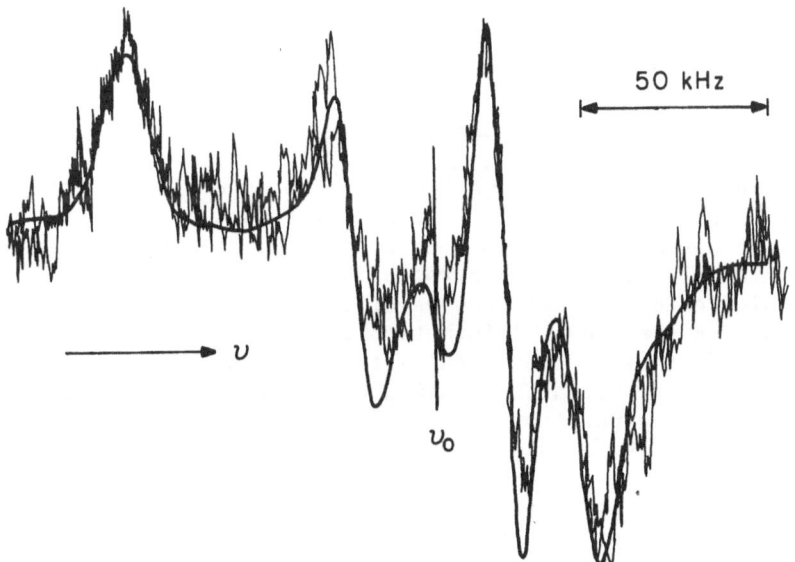

Fig. 5 O^{17} NMR derivative spectrum and computer simulation of B_2O_3 glass. ν_o = 7 MH$_z$.

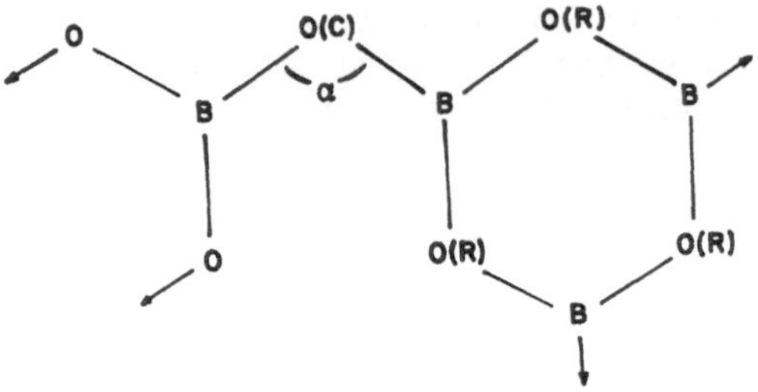

Fig. 6   Boroxol model for B_2O_3 glass showing the positions of the
         ring oxygen site O(R) and the connecting oxygen site O(C).

O(R) site in a boroxol ring because the boroxol rings are expected
to have little distortion in the rings and, consequently, only a
small distribution in the asymmetry parameter η.   Site 2: Q_{cc}^o = 5.75
MHz, $σ_{Q_{cc}}$ = 0, η = 0.4, $σ_η$ = 0.2.   This is identified with the O(C)
site outside the boroxol rings.   Since the connecting oxygens O(C)
will be subject to a distribution in the B-O-B bond angle, it is
expected that η for the O(C) will display a distribution in values.
A Townes-Dailey calculation using $σ_η$ = 0.2 indicates that the distribution in the bonding angle is about 1.7°.   The value of the weighting
factor of (Site 1)/(Site 2) is 1.2, considerably smaller than 2.
This indicates that some of the boroxol rings are linked not just by
connecting oxygens O(C), but that there may exist a variety of different combinations of BO_3 triangles linking some of the boroxol rings
The oxygen sites in these connecting triangles will be similar to the
O(C) site.

B^{10} NMR IN ALKALI-BORATE GLASSES

Both B^{11} and B^{10} NMR have been used to determine the structure
of alkali borate glasses. B^{11} NMR has been used to measure N_4, the
fraction of boron atoms in four-coordination, as a function of R,
the ratio of alkali oxide to boron oxide in the glasses [14,15].   As
has been indicated for the studies of vitreous B_2O_3, B^{10} NMR spectra
can yield more structural information about the glasses than B^{11}
NMR because features of the richly-structured B^{10} spectrum are
quite sensitive to variations in Q_{cc} and η, and the smaller B^{10}
dipolar broadening permits better resolution of the features.
Matching of computer-simulated and experimental spectra yields
quadrupole parameters which can be identified as arising from boron
atoms in particular structural groupings, and the relative abundances of these groupings can also be determined.

B^{10} NMR studies of solium borate glasses included glasses with
sodium oxide content up to 35 mol% [16] while those of lithium
borate glasses went up to 65 mol% lithium oxide content [17].   The
data from the two systems agree quite well. B^{10} NMR experimental

results for lithium borate glasses are presented in Figure 7 a and b.
The spectra are analyzed in general accordance with Krogh-Moe's
proposal [18] that the glasses of a system will contain the structural groupings present in the crystalline compounds of that system.
Figure 8 shows the structural groupings that are expected in the
lithium borate system. The symbols for the structural groupings are
defined as follows: boroxol unit, B; diborate unit, D; tetraborate
unit, T; metaborate unit, M; pyroborate unit, P; orthoborate unit,
O; and loose BO_4 unit, L. Many of the structural groupings have
boron atoms in both three and four coordination. A grouping symbol
with a superscript 3 or 4 will denote the fraction of all the
three- or four-coordinated boron atoms in the glass that are found
in that grouping. These boron fractions will be related through
the following constraints.

(a) Site constraints:

$$T^3 = 3T^4, \quad D^3 = D^4 \tag{1}$$

(See the tetraborate and diborate groups in Figure 8)

(b) Charge conservation:

The sum of positive charges on all the alkali ions in the glass
must equal the sum of all the negative charges on the structural
units:

$$R = D^4 + T^4 + L^4 + M^3 + 2P^3 + 3O^3 \tag{2}$$

(c) Boron conservation:

$$1 = B^3 + T^3 + T^4 + D^3 + D^4 + L^4 + M^3 + P^3 + O^3 \tag{3}$$

(d) N_4 constraint:

$$N_4 = T^4 + D^4 + L^4 \tag{4}$$

In addition, N_4 has been found to be related to R as shown in
Figure 9 by B^{11} NMR [15]. The data have been fit by three straight-
line segments given by:

$$N_4 = R \qquad \text{for} \qquad 0 \le R < 0.4 \tag{5a}$$

$$N_4 = \frac{1}{3} + \frac{R}{6} \qquad \text{for} \qquad 0.4 \le R < 0.7 \tag{5b}$$

$$N_4 = \frac{5}{8} - \frac{R}{4} \qquad \text{for} \qquad 0.7 \le R \le 1.86 \tag{5c}$$

A structural model [17], which is a refinement of previous
models [16,19], is constructed under these constraints.

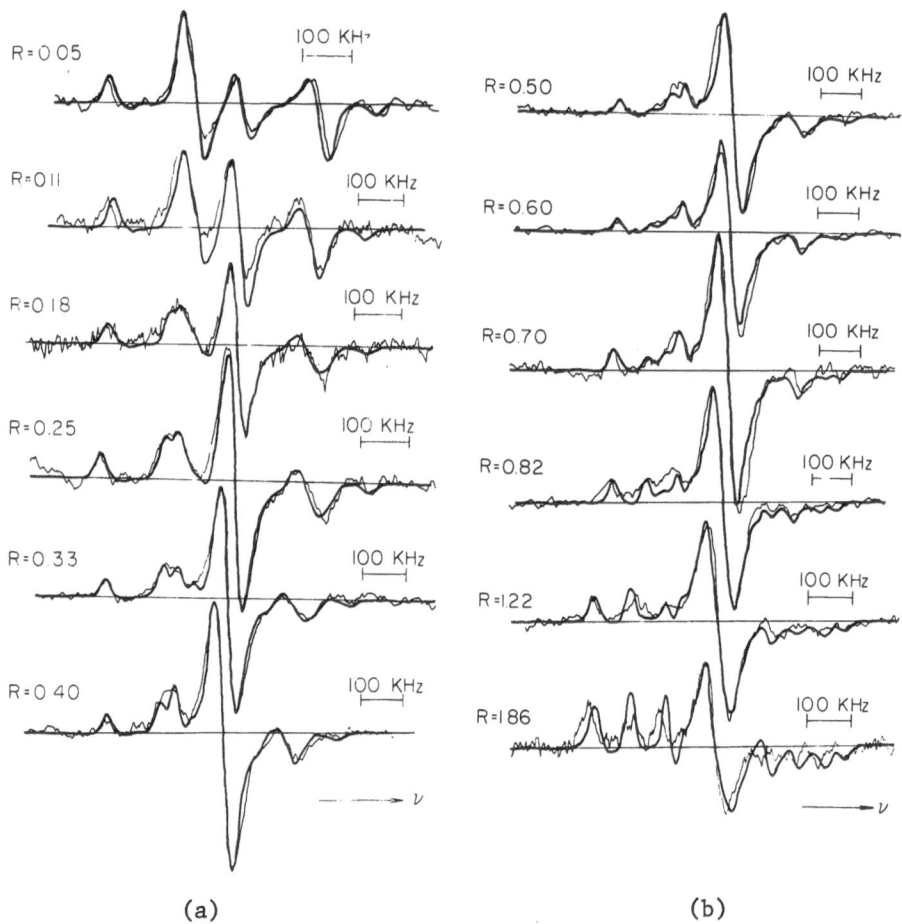

(a)   (b)

Fig. 7   Experimental B[10] NMR spectra for lithium borate glasses,
         where R = mol% Li_2O/mol% B_2O_3.   The smooth lines super-
         imposed on the spectra are computer-simulated lineshapes.

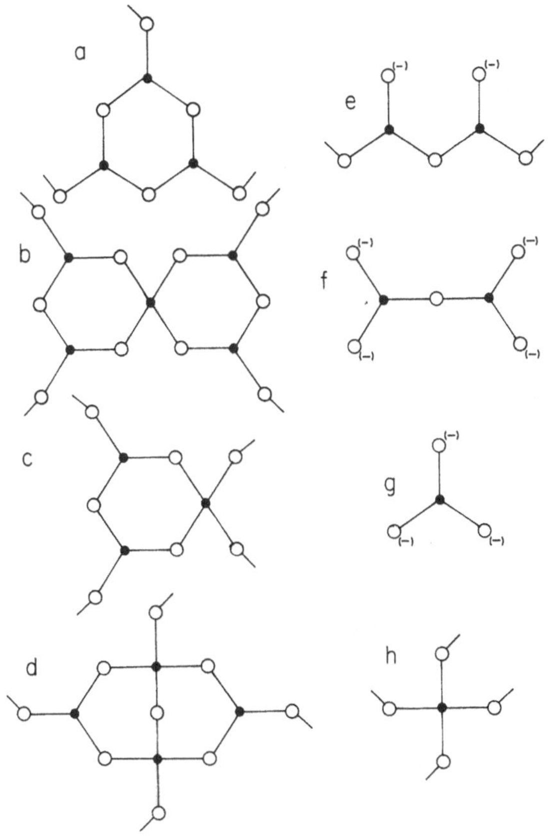

Fig. 8   Structural groupings in lithium borate glasses.   Solid
circles represent boron atoms, open circles oxygen atoms.
An open circle with a negative sign indicates a non-bridging
oxygen: (a) boroxol unit, (b) pentaborate unit, (c) triborate unit, (d) diborate unit, (e) metaborate unit, (f)
pyroborate unit, (g) orthoborate unit, (h) loose BO_4 unit.
A tetraborate unit is formed by connecting one oxygen of
the BO_4 configuration in the triborate unit to a BO_3 configuration of the pentaborate unit.

The glass-forming range is divided into three regions for
consideration, and structural groupings whose composition is far
removed from a given region are assumed not to exist in that region.
Then only one free parameter remains in Eqs. 1-4 when the constraints
(a) - (d) are applied to each of the three regions.   All the experimental spectra shown in Figure 7 appropriate for each region are
computer-simulated by varying the value of the only free parameter
for that region.   Overall, the fits achieved are quite good.   In
region I($0 \leq R \leq 0.4$) (see Figure 10) only boroxol, tetraborate,

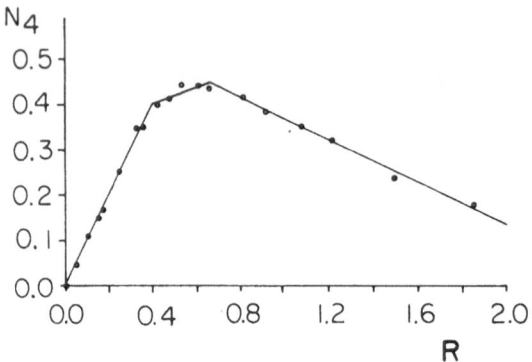

Fig. 9.   The fraction of four-coordinated borons in lithium borate
           glasses as determined by B^{11} NMR.

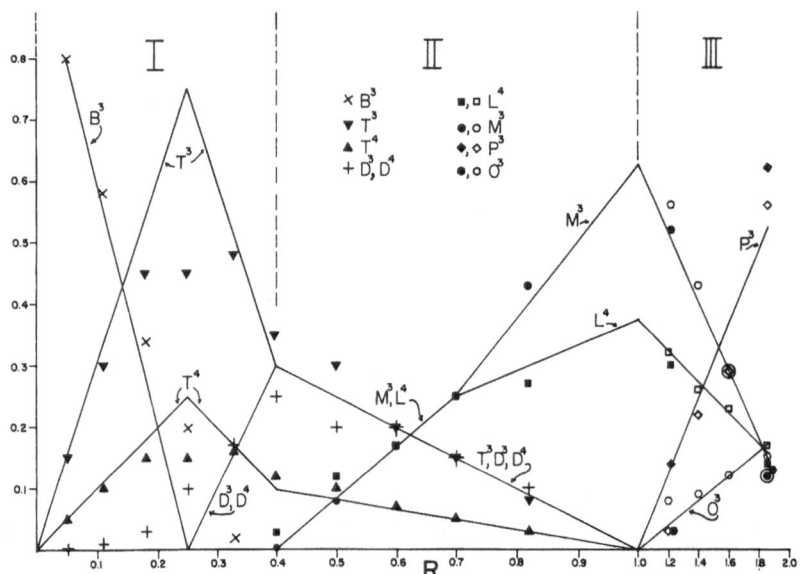

Fig. 10.   Weightings of the structural groupings used to calculate
           the computer-simulated lineshapes in Figure 7. The solid
           lines in Region I are from simple lever rule predictions;
           those in Region II and III are derived from the model
           summarized in the text.

and diborate units are assumed to exist. With the addition of Li_2O,
boroxol units are destroyed in favor of diborate units and tetraborate units. In region II ($0.4 \leq R < 1.0$), only diborate, tetraborate, metaborate, and loose BO_4 units are assumed to exist. Upon
addition of Li_2O, diborate and tetraborate units are proportionately

destroyed to form metaborate units and loose BO_4 units. In region
III ($1.0 \leq R \leq 1.86$), only metaborate, loose BO_4, pyroborate, and
orthoborate units are assumed to exist. Upon addition of Li_2O,
metaborate and loose BO_4 units are destroyed linearly to form pyroborate and orthoborate units. The relative abundances of the structural units used to calculate the computer-simulated lineshapes in
Figure 7 are plotted in Figure 10.

The sensitivity of computer simulation of B^{10} NMR spectra is
illustrated by data from the sodium borate system. Figure 11 displays for sodium borate glasses B^{10} NMR spectra similar to those of
Figure 7 for lithium borate glasses. Feature A of Figure 11 is shown
on a greatly expanded frequency scale in Figure 12. The computer
simulation (open circles) in Figure 12 is a "best fit" composed
entirely of a weighted sun of the spectra from the appropriate
structural groupings. Only one parameter is adjusted to determine
the relative weightings of the spectra from the various groupings.
For example, Figure 13 displays the significant changes in the
simulation caused by changing the parameter (which is δ, the excess
fraction of diborate units from the value predicted by the lever
rule) away from the "best fit" value of 0.08 to 0.06 and 0.10. The
departure from "best fit" for the latter case is clearly recognizable.

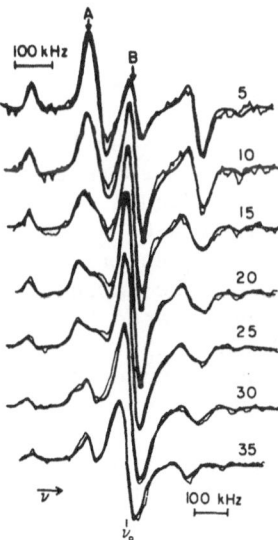

Fig. 11 B^{10} NMR spectra for sodium borate glasses. The molar %
          Na_2O content is indicated to the right of each trace. The
          darker line in each trace is a computer-simulated spectrum.

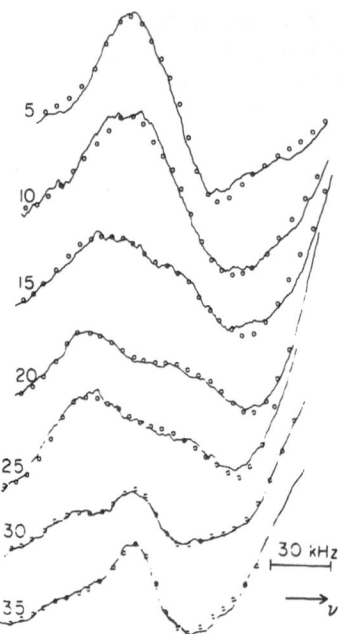

Fig. 12   Experimental B^{10} NMR spectra for the main feature (feature
A of Figure 11) for 3-coordinated boron atoms in sodium
borate glasses. The mol% Na_2O for each glass is given at
the left of each spectrum.

B^{11} NMR IN SODIUM BOROSILICATE GLASSES

Earlier NMR studies of this system [20,21] have been extended
to higher soda content [22,23]. A representative B^{11} NMR derivative
absorption spectrum is shown in Figure 14 for a K = 3, R = 2.5
sample, where K = mol% SiO_2/mol% B_2O_3 and R = mol% Na_2O/mol% B_2O_3.
Features of the spectrum can be identified with borons in three
different atomic environments. The narrow off-scale response
labeled (a) is due to four-coordinated borons which have small
values for the coupling constant Q_{cc}. Three-coordinated borons
have higher values for Q_{cc} and produce second-order features in the
spectrum. Features (b) are due to symmetric three-coordinated
borons (borons surrounded by either three bridging or three nonbridging oxygens) with low asymmetry parameter η. Features (c)
are due to asymmetric three-coordinated borons (borons with one or
two non-bridging oxygens) with high η. The fraction of four-
coordinated borons, N_4, is found by integrating the spectrum and
comparing the area of the integral due to feature (a) with the
total area. The fraction of boron in symmetric three-coordination,
N_{3S}, and the fraction of asymmetric three-coordinated borons, N_{3A},

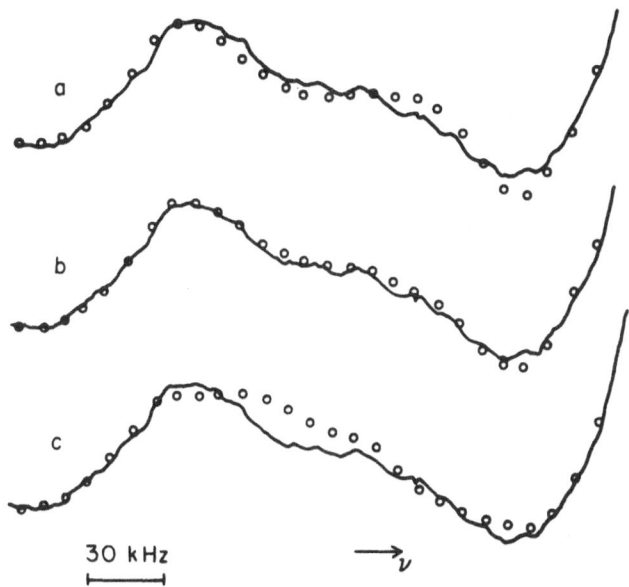

Fig. 13   A portion of the experimental B^{10} spectrum (feature A of
         Figure 11) for the sodium borate glass with 20 mol% Na_2O.
         (a), (b), and (c) show comparisons with computer-simulated
         spectra (open circles) using values of $\delta = 0.06$, 0.08, and
         0.10, respectively.

are obtained from computer simulation [24] of the features (b) and
(c). The experimental data for N_4 and N_{3A} are displayed in Figures
15 and 16, respectively. A structural model that fits the data is
suggested as follows:

    1) $R \leq 0.5$. In this region $N_4 = R$, the ternary system behaves
just like the binary alkali borate system. Diborate units are
formed with the addition of sodium oxide.

    2) $0.5 < R \leq 1/2 + 1/16$ K. In region, additional sodium oxide
combines with diborate units and silica tetrahedra to form reedmergnerite units [20]. A reedmergnerite unit is a boron tetrahedron with
each oxygen bridged to a silica tetrahedron. This process continues
until all the silica is used up at $R = R_{MAX} = 1/2 + 1/16$ K.

    3) $R_{MAX} < R \leq 1/2 + 1/4$ K. In this region the added sodium
oxide is used to form non-bridging oxygens on the reedmergnerite
units. This process ends at $R = R_{D1} = 1/2 + 1/4$ K.

    4) $R_{D1} < R \leq R_{D3} = 2 + K$. The added sodium oxide is shared
proportionately between diborate units and reedmergnerite units

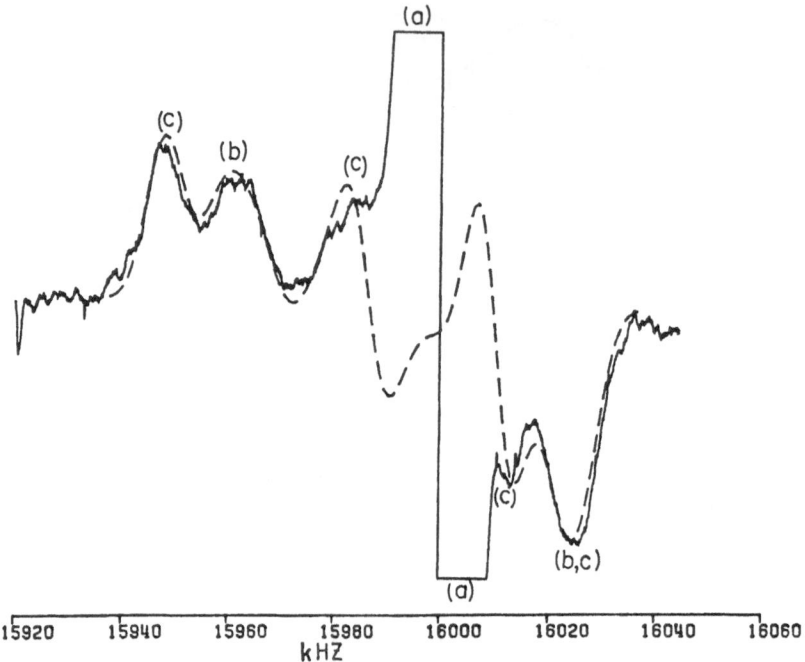

Fig. 14   Experimental B^{11} NMR spectrum for the K = 3, R - 2.5 sodium
borosilicate glass.   The dashed line is the computer-
simulated lineshape.

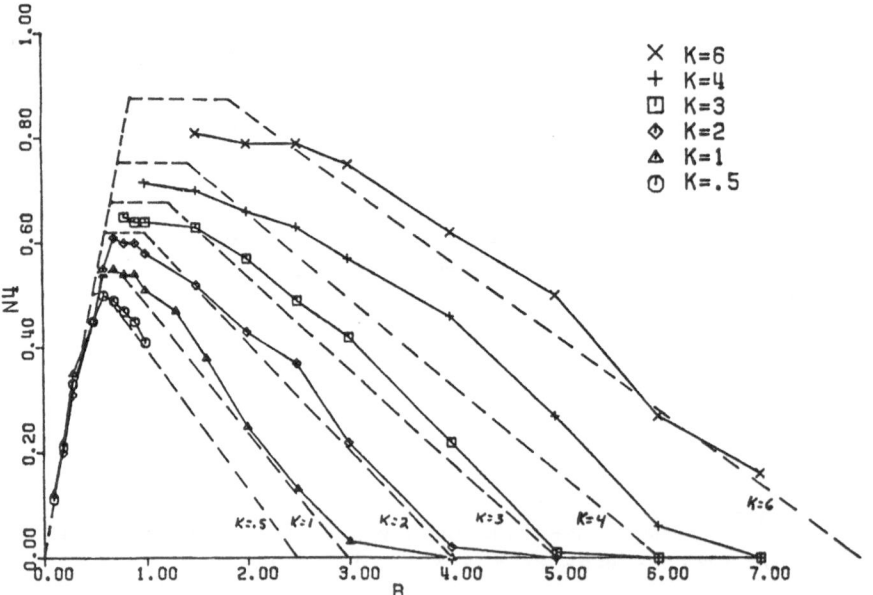

Fig. 15   Compositional dependence of N_4 for the sodium borosilicate
system.   Dashed lines are predictions based on the model
presented in the text.

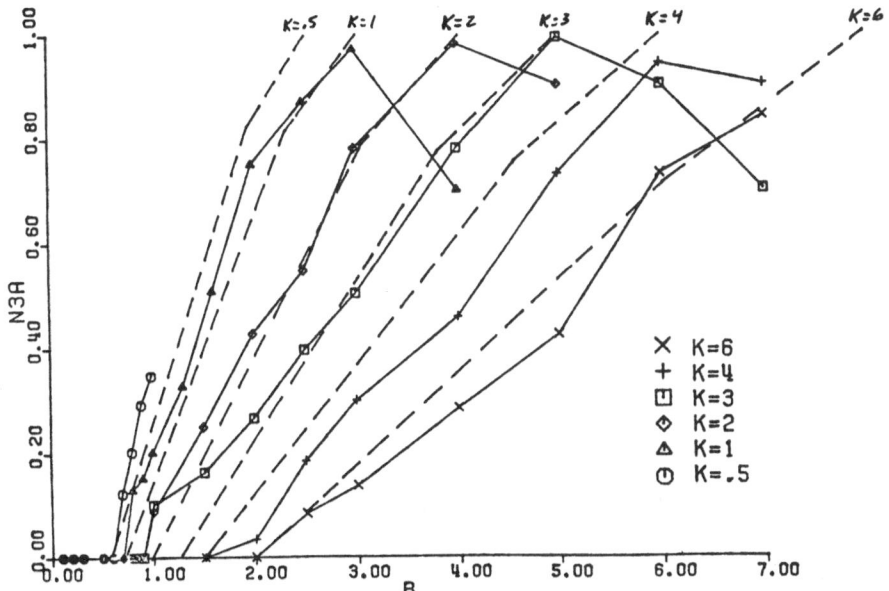

Fig. 16   Compositional dependence of N_{3A} for the sodium borosilicate
system. Dashed lines are predictions based on the model
presented in the text.

to form pyroborate units and silica tetrahedra with two non-bridging
oxygens per silicon. This process ends when both the diborate units
and the reedmergnerite units are exhausted at $R_{D3} = 2 + K$.

The compositional dependences of N_4 and N_{3A} predicted by this
model are plotted as dashed lines in Figures 15 and 16. The agreement between model and experiment is quite good. Quantitative
details of the model - including derivations of the expressions for
R_{MAX}, R_{D1}, and R_{D3} - are given in references [22] and [23].

B^{11} NMR AND LITHIUM BOROSILICATE GLASSES

The B^{11} NMR studies of alkali borosilicate glasses have recently
been extended [25] to glasses of the system $Li_2O-B_2O_3-SiO_2$. Figure
17 presents initial data for N_4 as a function of K and R where R =
mol% Li_2O/mol% B_2O_3. Comparison of Figures 15 and 17 reveals striking differences in the dependence of N_4 in R for the two systems
above R $\sim$ 1/2. This difference may be associated with the separate
presence of pentaborate and triborate groupings (Figure 8) in the
glasses containing Li_2O; these two groupings are believed to be
linked together to form tetraborate groupings in the Na_2O-based
glasses. Additional studies of ternary borosilicate glasses containing various alkali and alkaline earth oxides are in progress.

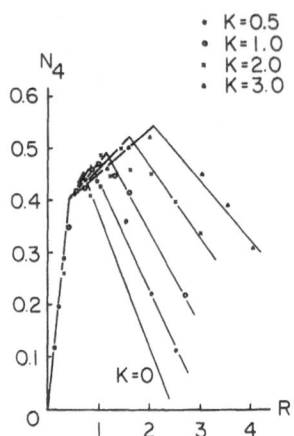

Fig. 17   Dependence of N_4 on R and K for lithium borosilicate
          glasses. The solid lines are least-squares fits to the
          the data.

REFERENCES

1. D. L. Griscom, in <u>Borate Glasses: Structure, Properties, Appli-
    cations</u>, eds. L. D. Pye, V. D. Frechette, and N. J. Kreidl,
    Plenum Press, New York (1978).
2. W. Muller-Warmuth and H. Eckert, Physics Reports, 88:92 (1982).
3. P. J. Bray, F. Bucholtz, A. E. Geissberger and I. A. Harris,
    Nuc. Inst. and Methods 199:1 (1982).
4. P. J. Bray, A. E. Geissberger, F. Bucholtz and I. A. Harris,
    J. Non-Cryst. Solids, 52:45 (1982).
5. A. H. Silver and P. J. Bray, J. Chem. Phys. 29:984 (1958).
6. G. E. Jellison, Jr., S. A. Feller, and P. J. Bray, J. Mag. Res.,
    27:121 (1977).
7. J. A. Abys, D. M. Barnes, S. Feller, G. Rouse and W. M. Risen,
    Jr., Mater. Res. Bull., 15:1581 (1980).
8. A. E. Geissberger and P. J. Bray, J. Non-Cryst. Solids, 54:121
    (1983).
9. R. L. Mozzi and B. E. Warren, J. Appl. Cryst., 2:164 (1969).
10. D. E. O'Reilly, J. Chem. Phys., 28:1262 (1958).
11. G. E. Jellison, Jr., L. W. Panek, P. J. Bray, and G. B. Rouse,
    Jr., 66:802 (1977).
12. J. Krogh-Moe, Phys. Chem. Glass, 6:46 (1965).
13. R. L. Mozzi and B. E. Warren, J. Appl. Cryst., 3:251 (1970).
14. P. J. Bray and J. G. O'Keefe, Phys. Chem. Glasses, 4:37 (1963).
15. G. E. Jellison, Jr., S. A. Feller and P. J. Bray, Phys. Chem.
    Glasses, 19:52 (1978).
16. G. E. Jellison, Jr., and P. J. Bray, J. Non-Cryst. Solids,
    29:187 (1978).
17. S. A. Feller, W. J. Dell and P. J. Bray, J. Non-Cryst. Solids,
    51:21 (1982).

18. J. Krogh-Moe, Phys. Chem. Glass, 6:46 (1965).
19. Y. H. Yun and P. J. Bray, J. Non-Cryst. Solids, 44:227 (1981).
20. Y. H. Yun and P. J. Bray, J. Non-Cryst. Solids, 27:363 (1978).
21. Y. H. Yun, S. A. Feller, and P. J. Bray, J. Non-Cryst. Solids, 33:273 (1979).
22. W. J. Dell, Ph.D. Thesis, Brown University, 1983.
23. W. J. Dell, P. J. Bray and S. Z. Xiao, B^{11} NMR Studies and Structural Modeling of $Na_2O-B_2O_3-SiO_2$ Glasses of High Soda Content, to be published in J. Non-Cryst. Solids (1983).
24. P. C. Taylor and P. J. Bray, J. Mag. Res., 2:305 (1970).
25. Xue-Wen Wu, M. L. Lui and P. J. Bray, to be published.

MAGIC-ANGLE SAMPLE-SPINNING NUCLEAR MAGNETIC RESONANCE SPECTROSCOPY

OF SILICATE GLASSES: A REVIEW

R. James Kirkpatrick[1], Todd Dunn[1,3], Suzanne Schramm[2], Karen Ann Smith[2], Richard Oestrike[1], and Gary Turner[2]

[1]Department of Geology, 1301 W. Green Street, Urbana, IL 61801; [2]School of Chemical Sciences, University of Illinois, 505 S. Mathews Street, Urbana, IL 61801 [3]Present address: Atomic Energy of Canada, Pinawa Man., ROE 1LO, Canada

INTRODUCTION

Nuclear magnetic resonance spectroscopy (NMR) using magnetic-angle sample-spinning (MASS) methods is a relatively new technique for examining the structural environment of atoms in both crystalline and amorphous solids.   In this paper we will briefly review the theory of MASS NMR and some of the data for crystalline silicates and silicate glasses. We will demonstrate that MASS NMR using various nuclides (primarily silicon-29, aluminum-27, and oxygen-17) can 1) determine the range of Si-O-Si bond angles in SiO_2 glass, 2) determine the presence or absence of Al(4) and Al(6) in aluminosilicate glasses, 3) estimate the range of polymerization states of SiO_4 tetrahedra in silicate glasses, 4) determine the presence of different types of oxygens in $CaO-MgO-SiO_2$ and Pb_2SiO_4 glasses, and 5) examine the structural similarities and differences between Pb_2SiO_4 glass and the various Pb_2SiO_4 crystalline polymorphs. Using the available data, we hope to illustrate both the potential application and some of the limitations of MASS NMR in examing amorphous solids.

THEORY

An elementary but complete introduction to NMR theory is given by Davis [1]. More advanced treatments are given by Farrar and Becker [2], Becker [3], and Akitt [4]. We will present here only a brief introduction using both the classical and quantum descriptions of the NMR phenomenon.

303

In the classical treatment, the NMR phenomenon is considered to arise from the interaction of the dipole moment of the nucleus, μ, and the applied magnetic field H_o. Because there is torque on the magnetic dipole, it precesses about H_o with a frequency, υ, given by

$$\upsilon = \frac{\gamma}{2\pi} \, H_o \qquad\qquad (1)$$

where γ is the magnetogyric ratio and is a property of the nucleus. υ is the Larmor frequency and for NMR is in the radio frequency range. The Larmor frequency can be measured by the absorption of radio frequency radiation oscillating in the plane perpendicular to the orientation of the H_o applied field.

In the quantum mechanical description of the NMR phenomenon, the nucleus is considered to have a quantized property called spin, I. The spin quantum number, m, assumes values from -I to I, giving $2I+1$ energy levels. I may take on integral and half-integral values. Most nuclides have spins in the range $I=0$ to $7/2$. Those with spin $I=0$ are not magnetically active. Those with $I=1/2$ behave as magnetic dipoles, and those with $I \geq 1$ as magnetic quadrupoles.

The energy levels of a nucleus are given by

$$E_m = \gamma h m H_o \qquad\qquad (2)$$

A nucleus in a lower energy spin state can be raised to a higher energy spin state by the absorption of a photon with an energy equal to the energy difference between the energy levels of the nucleus. The frequency of this radiation is given by

$$\upsilon = \frac{\gamma}{2\pi} \, H_o \qquad\qquad (3)$$

which is exactly the value of the Larmor frequency from classical theory.

Modern pulse-Fourier transform NMR spectrometers determine υ by detecting the radio frequency emission from a sample in which the population of nuclei in the higher energy spin states has been raised by absorption of a few microsecond-long burst of radio-frequency radiation. The emitted radiation is detected in the time domain and the signal is then Fourier transformed into the frequency domain to yield the spectrum. A much higher signal-to-noise ratio is obtained with this method than with the older continuous wave methods, which have also been used to examine glass structure [5], because the signals from up to several thousand pulses are added together.

NMR spectroscopy is chemically useful, because the electrons in the vicinity of the nucleus shield it from a small part of the applied magnetic field. This diamagnetic portion of the shielding arises from the motion of the electrons in the vicinity of the nucleus, which gives rise to a magnetic field which partially counteracts the applied field. In a molecule or crystal, some of the electrons are involved in chemical bonding, which restricts their motion and reduces their shielding effect. The latter is the paramagnetic portion of the shielding. Thus, a bare nucleus should be unshielded, an unbound atom the most shielded, and a bound atom shielded to an intermediate amount depending on the local structural and compositional environment. The shielding is described by the shielding tensor, with the isotopic shielding σ_i, given by $1/3 \ Tr(\sigma_{11} + \sigma_{22} + \sigma_{33})$, where the σ_{jj} are the diagonal elements of the shielding tensor.

Because the absolute value of the resonance frequency is difficult to measure accurately enough to detect the effect of shielding, the NMR data are reported as the parts per million (ppm) difference between the resonance frequency of the sample and that of an experimentally practical standard, tetramethylsilane in the case of silicon-29. This difference is called the chemical shift and is the primary NMR parameter discussed in this paper.

In a solid at rest there are a wide variety of phenomena which can cause the resonance frequency of an individual nucleus to be different than the isotropic value. These interactions include the chemical shift anisotropy (due to the tensor properties of the shielding), dipole-dipole interactions between nuclei, electron-coupled interactions (J-coupling), and, for nuclei with spin $I>1$, interactions between the magnetic quadrupole moment of the nucleus and any anisotropic electric field gradient at the nucleus. These interactions are best treated as perturbations to the Zeeman interaction described above.

Because of these interactions, the nuclei of a particular nuclide in a solid sample will have a wide range of energy levels, giving rise to very broad peaks, which provide little information. In liquids, molecular motion occurs faster than the Larmor frequency and averages the interactions, giving rise to very narrow peaks and allowing determination of the chemical shifts to an accuracy of 0.1 ppm. Thus, NMR found its first chemical application to the study of molecules in solution.

In solids, the peak-broadening effects of the chemical shift anisotropy and dipole-dipole interaction as well as the symmetric part of the electron-coupled interaction and the first order effect of the electron-quadrupole interaction can be removed by mechanically spinning the sample at frequencies of the order of the peak breadth in frequency units (usually in the kilohertz

range) at an angle of 54.7° to the orientation of H_O applied
magnetic field. 54.7° is the magic-angle and this process is called
magic-angle sample-spinning (MASS). The antisymmetric part of the
electron-quadrupole interaction appears to be small for most solids
and will be ignored here. Second-order quadrupolar interactions
can be significant and will be discussed below.

The quantum mechanical basis for MASS has been discussed by
Andrews [6,7]. We will note here only that the Hamiltonians for
the interactions that are averaged contain terms of $(3\cos^2\theta-1)$,
where θ is the angle of rotation relative to the orientation of
the H_O magnetic field. At 54.7° these terms reduce to zero.

Figure 1A [8] illustrates the energy levels, with Zeeman and
chemical shift anisotropy contributions, for a spin I=1/2 nuclide
such as silicon-29. Figure 1B shows the broad static spectrum
of silicon-29 in crystalline sodium disilicate. On MASS (Figure 1C)
the broad peak breaks up into a true resonance (CB=center band) and
spinning side-bands (SSB), which arise from the time dependent terms
in the various Hamiltonians. These spinning side-bands move outward from the true resonance and become smaller as the spinning
speed increases. Their spacing in frequency units is at the
spinning speed. The center band is usually larger than the side-
bands, although here it is not. Spin I=1/2 nuclides have no quadrupolar moments and thus do not suffer from second order quadrupolar line broadening. The peaks are, therefore, quite narrow and
do not change position with the magnitude of the applied field.

Figure 1D shows a nuclear energy level diagram for a spin
I=3/2 nuclide, such as sodium-23, including the Zeeman and second-
order quadrupolar interactions. There are four energy levels and
three allowed transitions, each with a different energy difference
and different frequency. Figure 1E shows the contributions to a
static spectrum from each allowed transition. On MASS this spectrum
breaks up into a center band and side-bands. Until recently it was
thought that quadrupolar nuclei could not be effectively examined
in solids because the peak breadth is hundreds of kilohertz, far
wider than samples can be spun. Recently, however, it has been
shown that it is possible to monitor only the center (1/2, -1/2)
transition for half-integral nuclides (including boron-11,
oxygen-17, sodium-23 and aluminum-27). The theoretical and
experimental aspects of solid-state NMR of quadrupolar nuclides
have been described by Abragam [9], Kundla, et al. [10], Meadows,
et al. [11], Ganapathy, et al. [12], and Schramm and Oldfield [13].
Figure 1G shows the peak for the (1/2, -1/2) transition of
sodium-23 in Na_2MoO_4. Note that the peak breadth in kHz is about
an order of magnitude less than that of the full spectrum.
Although there is only one resonance, the peak is broadened and
split by a large second-order quadrupolar interaction. The shapes
of the peaks for quadrupolar nuclei can be quite complex [12].

Under MASS the shapes and breadths of the peaks are controlled by the asymmetry parameter, which is a measure of the deviation of the electric field gradient at the nucleus from cylindrical symmetry, and the quadrupole coupling constant (e^2qQ/h).

When quadrupolar interaction is the dominant peak broadening mechanism, spectral resolution increases as $(H_0)^2$. Thus, the largest possible H_0 field should be used. For quadrupolar nuclides with large e^2qQ/h, the peak maximum does not correspond to the isotropic chemical shift, and simultaion of the spectrum is needed to obtain these values. At the highest field presently available commercially (11.7T), however, the peaks for sites with $e^2qQ/h \leq 2kHz$ are within about 1 to 5 ppm of the isotropic values.

For quadrupolar nuclei, better peak narrowing can often be obtained by spinning at an angle other than the magic-angle if second-order quadrupolar effects are the dominant line-broadening mechanism [12]. Thus, for sodium-23 in Na_2MoO_4 the best narrowing is obtained with a spinning angle of 75° (Figure 1H).

MASS NMR OF CRYSTALLINE SILICATES

It is not presently possible to calculate NMR chemical shifts for known or assumed structures. MASS NMR can, however, yield information about the structures of poorly known substances, such as glasses, by comparison with MASS NMR data for well-known crystalline materials. We will, thus, briefly review the MASS NMR data for silicate crystals.

Silicon-29

The first systematic examination of the silicon-29 MASS NMR behavior of crystalline silicates was by Lippmaa, et al. [14,15]. Since then, there have been a large number of studies of the silicon-29 MASS NMR behavior of zeolites, clays, and natural minerals. Much of this work has been reviewed by Thomas, et al. [16], Smith, et al. [17], Kirkpatrick, et al. [8,18], and Magi, et al. [19]. The following discussion is taken from these papers. Figure 2 presents typical silicon-29 MASS NMR spectra of crystalline silicates. Note that the full-width at half-height (FWHH) for each peak is typically 1-4 ppm for aluminum-free silicates or Al/Si ordered aluminosilicates.

One of the most important sets of observations is that the silicon-29 chemical shift becomes more shielded (more negative) as the polymerization of the phase increases (Figures 3A and 3B). The ranges of silicon-29 chemical shifts for each polymerization type are best resolved for silicates containing no aluminum (Figure 3B). For framework silicates and sheet silicates, the silicon-29 chemical shift becomes less shielded (less negative)

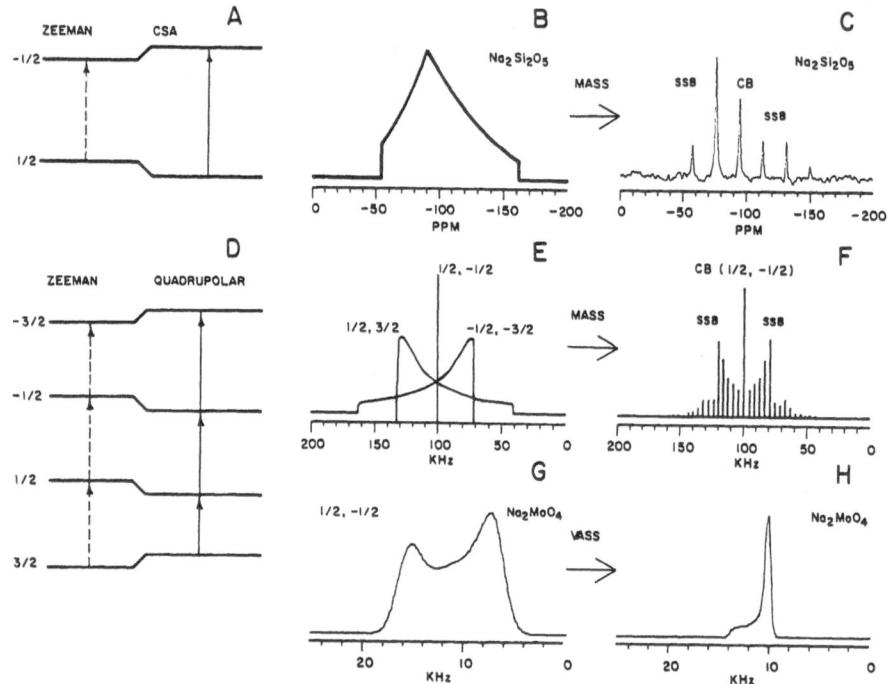

Fig. 1   A) Energy level diagram for an I=1/2 nuclide.   B) Static
        silicon-29 (1=1/2) spectrum for sodium disilicate.   C)
        Silicon-29 MASS NMR spectrum of sodium disilicate (SSB –
        spinning sideband, CB – center band).   D) Energy level
        diagram for an I=3/2 nuclide, including the Zeeman and
        quadrupolar interactions.   E) Static spectrum of an I=3/2
        nuclide, showing the contributions from the three allowed
        transitions.   F) MASS spectrum of an I=3/2 nuclide, showing
        sideband development.   G) NMR spectrum of the (1/2, – 1/2)
        transition for sodium-23 in Na_2MoO_4 at a spinning angle of
        54.7° (magic-angle).   H) Same as (G), but at spinning angle
        of 75° (VASS).

as the number of next-nearest neighbor (NNN) tetrahedral aluminum
atoms increases (Figs. 3C and 3D). Thus, it appears that the silicon-
29 chemical shift is very sensitive to NNN effects.

    One important result of this is that for Al/Si disordered
aluminosilicates it is possible to resolve signals from silicon atoms
with different numbers of aluminum NNN in the same phase. For many
zeolite catalysts and sorbant materials, it is possible to resolve
five different types of silicon sites, Q^4(0Al-4Al), even though
crystallographically there is only one silicon site [16].

    There have been several attempts to correlate the silicon-29
chemical shift to structural and bonding parameters. Figure 4 [20]

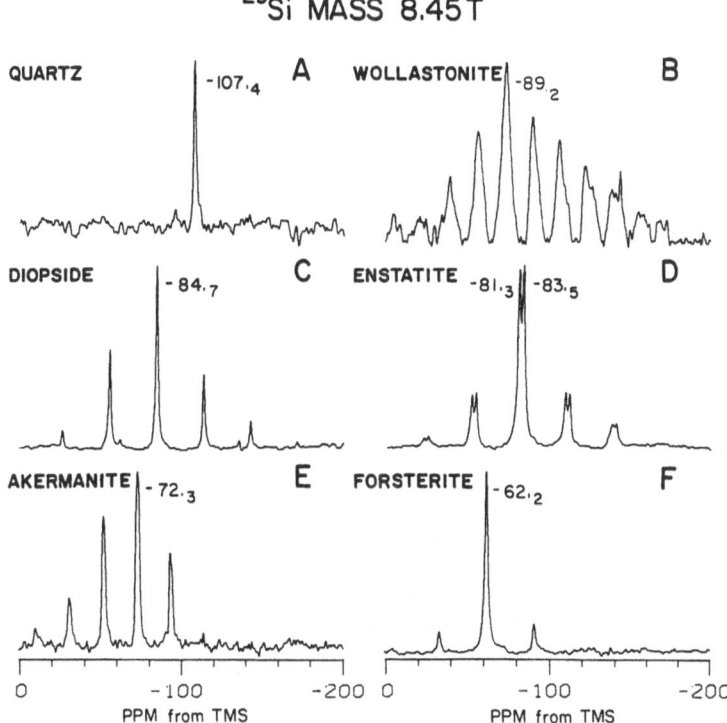

Fig. 2 Silicon-29 MASS NMR spectra for crystalline silicates
A) Quartz, SiO_2. B) Wollastonite, $CaSiO_3$, C) Diopside,
$CaMgSi_2O_6$, D) Clinoenstatite, $MgSiO_3$, E) Akermanite,
$Ca_2MgSi_2O_7$, F) Forsterite, Mg_2SiO_4.

presents the correlation with mean Si-O bond distance, mean secant
of the Si-O-T bond angle (T=Si or Al), and the sum of the total
Brown and Shannon [21] bond strengths to the four oxygens coordinated to the silicon of interest. Least squares fits of these
various correlations have been given by J. V. Smith and Blackwell
[22], J. V. Smith, et al. [23], and K. A. Smith, et al. [20]. The
very best correlations are for the framework silicates with the
secant of the mean Si-O-T bond angle. The best overall correlation
for all crystalline silicates is with the bond strength sum. Except
for the SiO_2 polymorphs, mean Si-O bond distance does not correlate
well with silicon-29 chemical shift. Using these correlations, it
is possible to obtain fair to excellent estimates of these bonding
parameters for materials with unknown structures.

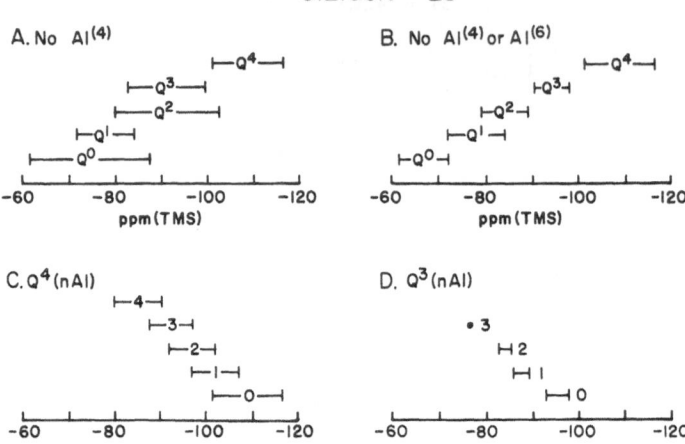

Fig. 3   Ranges of silicon-29 chemical shifts for A) silicates with
         no tetrahedrally coordinated Al, B) silicates with no
         aluminum, C) framework silicates with different members of
         NNN Al atoms to silicon, D) sheet silicates with different
         numbers of NNN tetrahedral Al atoms to silicon.

Aluminum-27

The aluminum-27 MASS NMR behavior of cyrstalline silicates has
also been extensively investigated. Interpretation of the spectra,
however, is complicated by its quadrupolar nature.

One of the major results of aluminum-27 MASS NMR is that it is
relatively easy to distinguish tetrahedrally coordinated from octahedrally coordinated aluminum [24,25]. At a H_o magnetic field
strength of 11.7 Tesla, the chemical shifts of tetrahedrally coordinated aluminum fall into the range +50 to +80 ppm, while those
of octahedrally coordinated aluminum fall into the range -10 to
+20 ppm. Figure 5 presents aluminum-27 MASS NMR spectra of typical
crystalline alumino-silicates.

Because of the quadrupolar behavior of aluminum-27, however,
signal from sites with large quadrupole coupling constants can be
lost. Large quadrupole coupling constants cause very broad peaks
which cannot be narrowed by MASS. For the Al_2SiO_5 polymorphs,
for instance, signal for the octahedral site of andalusite cannot
be resolved because of this problem. Large quadrupole coupling
constants appear to be due to large electric field gradients at
the nucleus, which are due to very distroted sites. In glasses,
we expect that structural and chemical disorder will cause a
range of electronic environments and, therefore, a range of a
quadrupole coupling constants for aluminum-27. Thus, in some

NESO-, SORO-, INO- and PHYLLOSILICATES

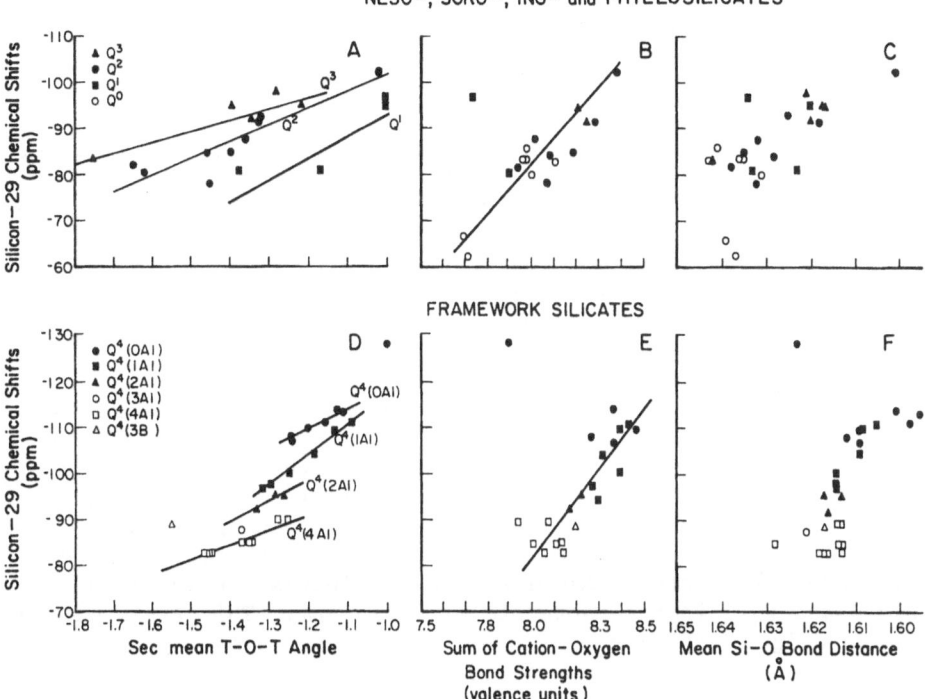

Fig. 4   Correlations of silicon-29 chemical shift with mean Si-O-T
         bond angle, cation-oxygen bond strength sum, and mean Si-O
         bond distance.

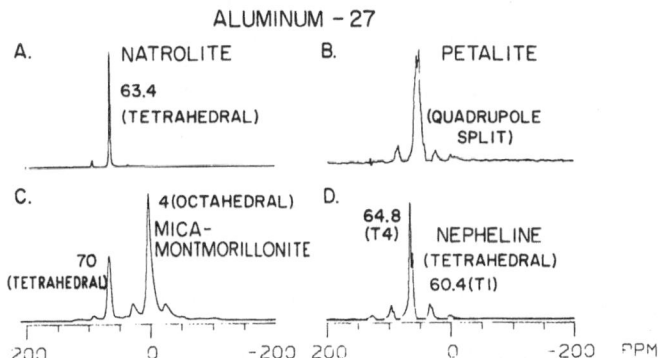

Fig. 5   Aluminum-27 MASS NMR spectra of crystalline aluminosilicates.
         A) Natrolite, a zeolite, B) Petalite, $LiAlSi_4O_{10}$, C) Mica-
         montmorillonite, a synthetic clay, as D) Nepheline,
         $(Na,K)AlSiO_4$.

cases some signal for quadrupolar nuclides in glasses may be lost.
For some types of sites, it is possible that all individual sites
are so distorted that they all have very large quadrupole coupling
constants, and therefore, yield no detectable signal.

The variations in aluminum-27 chemical shift with structure or
composition for tetrahedral and octahedral aluminum do not appear
to be as systemmatic as for silicon-29. Part of the problem may be
that because of quadrupolar behavior the peak maxima are displaced
a few ppm to less positive values from the isotropic values, even
for quite narrow peaks at 11.7 Tesla [18]. For tetrahedral aluminum
in framework silicates and aluminates and in sheet silicates there
appear to be trends of more positive chemical shift (less shielding)
with decreasing $Si/(Si + Al(IV))$. The ranges of aluminum-27 chemical shifts at 11.7 Tesla for framework and sheet silicates are well
defined and range from about +50 to +65 ppm for framework silicates
and from about +65 to +75 ppm for sheet silicates [18]. For the
position of the peak maxima of octahedral aluminum there appears
to be a trend of decreasing shielding with increasing polymerization
of the crystal structure from about -7 ppm for orthosilicates to +4
ppm for sheet silicates. Aluminas and aluminates are the least
shielded and fall in the range of +8 to +15 ppm. This trend may be
due to quadrupole effects.

Oxygen-17

Although oxygen-17 MASS NMR is potentially very useful in
examining solids, less is known about its behavior than silicon-29
and aluminum-27. This is primarily because it has a very low
natural abundance (0.04%) and must be examined in isotopically
enriched samples. Schramm, et al. [26] have shown that peaks for
bridging and nonbridging oxygens in crystalline silicates can be
resolved. Schramm and Oldfield [13] have shown that peaks for the
three nonbridging oxygens in forsterite (Mg_2SiO_4) can be resolved
and that simulation of the spectra can yield quadrupole coupling
constants and asymmetry parameters for each site. Oxygen-17 MASS
NMR spectra of crystal are shown below in conjunction with similar
spectra of glasses.

GLASSES

SiO$_2$ Glass

Silica glass, because of its compositional simplicity, is the
starting point for many discussions of silicate glass structure
and has been investigated using many methods.

Figure 6 presents silicon-29 MASS NMR spectra of a silicon-29
enriched glass made by fusion at about 1850°C for about 2 hours,
a silica gel dehydrated at 800°C for two hours, and simulations
of the spectra assuming Gaussian peak shapes. The spectrum of the

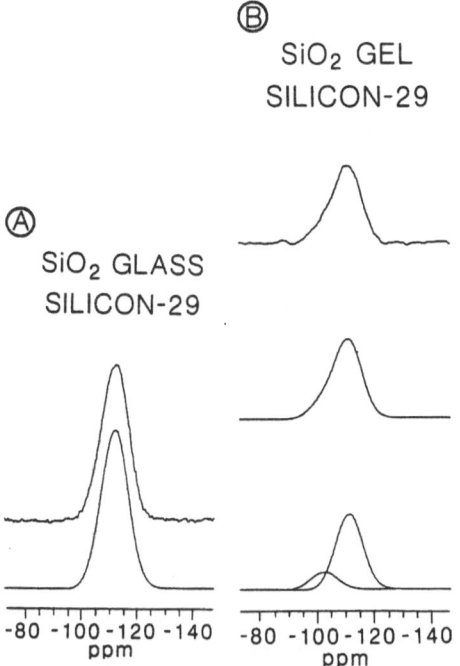

Fig. 6 Silicon-29 MASS NMR spectra and spectral simulations for
A) a silicon-29 enriched SiO_2 glass, B) a silica gel.

glass can be well fit with a single peak at -112.1 ppm and that of
the gel with two peaks, one at -112.1 ppm, the other at -103.3 ppm.
We interpret the single, broad peak for the glass to be due to Q^4
silicon sites with a range of mean Si-O-Si bond angles and a concomitant range of Si-O bond distances. We interpret the two peaks
of the gel to be due to Q^4 silicon sites (the large peak at -112.2
ppm) and Q^3 silicon sites with one OH group (the smaller peak at
-103.3 ppm). Silicon-29 MASS with cross-polarization to protons
(work in progress) greatly enhances the peak at -103.3 ppm,
clearly indicating that the silicon atoms giving rise to this peak
are bonded to an OH group. All these peaks are broadened relative
to crystalline silicates, because of a range of mean Si-O-Si bond
angles and mean Si-O bond distances. For SiO_2 glass the correlations of J. V. Smith and Blackwell [22] and K. A. Smith, et al. [20]
yield a mean Si-O-Si bond angle of 151°, with a range from 130°
to 180°, and a mean Si-O bond distance of 1.60 Å, with a range
from 1.56 Å to 1.66 Å. For the Q^4 peak of the gel, the mean Si-O-Si
bond angle is 151°, with a range from 136° to 180° and the mean
Si-O bond distance is 1.60 Å, with a range from 1.56 Å to 1.64 Å.

Dupree and Pettifer [27] and Murdoch, et al. [28] have also
published silicon-29 MASS NMR spectra of SiO_2 glass. Their samples,

however, were apparently commercially available glass with silicon-
29 at its 4.9% natural abundance. Silicon-29 spectra of very
silica-rich, anhydrous glasses with natural abundance silicon are
very difficult to obtain, primarily because of a very long relaxation time for the silicon-29 nucleus. The spectra in both papers
are quite noisy and appear to have been extensively smoothed
numerically. Because of potential differences is samples, the
numerical smoothing, and the difficulty in obtaining high signal-
to-noise spectra, the peaks in both of their spectra are considerably
broader than in ours. The full-width at half-height (FWHH) is 11.6
ppm for the spectrum of our glass, 13.5 ppm for that of Dupree and
Pettifer, and 14.9 ppm for that of Murdoch, et al. Figure 7 shows
our observed spectrum for SiO_2 glass, the observed spectrum of
Dupree and Pettifer, and the spectra calculated by Dupree and
Pettifer from x-ray radial distribution study and models of SiO_2
glass. The peak maximum of our spectrum is between that of the model
of Mitra [29] and that of the x-ray radial distribution analysis of
Mozzi and Warren [30]. The upper and lower limits, however, are
very close to those of Mitra. The model of Evans and Teter [3] compares least well.

        Recent Raman spectroscopy [32,33] as well as older physical
property data [34,35,36,37] indicates strongly that non-framework
silicate glasses are composed of SiO_2 tetrahedra with different
numbers of nonbridging oxygens. Silicon-29 MASS NMR data for
glasses in the system $CaO-MgO-SiO_2$ are consistent with these ideas
but cannot resolve individual peaks for the different types of
silicon sites [28]. Figure 8 shows silicon-29 spectra for glasses
along the join $SiO_2-CaMgSiO_4$ along with that of $Ca_2MgSi_2O_7$ glass.

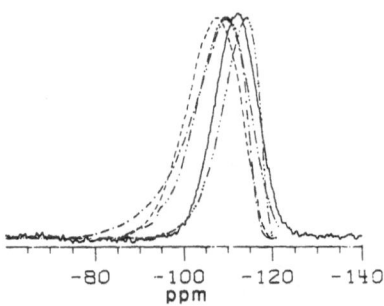

Fig. 7   Comparison of the silicon-29 spectrum of SiO_2 glass in
         Fig. 6a (solid line) with the observed silicon-29 spectrum
         for a natural abundance glass [27] (−·− line) and the
         spectra calculated from the models of Evans and Teter [31]
         (−·− line) and Mitra [29] (−····− line) and the x-ray
         radial distribution data of Mozzi and Warren [30] (−−− line),
         (calculations from Dupree and Pettifer [27]).

^{29}Si MASS 8.45T

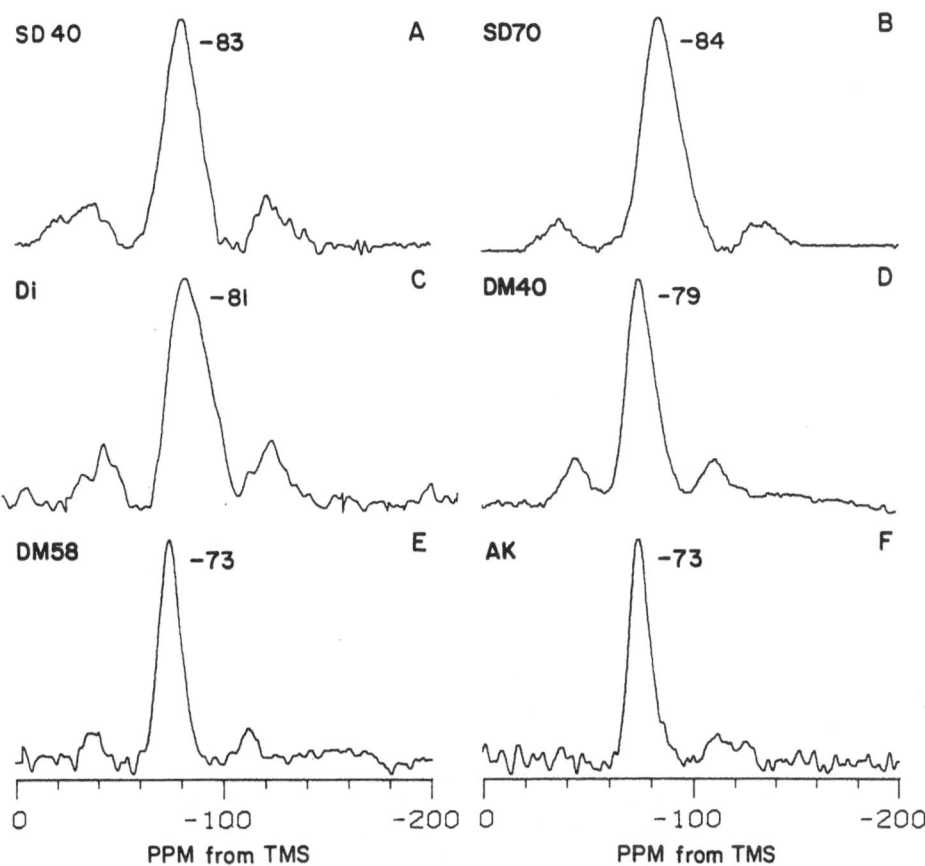

Fig. 8   Silicon-29 MASS NMR spectra of glasses in the system
CaO-MgO-SiO$_2$ obtained at a H$_o$ field strength of 8.45 T
(mole %). A) 60% SiO$_2$, 40% CaMgSi$_2$O$_6$ (Di), B) 30% SiO$_2$,
70% Di, C) Di, D) 60% Di, 40% CaMgSiO$_4$(Mo), E) 42% Di,
58% Mo, F) Ca$_2$MgSi$_2$O$_7$ (akermanite).

Figure 9 shows silicon-29 spectra for CaSiO$_3$ (Wo) and MgSiO$_3$ (En)
glasses. With decreasing silica content along SiO$_2$ - CaMgSiO$_4$,
the chemical shift of the peak maximum becomes less negative (less
shielded). This variation is the same as for crystalline silicates
(Fig. 3), which show, on an average, a decreasing shielding with
decreasing polymerization. As for SiO$_2$ glass, the peaks in these
spectra are much broader than those of crystalline phases. We
interpret this peak broadening as due to a range of electronic
environments at silicon due to both a range in the number of

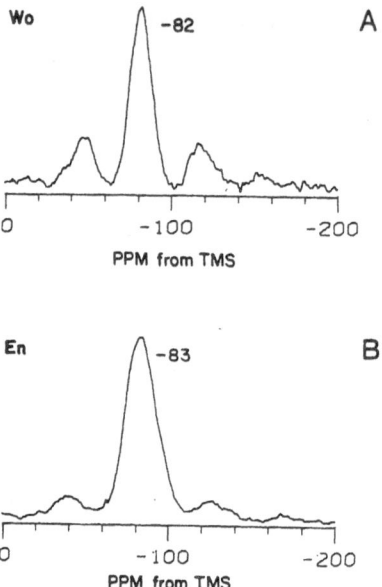

Fig. 9   Silicon-29 MASS NMR spectra of Wo (CaSiO$_3$) and En (MgSiO$_3$)
glasses obtained at an H$_o$ field strength of 8.45 T.   Note
the presence of more intensity at more negative (more
shielded) chemical shifts for En glass.

nonbridging oxygens per tetrahedron and a range of bond angles and
bond lengths for each type of silicon site. A fully polymerized
glass, such as SiO$_2$, experiences line broadening due to variable
bond angles and lengths only. Comparison of the range of chemical
shifts over which signal intensity occurs in glass spectra with the
ranges for different structural types in aluminum-free crystals
gives essentially the same range of polymerization states as
interpretations of Raman spectroscopy [32,22] (Fig. 10). The mean
number of nonbridging oxygens increases with decreasing SiO$_2$ content.
Because silicon-29 MASS NMR cannot resolve individual peaks for
different polymerization states in these glasses, the most we can say
is that the NMR data are consistent with the Raman data.

Like many spectroscopic methods, NMR examines the structural
and chemical environment only to about the first- and second-nearest-
neighbors [14,15,17,21] and cannot directly shed light on the question of the size of individual polymeric units in silicate glasses.

For the Ca-Mg metasilicate glasses Wo, Di, and En, the peak
maximum does not change significantly with Ca/Mg ratio, but the
full peak breadth increases with increasing Mg content (Figs. 8 and
9). This same result has been found by Murdoch, et al. [28]. The
increasing peak breadth with increasing Mg is consistent with a
larger concentration of very polymerized and very depolymerized

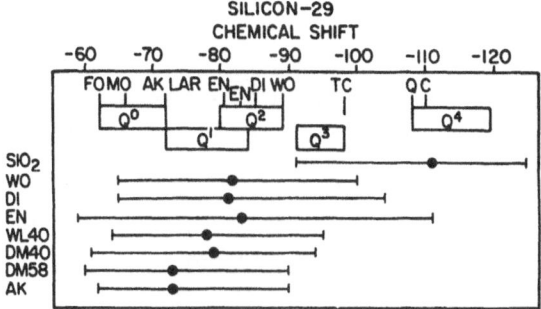

Fig. 10   Full breadths of silicon-29 MASS NMR peaks for glasses in
the system CaO–MgO–SiO$_2$. Dots indicate peak maxima.
Compositions are given in the caption to Figure 8.

silicon sites in the Mg-rich metasilicate glasses, as predicted by
the Raman spectroscopy and phase equilibrium data [33,38]. This
increase in the range of silicon sites is traditionally attributed
to the higher field strength of Mg relative to Ca.

The oxygen-17 MASS NMR spectra of crystals and glasses in the
system CaO–MgO–SiO$_2$ show that the structures of CaSiO$_3$ and MgSiO$_3$
glasses are similar to those structures of the isochemical crystalline places, but that the structures of intermediate compositions
glasses, such as CaMgSi$_2$O$_6$ and Ca$_2$MgSi$_2$O$_7$, are quite different from
those of the isochemical crystals [39]. Figure 11 presents the
oxygen-17 MASS NMR spectra of crystalline CaSiO$_3$ (parawollastonite),
MgSiO$_3$ (clinoenstatite), CaMgSi$_2$O$_6$ (diopside), and Ca$_2$MgSi$_2$O$_7$
(akermanite). Figures 12 and 13 present similar spectra for glasses
along the joins SiO$_2$ – CaMgSiO$_4$ and CaSiO$_3$ – MgSiO$_3$. For clinoenstatite crystals, there is a band from about 25 to 65 ppm due to
overlapping signal from the four NBO sites and two BO sites. For
En glass there is a broad peak that covers about the same chemical
shift range. Clearly, the oxygens in En glass are in electronic
environments similar to those in crystalline clinoenstatite. For
crystallite parawollastonite, there is a small band of peaks from
about 20 to 65 ppm and a larger band of peaks from about 75 to 140
ppm. We attribute the small band of peaks to the three BO sites
and the larger band to the six NBO sites. We make this assignment
because the small band covers essentially the same chemical shift
range as clinoenstatite, while the larger band is shifted significantly to less shielded values. The BO's are much less strongly
bonded to the large cations (Ca and Mg in this case) than the NBO's.
Thus, their chemical shifts should change much less with changing
Ca/Mg ratio than those of the NBO's. Apparently, Ca, with its
smaller electronegativity, causes the NBO's to be less shielded
than Mg. This interpretation is further supported by the oxygen-17
data for diopside crystals. This material contains one BO site
and two NBO sites, one of which is coordinated to one Mg and one Ca,

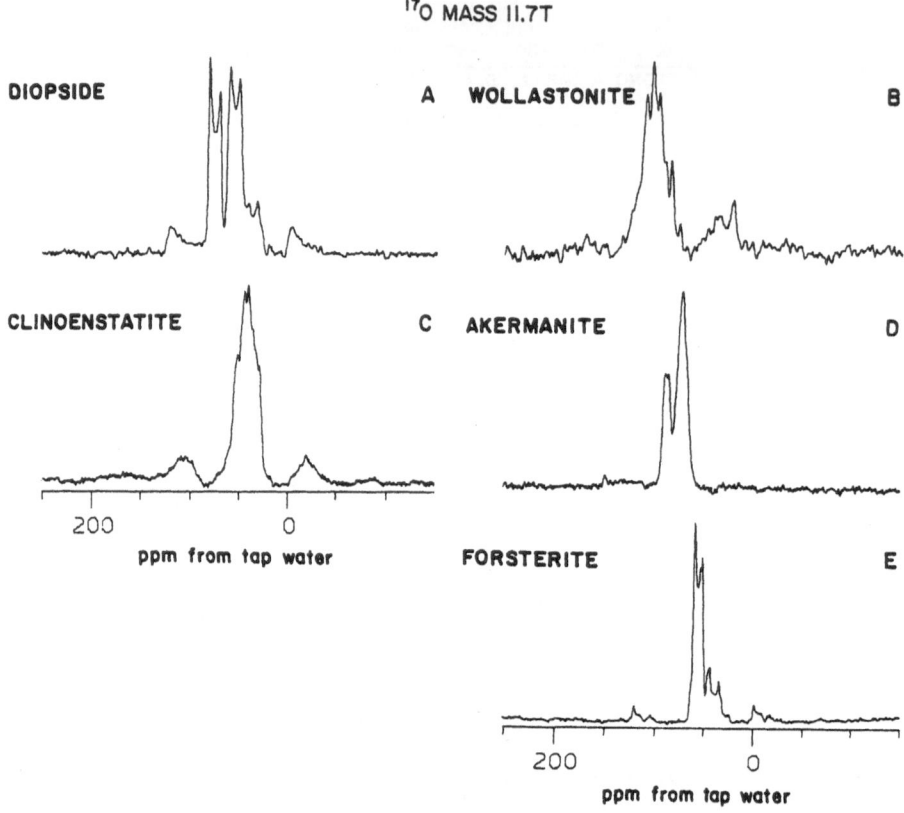

Fig. 11   Oxygen-17 MASS NMR spectra of crystalline silicates in the
          system CaO-MgO-SiO$_2$ obtained at a H$_o$ field strength of
          11.7 T.

the other of which is coordinated to two Mg's and one Ca.   Two
very similar quadrupolar split peaks with maxima at about 50 and
58 ppm, and about 70 and 80 ppm, are attributable to these two
NBO sites.   The positions of these two peaks are, as expected,
intermediate between the positions of the NBO sites in clinoensti-
tite and parawollastonite.

     The oxygen-17 spectra of CaMgSi$_2$O$_6$ and Ca$_2$MgSi$_2$O$_7$ composition
glasses are very different than those of the isochemical crystalline
phases.   Both contain peaks at about 107 ppm, attributable to NBO's
coordinated to only Ca ("wollastonite-like" sites), and broader
peaks at smaller chemical shifts, attributable to overlapping
signal from NBO's coordinated to Ca and Mg ("diopside-like" sites)
or to only Mg ("enstatite-like" sites) and all the BO's.   As
expected, the peak we assign to NBO's coordinated to only Ca is
much larger for Ca$_2$MgSi$_2$O$_7$ glass than for CaMgSi$_2$O$_6$ glass.   Along

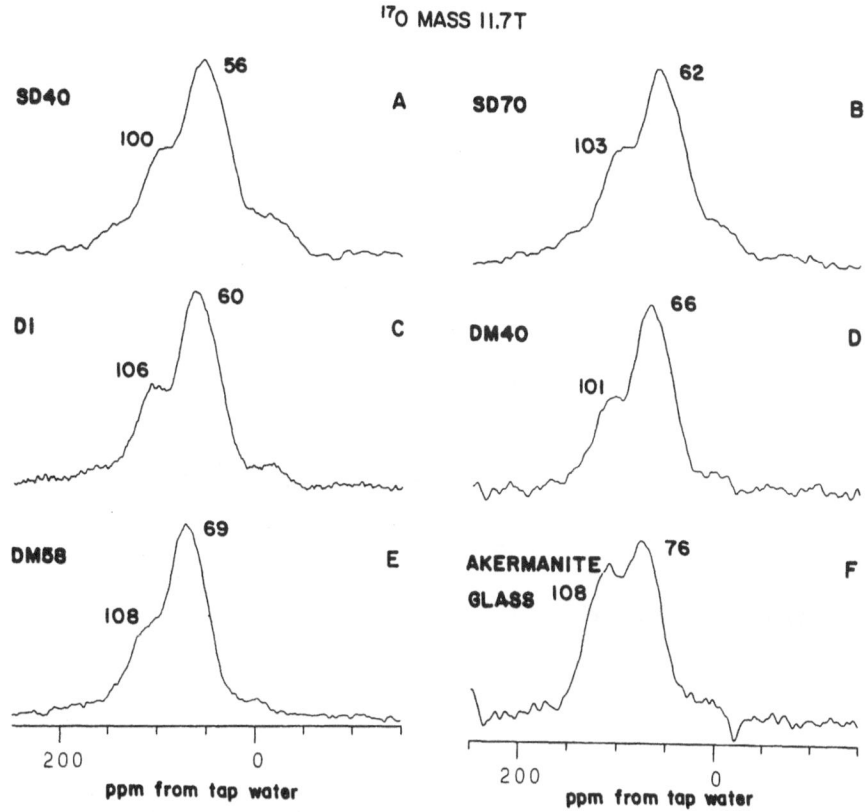

Fig. 12   Oxygen-17 MASS NMR spectra of glasses in the system
CaO–MgO–SiO$_2$ obtained at a H$_o$ field strength of 11.77 T.
Compositions are given in the caption to Figure 8.

the join CaSiO$_3$ – MgSiO$_2$ the intensity of the peak at about 107 ppm
for "wollastonite-like" sites decreases as the Wo content of the
glass decreases. Although we have used the expressions "wollastonite-
like", "enstatite-like", and "diopside-like" for nonbridging oxygens,
because there are likely silica tetrahedra in the glasses with differing number of NBO's, these expressions refer to the types of
modifier cations coordinated to the oxygen and not the polymerization of the silicon.

Aluminosilicate Glasses

Aluminum Coordination: Both aluminum-27 and silicon-29 MASS
NMR are potentially powerful tools for determining the structural
environment of aluminum in aluminosilicate glasses. Aluminum-27
is useful because the chemical shifts of tetrahedral and octahedral
aluminum fall into distinct ranges, because the chemical shifts of
Q^4 and Q^3 tetrahedral aluminum in aluminosilicates fall into

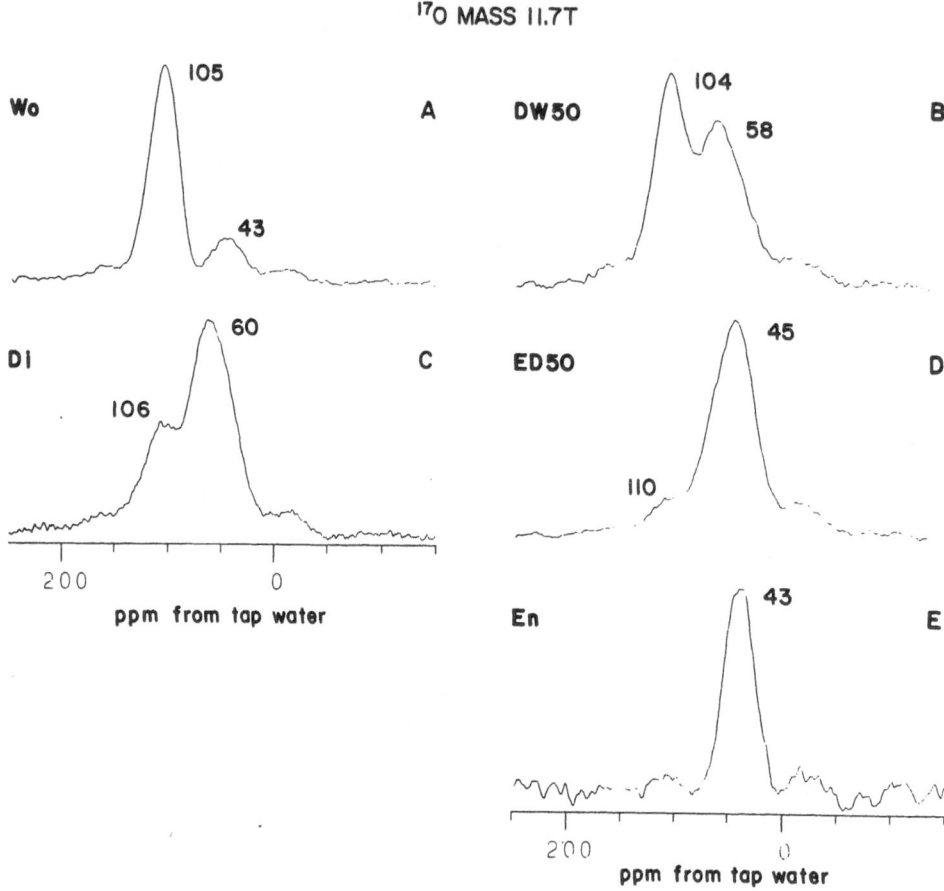

Fig. 13   Oxygen-17 MASS NMR spectra of glasses along the Ca-Mg
metasilicate join (mole %). A) Wo, B) 50% Wo, 50% Di,
C) Di, D) 50% Di, 50% En, E) En.

distinct ranges, and because the chemical shift of tetrahedral
aluminum becomes more positive (less shielded) with decreasing
Si/(Si+Al(IV)) in aluminosilicate glasses. Silicon-29 is useful
because, for less polymerized silicon sites, octahedral aluminum NNN
cause more negative (more shielded) silicon-29 chemical shifts,
while tetrahedral aluminum NNN cause less negative (less shielded)
chemical shifts.

Figure 14 shows aluminum-27 MASS NMR spectra of $CaAl_2Si_2O_8$
(anorthite) and $NaAlSi_3O_8$ (albite) glasses and crystals [8]. In
general, the peak maxima for the glasses are at nearly the same
chemical shifts as those of the isochemical crystals but are somewhat broadened. The peaks for the crystals themselves are somewhat
broadened due to quadrupolar effects and, for crystalline anorthite,

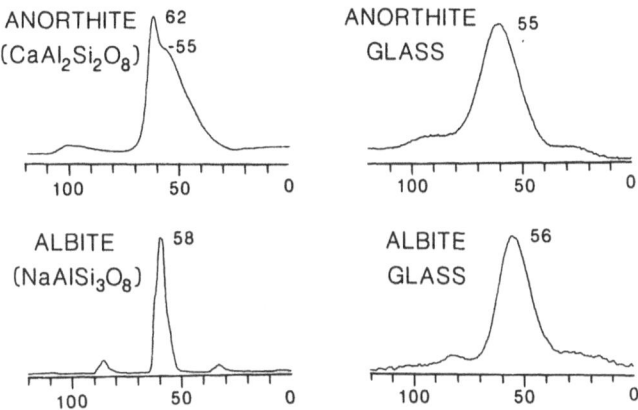

Fig. 14 Aluminum-27 MASS NMR spectra of $CaAl_2Si_2O_8$ and $NaAlSi_3O_8$ crystals and glasses obtained at a H_o field strength of 11.7 T.

by the fact that there are eight structurally distinct aluminum sites. For these glasses, there is no aluminum-27 signal near 0 ppm which would indicate the presence of octahedral aluminum. Signal from octahedral aluminum may be lost because of large quadrupole coupling constants. The apparent lack of octahedral aluminum, however, is consistent with x-ray radial distribution [40] and Raman spectroscopic [33] data for similar glasses.

Almost all of the glasses we and others have examined to date have molar alkali or alkaline-earth to alumina ratios of one or greater. All such glasses including $NaAlSi_3O_8$ and $CaAl_2Si_2O_8$ show peaks for only tetrahedral aluminum. de Jong, et al. [41] also found only peaks for tetrahedral aluminum in a wide variety of aluminosilicate glasses examined at a lower H_o magnetic field strength of 4.7 Tesla.

Evidence that octahedral aluminum can be detected in non-crystalline solids comes from aluminum-27 MASS NMR data for x-ray amorphous gels prepared by precipitation from aqueous solution at room temperature and then dehydration by heating at a few hundred degrees C. For Al_2O_3–SiO_2 gels dehydrated at 450°C, Thomas, et al. [42] have found signals for octahedral aluminum at a chemical shift of about +10 ppm and for tetrahedral aluminum at a chemical shift of about +60 ppm. Clearly, at least in some amorphous materials, both octahedral and tetrahedral aluminum can be detected.

Pb_2SiO_4

One of the most useful applications of MASS NMR has been the silicon-27 and oxygen-17 examination of Pb_2SiO_4 glass and the various Pb_2SiO_4 crystalline polymorphs [43]. Figures 15 and 16

Pb_2SiO_4 Silicon-29

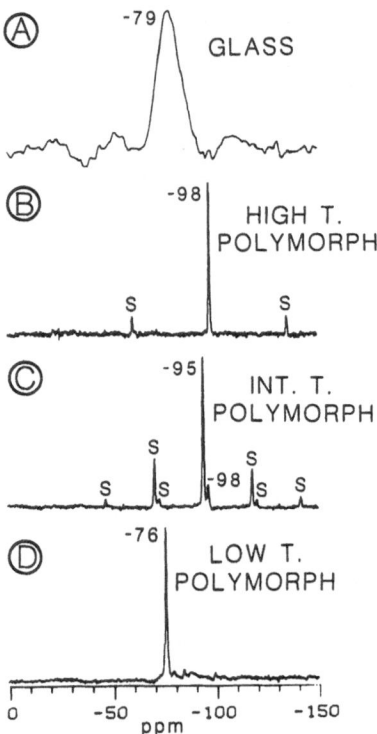

Fig. 15 Silicon-29 MASS NMR spectra of Pb_2SiO_4 glass and various
        Pb_2SiO_4 crystalline polymorphs obtained at a H_o field
        strength of 8.45 T. The sample in C contains a small
        amount of the high-T polymorphs. S indicates a spinning-
        sideband.

show silicon-29 and oxygen-17 spectra for these materials. Pb_2SiO_4,
although chemically an orthosilicate, does not crystallize with an
orthosilicate structure. The high and intermediate temperature
polymorphs have ring tetramer structures; the low temperature
polymorph has a dimer structure [44,45] Figures 15B and 15D show
that these crystalline phases have silicon-29 chemical shifts
(-98, -95, and -76 ppm, respectively) in the correct range for these
type of structures.

    Because the Pb_2SiO_4 polymorphs have an orthosilicate composition,
but more polymerized structures, they must contain oxygens that are
coordinated to only lead (PBO's) as well as bridging and nonbridging
oxygens (BO's and NBO's, respectively). The oxygen-17 spectra of
the low and intermediate temperature polymorphs (Figures 16B and
16C) show that all three types of oxygens can be resolved. We
assign the peaks at about 45 and 65 ppm to BO's (in the same range

Pb$_2$SiO$_4$ Oxygen-17

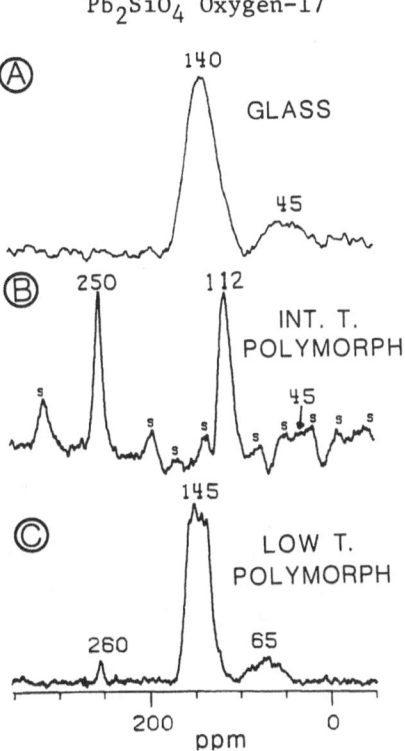

Fig. 16   Oxygen-17 MASS NMR spectra of Pb$_2$SiO$_4$ glass and crystalline polymorphs obtained at a H$_o$ field strength of 11.7 T.

as the Ca- and Mg-metasilicates), those at about 112 and 145 ppm to NBO's, and those at about 250 and 260 ppm to PBO's. The PBO chemical shift is consistent with the value of 289 ppm for crystalline PbO (Turner, unpublished data). The relative height of the peak for the PBO's is larger for the intermediate temperature polymorph than for the low temeprature polymorph, in agreement with the larger number of oxygens of this type in the more polymerized intermediate tempera- ture polymorph. The oxygen-17 spectrum for the high-temperature Pb$_2$SiO$_4$ polymorph was not obtained because of rapid oxygen exchange with the furnace atmosphere.

The silicon-29 and oxygen-17 spectra of Pb$_2$SiO$_4$ glass (Figs. 15A and 16A) show clearly that the structure of the glass is more like that of the low temperature, dimer-structure polymorph than those of the high and intermediate temperature, ring-tetramer- structure polymorphs. The peak maximum of the silicon-29 spectrum of the glass is at -79 ppm, in the range for dimers, and there is little signal intensity in the -95 to -100 ppm range, which would indicate the presence of ring-tetramer units. The peak does, however, extend into the range of -83 to -92 ppm, where the peaks for Pb-metasilicate (with a chain structure) fall (Dunn, unpublished data). The oxygen-17 spectrum of the glass is also much more similar to that of the low temperature, dimer-structure polymorph. There are broad peaks at about 60 and 140 ppm that we attribute to BO's and NBO's, respectively. There is no peak in the range of 260 ppm for PBO's which could be due to loss of signal due to a large Pb_2SiO_4 glass having a monomer structure. The latter is unlikely, however, because the silicon-29 chemical shift is similar to that of the dimer-structure polymorph and because the presence of bridging oxygens in an orthosilicate composition glass (the 60 ppm peak) requires stoichiometrically that some oxygens coordinated to only Pb be present. These data indicate that Pb_2SiO_4 glass has a structure dominated by silica tetrahedra with one bridging oxygen per silicon. As for most silicate glasses, however, it seems likely that some concentration of other types of silicon sites (probably including Q^0 and Q^2 sites) are also present.

PROGNOSIS

We have attempted in this paper to illustrate both some of the potential applications and some of the limitations of MASS NMR spectroscopy for the examination of glass structures. Limitations include, for instance, lack of resolution of separate silicon signals for different types of silicon sites for many compositions and the potential loss of signal from large e^2qQ/h sites of quadrupolar nuclides. When used for specific and well defined problems (for instance, the comparison of Ca-Mg metasilicate and Pb_2SiO_4 crystal and glass structures) MASS NMR does, however, appear to yield information that cannot be obtained by other methods. As the MASS NMR behavior of more nuclides (e.g., boron-11 [46], sodium-23, magnesium-25, and phosphorous-31) becomes better understood, it seems likely that this method will offer attractive possibilities for examining the local structural and chemical environment of a wide variety of species in amorphous phases. The pulse-Fourier-transform NMR technique in conjunction with magic angle sample spinning offers significant advantages over traditional continuous wave NMR because of greatly increased sensitivity and resolution.

REFERENCES

1. J. C. Davis, Advanced Physical Chemistry, Ronald Press, New York, p. 632 (1965).
2. T. C. Farrar, and E. D. Becker, Pulse and Fourier Transform NMR: Introduction to theory and methods, Academic Press, New York, (1971).
3. E. D. Becker, High Resolution NMR, Theory and Applications, 2 Ed., Academic Press, New York, p. 354 (1980).
4. J. W. Akitt, NMR and Chemistry: An Introduction to the Fourier Transform-Multinuclear Era, 2 Ed., Chapman and Hall, London, p. 263 (1983).

5. P. J. Bray, F. Bucholtz, A. E. Geissberger, and I. A. Harris, Nucl. Inst. and Methods, 199:1-15 (1982).

6. E. R. Andrew, Progress in NMR Spectroscopy, 8:1-39 (1971).

7. E. R. Andres, Int. Rev. Physical Chem. 1:195-224 (1981).

8. R. J. Kirkpatrick, R. A. Kinsey, K. A. Smith, D. M. Henderson, and E. Oldfield, Am. Min. 70:106-123 (1985).

9. A. Abragam, The Principles of Nuclear Magnetism, Clarendon Press, Oxford.

10. E. Kundla, A. Samoson, and E. Lippmaa, Chem. Phys. Lett. 83:229-232 (1981).

11. M. D. Meadows, K. A. Smith, R. A. Kinsey, T. M. Rothgeb, R. P. Skarjune, and E. Oldfield, Proceedings of the National Academy of Science of the United States, 79:1351-1355 (1982).

12. S. Ganapathy, S. Schramm, and E. Oldfield, Chem. Physics, 77:4360-4365 (1982).

13. S. Schramm, and E. Oldfield, J. Amer. Chem. Soc. 106:2502-2506 (1983).

14. E. Lippmaa, M. Magi, A. Samoson, G. Engelhardt, and A. -R. Grimmer, J. Amer. Chem. Soc. 102:4889-4893 (1980).

15. E. Lippmaa, M. Magi, A. Samoson, M. Tarmak, and G. Engelhardt, J. Amer. Chem. Soc. 103:4992-4996 (1981).

16. J. M. Thomas, J. Klinowsky, S. Ramdas, M. W. Anderson, C. A. Fyfe, and G. C. Gobbi, American Chemical Society Symposium Series 218, Intrazeolite Chemistry, G. D. Stuckey and F. G. Dwyer, Eds., American Chemical Society, Washington, DC p. 159-180 (1983).

17. K. A. Smith, R. J. Kirkpatrick, E. Oldfield, and D. M. Henderson, Am. Min. 68:1206-1215 (1983).

18. R. K. Kirkpatrick, K. A. Smith, S. Schramm, G. Turner, and W. -H. Yang, Ann. Rev. Earth Planet. Sci. 13:29-47 (1985).

19. M. Magi, E. Lippmaa, A. Samoson, G. Engelhardt, and A. -R. Grimmer, J. Phys. Chem. 88:1518-1522 (1984).

20. K. A. Smith, R. J. Kirkpatrick, D. M. Henderson, R. Oestrike, and E. Oldfield, in preparation.

21. I. D. Brown, and R. D. Shannon, Acta Cryst., A29:266-282 (1973).

22. J. V. Smith, and C. S. Blackwell, Nature, 303:223-225 (1983).

23. J. V. Smith, C. S. Blackwell, and G. L. Hovis, Nature, 309:140-142 (1984).

24. D. Muller, W. Gessner, H. J. Behrens, and G. Scheler, Chem. Phys. Lett. 79:59-62 (1981a); D. Muller, D. Hoebbel, and W. Gessner, Chem. Phys. Lett. 84:25 (1981b).

25. R. A. Kinsey, R. J. Kirkpatrick, J. Hower, K. A. Smith, and E. Oldfield, Am. Min. 70:537-548 (1985).

26. S. Schramm, R. J. Kirkpatrick, and E. Oldfield, J. Amer. Chem. Soc., 105:2483-2485 (1983).

27. E. Dupree, and R. F. Pettifer, Nature, 308:523-525 (1984).

28. J. B. Murdoch, J. F. Stebbens, and I. S. E. Carmichael, Am. Min. 70:332-343 (1985).

29. J. K. Mitra, Phil. Mag., 45B:529-548 (1982).

30. R. L. Mozzi, and B. E. Warren, J. Appl. Cryst., 2:164-172 (1968).

31. D. L. Evans, and M. Teter, The Structure of Non-Crystalline Solids, P. H. Geskell, Ed., Taylor and Francis, London, (1977).

32. S. A. Brawer, and W. B. White, J. Non-Cryst. Solids, 23:261 (1977).

33. B. O. Mysen, D. Virgo, and F. A. Seifert, Rev. Geophys. Space Phys., 20:353 (1982).

34. J. O. Bockris, J. D. MacKenzie, and J. A. Kitchner, Trans. Faraday Soc., 52:1734 (1955).

35. J. O. Bockris, J. W. Tomlinson, and J. L. White, Trans. Faraday Soc., 52:299 (1956).

36. J. O. Bockris, and D. C. Lowe, Trans. Faraday Soc. 51:1734 (1954).

37. J. O. Bockris, and A. K. N. Reddy, Modern Electrochemistry, Vol. 1, Plenum Press, New York (1970).

38. A. L. Boettcher, C. W. Burnham, K. E. Windom, and S. E. Bohlen, J. Geol. 90:127 (1982).

39. S. Schramm, Ph.D. Thesis, Department of Chemistry, University of Illinois at Urbana-Champaign (1984).

40. M. Taylor, and G. E. Brown, Geochim. Cosmochim. Acta. 43:61-76 (1984).

41. B. H. W. S. de Jong, C. M. Schramm, and V. E. Parziale, Geochim. Cosmochim. Acta, 47:1223 (1983).

42. J. M. Thomas, J. Klinowski, P. A. Wright, and R. Roy, Angew. Chem. Int. Ed. Engl., 22:614-616 (1983).

43. T. Dunn, R. J. Kirkpatrick, and E. Oldfield, in preparation; E. Lippmaa, A. Samoson, M. Magi, R. Teeaar, J. Schraml, and J. Gotz, J. Non-Cryst. Solids, 50:215-218 (1982).

44. J. Gotz, D. Hoebbel, and W. Wieker, Zeit. Anorg. Allg. Chem., 418:29 and 416:163 (1975).

45. R. M. Smart, and F. P. Glasser, Phys. Chem. Glasses, 19:95 (1978).

46. S. E. Schramm, and E. Oldfield, J. Chem. Soc., Chem. Comm., 980-981 (1982).

47. C. M. Schramm, B. H. W. S. de Jong, and V. E. Parziale, J. Am. Chem. Soc., 106:4391-4402 (1984).

48. R. Dupree, D. Holland, P. W. McMillan, and R. F. Pettifer, J. Non-Cryst. Solids, 68:399-410 (1984).

49. A. -R. Grimmer, M. Magi, M. Hahnert, H. Stade, A. Samosen, W. Wieker, and E. Lippmaa, Phys. Chem. Glasses, 24:105-109 (1984).

50. U. Selvary, K. J. Rao, C. N. R. Rao, J. Klinowski, and J. W. Thomas, Chem. Phys. Lett. 114:24-27 (1985).

51. R. Oestrike, Ph.S. Thesis, University of Illinois, Urbana-Champaign (1985).

52. R. J. Kirkpatrick, K. A. Smith, R. Oestrike, C. A. Weiss, Jr., and E. Eldfield, Am. Min. 71:in press (1986).

53. G. L. Turner, K. A. Smith, R. J. Kirkpatrick, and E. Oldfield, J. Mag. Res. (in press).

54. G. L. Turner, R. J. Kirkpatrick, S. Risbud, and E. Oldfield, J. Am. Cer. Soc. (submitted).

55. W. -H. Yang, and R. J. Kirkpatrick, in preparation.

NOTE ADDED IN PROOF

Since this paper was written, there have been several published studies of glasses using MASS NMR techniques. These include silicon-29 studies of Na_2O-SiO_2 and Li_2O-SiO_2 glasses which demonstrate that peaks for Q^0-Q^4 silicon sites can be resolved (in contrast to $CaO-SiO_2$ and $MgO-SiO_2$ glasses) and that, as expected, the average polymerization decreases with increasing alkali content [47,48,49,50]. For these glasses, there is also a systemmatic deshielding of the peaks for Q^4 and Q^3 silicon sites, indicating a decrease in the mean Si-O-Si bond angle with increasing alkali content. For many of the spectra in these papers, however, the signal/noise is poor, and the exact distribution of sites is difficult to determine.

Additional aluminum-27 work has shown that for glasses along the joins $SiO_2-MAl_2O_4$ (M = one alkaline earth or two alkalis) the aluminum-27 chemical shift becomes progressively deshielded, paralleling the trend for framework silicates [51]. New quanitation techniques have shown that for many glasses at 11.7 T, all the aluminum-27 is being detected [52]. One important conclusion from this result is that at least some of the glasses along the joins $SiO_2-MAl_2O_4$, which all have the composition of framework silicates, in fact have a framework structure with no nonbridging oxygens.

Additional studies include the following: boron-11 [53], which can resolve multiple trigonal and tetrahedgral boron sites at 11.7 T; nitrogen-15 [54], which demonstrates that nitrogen in Mg-Si-Al-O-N glasses is in a nitride environment, and phosphorous-31 [54,55], which demonstrates that in phosphate glasses different phosphorus sites can be resolved and that in alkaline earth silicate glass phosphorus is present only as monomeric units.

INVESTIGATION OF THE "GLASSY" PROPERTIES OF NEUTRON AND ELECTRON

IRRADIATED CRYSTALLINE QUARTZ RELATED TO THE ANOMALOUS LOW

TEMPERATURE DYNAMICS OF NON-CRYSTALLINE SOLIDS

C. Laermans

Kath. University Leuven, Physics Department, VSHD
3030 Leuven, Belgium

I. INTRODUCTION

At low temperatures, non-crystalline solids exhibit properties
which are unexpectedly different from these in their crystalline
counterparts [1]. The dynamic properties, as there are specific
heat, thermal conductivity, ultrasonic attenuation and velocity
show a surprising behavior which is similar for most of the glasses
studied until now: below 1K the specific heat is linear with
temperature and much larger as in crystals, the thermal conductivity
behaves as T^2 and is much smaller as in crystals and as far as
the ultrasonic properties are concerned the main feature is a nonlinear behavior of the ultrasonic attenuation, also called ultrasonic saturation by analogy with NMR spectroscopy. These properties
are unexpected, because at low temperatures the phonon wavelength
is much higher than the scale of the structural disorder in the
non-crystalline solid and therefore the thermal properties should
be similar as in the crystal; the same is also true for the ultrasonic wavelength.

While in crystals the thermal and ultrasonic properties can be
explained by the Debye theory and the presence of phonons, in non-
crystalline solids the existence of some kind of localized low
energy excitations has to be assumed, which are present in addition
to phonons. A phenomenological model which can describe almost all
the data until now, even obtained on glasses of very different
structure was already put forward in 1972 [2]. The model assumes
the existence, in the amorphous network, of configurational
tunneling states, more simply described as two-level systems (TLS),
an elastic equivalent of a spin 1/2 system. The basic problem
however is still unresolved: it is still unknown what the microscopic

329

nature is of these excitations, or with other words of the tunneling
particle.

   In an attempt to contribute to a better understanding of the
microscopic origin of the dynamical anomalies in non-crystalline
solids, another approach resulted from our study of neutron
irradiated quartz crystals. After a relatively low fast neutron
dose (8×10^{19} n/cm^2), the quartz was slightly defective, although
still long-range ordered, and showed a 9 GHz hypersonic attenuation
which was nonlinear, similar to previous findings in non-crystalline
solids [3]. It turned out that neutron irradiation induces excitations similar to these in non-crystalline solids and that their
density of states can be altered in a controlled way by changing
the neutron dose. Therefore the study of an irradiated crystal
was found to be a tool in the investigation of the low temperature
anomalies of amorphous solids. Also studies in electron-irradiated
quartz were carried out and the thermal conductivity data are very
similar to those in neutron irradiated quartz and therefore gave
evidence for "glassy" dynamic properties.

   Upon neutron irradiation the damage is mainly due to elastic
collisions. Because of the high energy of the fast neutrons these
collisions cause displacement cascades and as a consequence damaged
regions. For doses smaller than 3×10^{19} n/cm^2 long range order is
maintained while it gradually disappears for higher doses. At doses
of 2×10^{20} n/cm^2 the quartz is completely vitrified [4]. More recent
diffuse X-ray studies gave evidence for the presence of 20 Å diameter disordered clusters for doses as low as $3^* \times 10^{18}$ [5]. From
these studies a possible conclusion could be put forward that in
neutron-irradiated quartz the dynamical anomalies might be due to
a corresponding degree of amorphization of the crystals, a conclusion
which is however not completely followed by other authors. For
electron-irradiated quartz however, the situation is quite different.
High energy electrons are not expected to cause displacement cascades
like neutrons and large damage regions are not expected for the
doses used.

   The aim of this paper is to review the dynamical studies in
irradiated quartz related to the low temperature anomalous dynamics
in non-crystalline solids and also to review related damage studies.
In order to be able to compare to similar experiments in non-
crystalline solids first the anomalies in non-crystalline solids
will be summarized as well as the phenomenological TLS model. Then
the "glassy" properties of neutron and electron irradiated quartz
will be reviewed and discussed. Finally, after a summary of the
general knowledge of radiation induced disorder in quartz, recent
related damage studies will be discussed.

---

* in their dose units which differ from the ones we usually use;
  for more details see Section III.

II.   ANOMALOUS THERMAL AND ACOUSTIC BEHAVIOR OF NON-CRYSTALLINE
      SOLIDS AT LOW TEMPERATURES

A.   Experiments

The extensive research on the dynamical properties of non-crystalline solids, the last decade was caused by the discovery in 1971, by Zeller and Pohl [6] that non-crystalline solids exhibit a specific heat which varies almost linearly with temperature below 1K instead of the usual T^3 dependence expected on account of the Debye theory (see Fig. 1).   As already mentioned in the introduction a similar (Debye) behavior as in a crystal was expected because the phonons at low temperatures have a wavelength (of the order of 1000 Å) which is several hundred times larger than the interatomic distance and there is no evidence for structural fluctuations over a phonon wavelength.   In addition, it was found that the absolute value of the specific heat in vitreous silica could be up to a factor of 1000 larger than the Debye specific heat in quartz.

Not only the specific heat was expected to vary as T^3 below 1K, also the thermal conductivity.   And again the non-crystalline solids behave quite different from expectation, showing a nearly T^2 behavior and an absolute value of the thermal conductivity up to 1000 times smaller in silica as in quartz (see Fig. 2 for a typical example).   In addition, it has been found that not only silica glasses

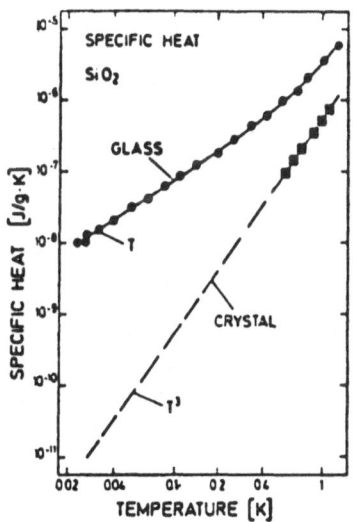

Fig. 1   Specific heat of vitreous silica and crystalline quartz as
         a function of temperature (from /6/ and /7/).   The specific
         heat of a pure quartz crystal is proportional to T^3 between
         0.6K and 5K.   Below 0.6K no experimental data exists so far.

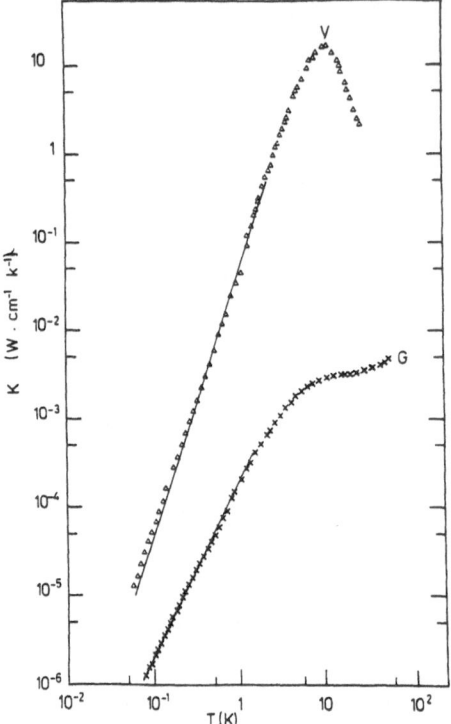

Fig. 2   Thermal conductivity of crystalline quartz (V) and of glass
         (G) as a function of temperature (from /8/).   Full lines,
         see text.

show these anomalies but that almost all non-crystalline solids
investigated until now have a thermal conductivity which is very
similar in magnitude and which depends very little on impurities;
this again is very different from crystals.

   Since it had been shown that in the temperature range in
question heat is carried by phonons [9], the large extra specific
heat and the small thermal conductivity were supposed to be due to
a new type of (low energy) excitations which would be strong phonon
scatterers. For instance, if the low energy excitations were of a
two-state nature, namely two-level systems, resonant phonon scattering would reduce the thermal conductivity. Strong phonon scatterers
should however also influence ultrasonic experiments and in addition
resonant phonon scattering would cause a saturation of the ultrasonic
attenuation. Indeed this was observed: the ultrasonic attenuation
was found to show such a nonlinear behavior [10,11]. A typical
example is given in Fig. 3: it is seen that increasing the intensity
of the ultrasonic beam which travels in the non-crystalline solids
causes a decrease in attenuation, which can be understood in terms
of an increasing population of the upper level of the TLS.

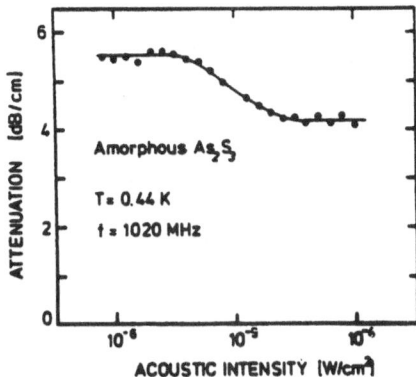

Fig. 3   Ultrasonic attenuation as a function of intensity in amorphous As_2S_3 (from /13/).

    Besides the ultrasonic saturation, the temperature dependence
of the ultrasonic attenuation and velocity is also very typical
for non-crystalline solids. While in perfect crystals the three
phonon processes, causing the attenuation at low temperatures are
frozen out below 20K and too small to give an observable attenuation, in amorphous solids the attenuation is rather high and shows
a T^3 behavior. Figure 4 shows typical attenuation data for a glass,
in this case neutron irradiated vitreous silica. Neutron irradiation
of a glass does not influence the "glassy" properties qualitatively;
there is however a small change in the quantitative behavior which
we will not discuss here. Figure 5 gives a typical example of the
temperature dependence of the sound velocity which is also "anomalous".
Both effects can also be understood in terms of the TLS model as we
will see later.

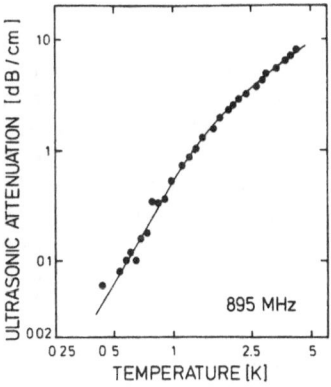

Fig. 4   Ultrasonic attenuation as a function of temperature for
longitudinal ultrasound in neutron irradiated vitreous
silica after substraction of the residual loss (from /13/).

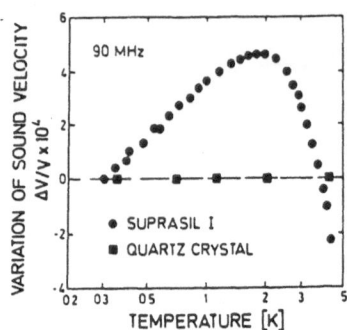

Fig. 5   Relative variation of longitudinal sound velocity in vitreous
         silica as a function of temperature (from /14/). Reference
         temperature T_0 = 0.3K. Full squares show for comparison the
         independent velocity in a quartz crystal.

The experiments mentioned so far could be explained taking into
account the existence of TLS and the population changes of the two
states of the TLS. We may expect a new phenomenon if we consider the
TLS- in analogy with NMR- as "pseudo-dipoles" which cause a "pseudo-
magnetization" and as a consequence coherent effects similar to the
photon echo, should be possible. Indeed, the first phonon echo was
observed by Golding and Greabner [15] at 20 mK in vitreous silica.
If two ultrasonic pulses are applied to the sample, the first pulse
creates a transverse pseudo-magnetization of the TLS and the second
pulse reverses time for the excited TLS. After a certain time,
rephasing occurs and the TLS coherently emit phonons. The acoustic
pulse which is generated is called a phonon echo. The experiments
can only be done at millikelvin temperatures in order to have
sufficiently long relaxation times for the echo to be observable.
If the glass is excited by a three pulse sequence, even a "stimulated
echo" can be observed [16].

Besides thermal and acoustic anomalies other related observations
were found as there are the saturation of a microwave [17] and electric echoes [18]. Also, so called "cross" effects were found, namely
the saturation of the ultrasonic attenuation by a microwave [19] and
vice versa [20]. Similar findings occurred in a variety of non-
crystalline materials, including amorphous semiconductors (for
instance, see Fig. 3) and glassy metals. Recently, for example, it
has been demonstrated that the specific heat shows a linear temperature dependence in a superconducting amorphous metal well below T_C
where the electronic specific heat can be neglected.

B.   Theoretical Model: TLS Model

The theoretical model which describes best the low temperature dynamical anomalies in non-crystalline solids is the so-called
tunneling model, originally put forward by Anderson et al. [2] and

at the same time by Phillips [2]. This theory which is phenomenological, describes both the thermal properties and the acoustic
properties, unlike some of the other proposed models. In this
model they assume the existence, in the non-crystalline solids, of
low energy excitations which are present in addition to phonons.
The elastic potential of the low energy excitation is represented
by a double well (see Fig. 6). The theory assumes that only the
lowest level of both wells is relevant and that because of a small
asymmetry, a particle (of unknown nature) tunnels between two nearly
degenerate energy levels. They suggested that these excitations in
amorphous solids are a natural consequence of the metastable state
of the glassy materials, which have many configurational minima of
the free energy, being almost but not exactly equivalent. In
Fig. 6, the energy is plotted for two neighboring minima as a
function of some (unknown) configurational coordinate d. The assumption is that there exists a strong coupling between long wavelength
phonons (as present at low temperatures) and these tunneling states.
If ε is the asymmetry of the potential and Δ the overlap, the
Hamiltonian can be diagonalized and the energy difference between
the two lowest energy levels is $E = \sqrt{\varepsilon^2 + \Delta^2}$ X and the model can be
simplified to the so-called two-level system model, which is similar
to the spin 1/2 problem [21]. To explain the linear specific heat,
a broad distribution of the energies E is needed and that may arise
in the model from the variation of ε and Δ from site to site in the
amorphous network. From the specific heat, a density of states of
the TLS is obtained in vitreous silica of $n_0 \cong 8 \times 10^{32}$/erg cm^3 or
$n_0 \cong 1 \times 10^{17}$/kcm^3.

Since in this paper the focus is mainly on the thermal conductivity and ultrasonic properties, as far as the dynamical properties are concerned, it is the interaction between phonons and
the TLS which will interest us. The dominant interaction between
the TLS and the phonons is the so-called <u>direct</u> <u>process</u>: namely
a TLS resonantly scatters phonons. Such a process gives rise to a
much smaller mean free path of the phonons in non-crystalline solids
as in crystals and therefore a higher relaxation rate. This
relaxation rate is in the TLS model given by [21]:

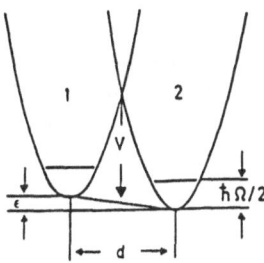

Fig. 6 Double well potential with barrier height V, asymmetry ε,
and distance d between the two minima. $\hbar\Omega/2$ is the ground
state energy.

$$\tau_{res}^{-1} = \frac{\pi n_o M^2}{\rho v^2} \omega \tanh \frac{\hbar \omega}{2kT} \frac{1}{\sqrt{1 + J/J_c}} \tag{1}$$

Here n_o is the density of states of the TLS, M is the coupling
constant for this process between the phonons and the TLS, $\tanh \hbar\omega/2kT$
describes the thermal occupation of the TLS, J is the intensity
of the phonon beam and J_c is a critical intensity, it is the
intensity at which ultrasonic saturation effects will become
noticeable. Other parameters have usual meanings.

The resonant interaction between phonons and TLS strongly
influences the thermal conductivity in non-crystalline solids below
1K, in fact it is the main scattering mechanism of phonons and thermal resistivities up to a factor of 1000 higher as the boundary
resistance can be found. (In pure crystals the boundary scattering
becomes dominant for T < 1K). If we insert the relaxation time of
expression [1] in the Debye expression for the thermal conductivity,
in the dominant phonon approximation [22] we find for the thermal
conductivity in non-crystalline solids:

$$K \quad \alpha \quad \frac{T^2}{n_o M^2} \tag{2}$$

It gives us the well-known T^2 law which is observed in many glasses
until now. (As a matter of fact, experimentally, one finds that
the exponent is only approximately equal to 2). The intensity
factor reduces to 1 in the case of thermal conductivity measurements
since $J \ll J_c$. In ultrasonic attenuation experiments, the intensity
of the ultrasonic beam can be of the order of or higher than J_c
and the mean free path of the phonons can become energy dependent,
the population of the levels can be equalized leading to the so-
called saturation of the ultrasonic attenuation. For $J \simeq J_c$ one
finds from [1] for the u.s. attenuation:

$$\alpha_{res} \quad \alpha \quad \frac{n_o M^2}{\sqrt{J}} \frac{\omega^2}{T} \tag{3}$$

Apart from the resonant interaction between TLS and phonons
another process will be observable in the ultrasonic experiments,
the relaxation process. When an ultrasonic wave travels through a
non-crystalline solid, the thermal equilibrium of the TLS will be
disturbed. The systems try to relax into a new equilibrium which
is reestablished via the emission and the absorption of thermal

phonons. The result is an ultrasonic absorption which is intensity
independent since now all TLS that are thermally activated take part
in this relaxation process. At low temperatures the condition
$\omega\tau \gg 1$ holds, where $\omega/2\pi$ is the frequency of the sound wave and τ
is the relaxation time necessary to re-establish thermal equilibrium.
In this regime the relaxation attenuation can be shown to be given
by [21]

$$\alpha_{rel} \quad \alpha \quad n_o M^2 \, D^2 \, T^3 \, \omega^o \tag{4}$$

In this expression, D is the coupling constant for this relaxation
process. This gives the experimentally found T^3 dependence of the
u.s. attenuation. At higher temperatures where $\omega\tau \ll 1$ the relaxation process gives rise to a T^o ultrasonic attenuation [23] which
can only be explained using the original tunneling mode [2].

The strong and temperature dependent absorption leads via the
Kramers-Kronig relation to a temperature dependence of the velocity
of sound. With temperature, the velocity first increases due to the
resonant interaction between the TLS and the u.s. phonons, passes
a maximum, and decreases because of the predominance of the relaxation process. Applying the Kramers-Kronig relation to Eq. 1, and
integrating over the energy distribution of the TLS, one obtains
for the variation of the velocity of sound $\Delta v = v(T) - v(T_o)$ where
T_o is an arbitrary reference temperature [21]:

$$\frac{\Delta v}{v} = \frac{n_o M^2}{\rho v^2} \, \ell n \, \frac{T}{T_o} \tag{5}$$

leading to a logarithmic temperature dependence of the sound
velocity as observed for T < 1K.

III. "GLASSY" DYNAMICS OF NEUTRON IRRADIATED QUARTZ

A. Ultrasonic properties

The first dynamic experiment to be reported on irradiated
quartz, related to the low temperature in anomalies in non-crystalline
solids, dates back to 1979 where it was shown that a saturation of
the 9 GHz hypersonic attenuation could be seen in slightly neutron
irradiated quartz, which was still long range ordered [3]. (Hypersonic is the word used for ultrasonic at frequencies above 1 GHz).

The used irradiation dose was 8×10^{18} n/cm^2 (E $\sim$ 0.3 MeV).*

Figure 7 shows the hypersonic attenuation as a function of the intensity of the acoustic beam at 1.4 K. The nonlinear behavior is very clear: at high intensities the attenuation is constant; this is the regime where the upper level of the TLS is fully populated and therefore the resonant interaction is fully saturated. At decreasing intensities the saturation effect decreases and the attenuation increases. Because of sensitivity limitations of the GHz equipment, lower intensities could not be reached and the fully unsaturated regime is not seen. In Fig. 8 the same data are seen but now the

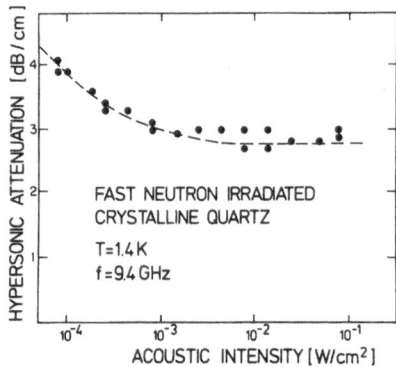

Fig. 7 Hypersonic attenuation as a function of intensity for the
longitudinal mode (from /3/).

---

1. In the original paper [3] the dose given was 6×10^{18} n/cm^2
(E $\gtrsim$ 1 MeV), in the mean time it has become clear that a more correct way to give this dose is 8×10^{18} n/cm^2 (E $\gtrsim$ 0.3 MeV). It has also to be born in mind that fast neutron doses are known generally only with an accuracy of 30%.
2. Neutron doses are determined in a different way in different irradiation sites. As a consequence, different reactor sites may give their doses in different units. In this work always the so-called fission dose will be given, which corresponds mainly to the dose of the neutrons with energy $\gtrsim$ 0.3 MeV. Where we compare the results of other authors, we will calculate an equivalent dose in our dose units. However, it has to be taken into account, when comparing the data of different authors that there are flux calibration differences and differences in energy spectrum of the neutrons at the site of the sample. In our work, the goal is to compare mainly data taken on samples irradiated at the same reactor site and if possible, in the same reactor cycle.

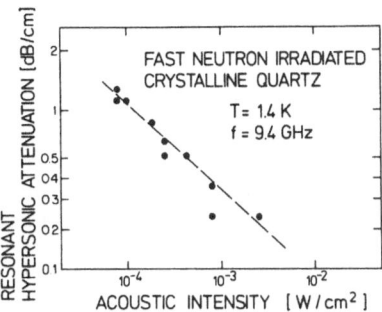

Fig. 8 Hypersonic attenuation as a function of intensity after
        subtraction of the intensity independent part (from /3/).

constant high energy attenuation is subtracted. It can be seen
that the $1/\sqrt{J}$ law found for the resonant interaction between phonons
and TLS in glasses, can fit the data (see Eq. (3)). Figure 9 shows
the temperature dependence of the attenuation, the temperature
independent part at low temperatures being subtracted and it can
be seen that T^3 can fit the data, indicating a relaxation attenuation
of the hypersonic beam due to interaction with TLS (see Eq. (4)).
In unirradiated quartz crystals it is known that no temperature
dependence is found for the 9 GHz attenuation in this temperature
range. Both sets of data give convincing evidence for the presence
of TLS, as in non-crystalline solids, in neutron-irradiated quartz,
which is defective but still long range ordered [3].

     Figure 10 shows more recent data of the 0.5 GHz ultrasonic
attenuation as a function of temperature for different neutron
doses. For this data a residual attenuation has been subtracted
so that the data would fit T^3 [24]. The subtracted value is about

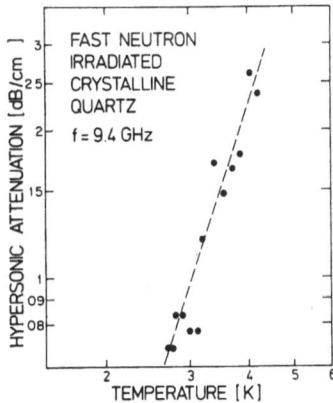

Fig. 9 Hypersonic attenuation as a function of temperature for the
        longitudinal mode after subtraction of the residual loss
        (from /3/).

the attenuation at 1.5 K. From data of Nava [31], it can be seen
that α does not vary very much below 1.5 K. For comparison, similar
data for vitreous silica [25] and neutron-irradiated vitreous silica
[13] are given. It is seen that the coefficient of T^3 increases
with increasing neutron dose, which is most probably due to an
increase of the density of states with dose since it is very likely
that the coupling parameters D and M are independent of dose (see
Eq. (3)). In fact, as we will see later, it has been found that M
in irradiated quartz, is similar as in vitreous silica [26]. Figure
11 shows the coefficient of T^3 relative to the value of that in
vitreous silica as a function of dose, indicating a linear increase
with dose, for the doses used, and also a linear increase of the
density of states with dose.

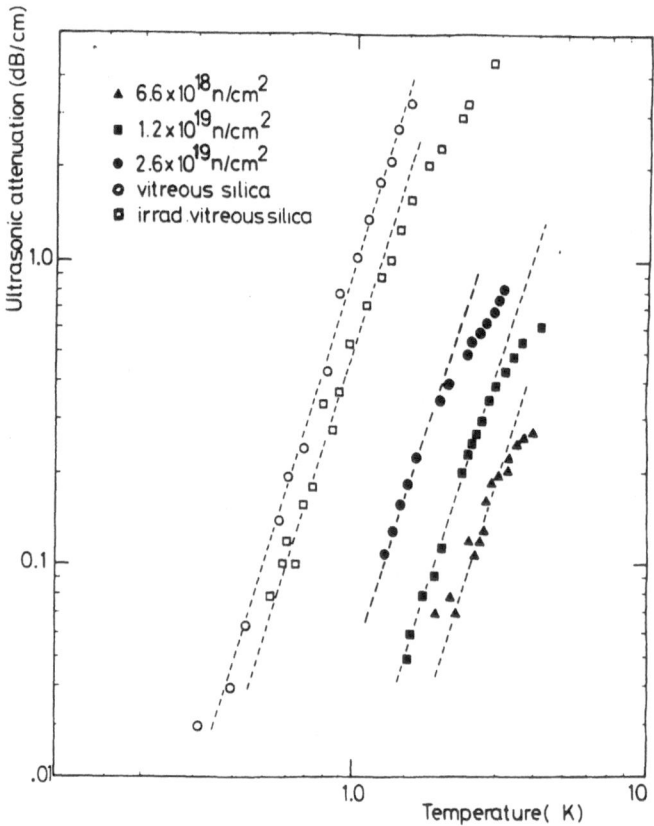

Fig. 10   Ultrasonic attenuation as a function of temperature and
          neutron dose (490 MHz longitudinal mode) after substraction
          of the residual loss (from /24/). Neutron irradiated
          quartz: (from /24/): ▲6.6×10^{18} n/cm^2 (N$_3$); ■ 1.2×10^{19}
          n/cm^2 (N$_4$); ● 2.6×10^{19} n/cm^2 (N$_5$). ○ Vitreous silica
          (570 MHz) (from /25/). ▫ Neutron irradiated vitreous
          silica (895 MHz) (from /13/).

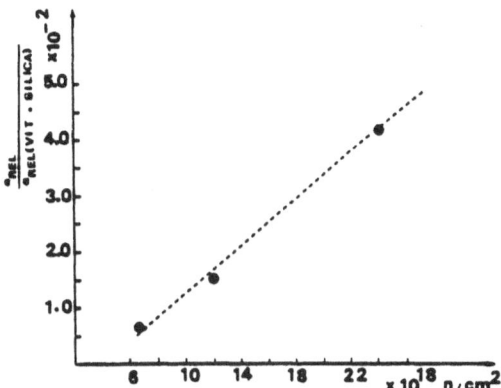

Fig. 11   Coefficient of T^3 in Fig. 10 for neutron irradiated quartz
as a function of dose relative to that of vitreous silica
(from /13/).

The ultrasonic attenuation studies given above, give no
independent value of n_0, M or D^*. An interesting study done by
Golding et al. [26] however, made it possible to determine M and n_0
in the sample, used for the previous hypersonic studies. As in
glasses, phonon echoes could be generated and an independent value
for M was found: 1.2 eV which is within the accuracy of the experiment similar as the value found in vitreous silica. From additional
velocity measurements, as a function of temperature, the product
$n_0 M^2$ was also determined (see Eq. (5)) and found to be 13 times
smaller as in vitreous silica. From these results, it has to be
concluded that there are TLS present in neutron-irradiated quartz
and that they are very similar in nature to these in vitreous
silica.

B.   Thermal Conductivity and Specific Heat

Figure 12 shows the thermal conductivity K as a function
of temperature for different neutron doses [8]. The slope of K
below 1K for the neutron-irradiated sample is neither 3 or 2 indicating a behavior between the pure crystal V and the non-crystalline
solid G. The thermal conductivity is smaller as in a pure crystal
V, indicating an additional scattering mechanism apart from the
boundary scattering (dominant in a pure crystal for T < 1K). On
the other hand, it is higher than in a glass, indicating that the
TLS are not the dominant scattering mechanism as in non-crystalline
solids. The thermal conductivity due to boundary scattering can be
calculated from the sound velocities in the crystal, taking into

---

*   Later on, using other determinations of the parameters n_0 and M,
we will be able to estimate D from these experiments.

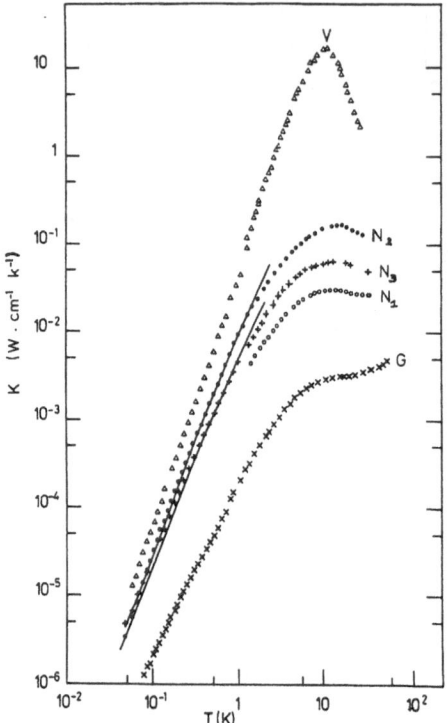

Fig. 12   Thermal conductivity as a function of temperature.   V:
          unirradiated quartz; neutron irradiated quartz: N_1 :
          8×10^{18} n/cm^2; N_2 : 1×10^{18} n/cm^2; N_3 : 6.6×10^{18} n/cm^2;
          G : soda=silicate glass.   Full lines are calculated
          curves (see text).   From /8/.

account the phonon focusing.   The calculated value corresponds very
well to the experimental data.   This has been represented in Fig. 2
where the straight line represents the calculated values.   (For
higher temperatures T > 1K, where point defect scattering and
umklapp processes become dominant, a fit was made, see also Section
IV).   Then, introducing <u>one</u> parameter, the data for the neutron
irradiated samples N_2 and N_3 are fitted, starting from the full
Debye expression [22]

$$K = \frac{k}{2\pi^2 v} \left(\frac{kT}{\hbar}\right)^3 \int_0^{\theta/T} \frac{x^4 \ell^x / \ell^x - 1)^2}{\tau^{-1}(x,T)} \qquad (16)$$

and using, apart from the calculated value of the relaxation time
for the boundary scattering τ_B, one more relaxation time, namely
for the TLS.

$$\tau^{-1} = \tau_B^{-1} + \tau_{TLS}^{-1}$$

With τ_{TLS} from Eq. (1). From this fit we could obtain values for n_0M^2 as seen in Table I[8].

TABLE I [8]

Sample	dose** (n/cm³)	$n_0\overline{M}^2$ (erg/cm³)	$\dfrac{(n_0\overline{M}^2)}{(n_0\overline{M}^2)^*\text{vit.sil.}} \underset{\cong}{\sim} \dfrac{n_0}{n_0},\text{vit.sil.}$
N_2	1×10^{18}	2.7×10^6	0.023
N_3	6.6×10^{18}	5.8×10^6	0.048
N_1	8×10^{18}		0.065 from long. ultrasound [26]

* Taking $n_0\overline{M} = 1.2 \times 10^8$ erg/cm³ a calculated mean value for vitreous silica.

**See also the note in Section III.A.

Taking into account that M is similar in neutron-irradiated quartz as in vitreous silica, as seen in Section III.A, the last column of Table I gives a relative density of states of TLS in neutron irradiated quartz compared to vitreous silica. As a matter of fact, it was only experimentally found that M_L, namely the coupling constant for the longitudinal phonons is similar as in vitreous silica, M involves also M_T, with T for transverse.

The above results are on samples irradiated in S.C.K. Mol. Another set of samples has been irradiated in Grenoble and studied recently [27]. The results are given in Fig. 13 together with previous results and in Table II. The doses are converted to fission doses to be able to compare to the doses given previously. (In CEN Grenoble, the doses are given two ways: $E \gtrsim 0.1$ MeV and $E \gtrsim 1$ MeV; since we know the conversion factor for $E \gtrsim 0.1$ MeV to fission dose, we can calculate a fission dose.)

There is a slight difference in the numbers for the TLS density as calculated here and in the paper [27] because here a calculated mean value for n_0M^2 for the vitreous silica, as in Table I, has been used while in Ref. 27 the n_0M^2 from the fit of the measured (soda)-silicate glass has been used.

A very interesting thermal study (thermal conductivity and specific heat) was published by Gardner and Anderson, also in 1981 [28] for four neutron-irradiated samples, the irradiation doses being always a decade apart. Before coming to conclusions about the presence of TLS, the contribution of three different excitations

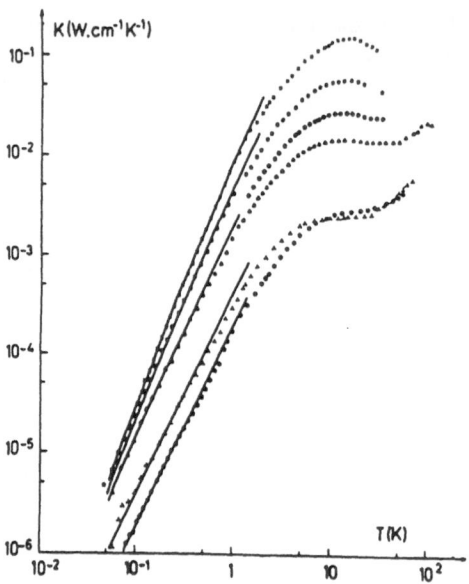

Fig. 13 Thermal conductivity versus temperature. Same curves as
         Fig. 12; additional curves: Pony I:Δ ; Pony II:Δ ; Pony III:
         λ (fourth and fifth curve from the top) /27/.

TABLE II [27]

Sample	dose (n/cm^2)	$n_o \overline{M^2}$ (erg/cm^3)	$\dfrac{n_o}{n_o, \text{ vit.sil.}}$ x
Pony I	1.1×10^{19}	1.56×10^7	0.13
Pony II	3.4×10^{19}	7.8×10^7	0.65
Pony III			

x
 As previously, a calculated mean value for $n_o \overline{M^2} = 1.2 \times 10^{18}$ erg/cm^2
has been used.

had first to be taken into account but could be subtracted: the
impurities gave rise to excitations which influence the specific
heat (because neutron irradiation is always accompanied by
γ-irradiation); furthermore, there were some unknown excitations
causing a Schottky anomaly in the specific heat and finally inclusions
which reduce the thermal conductivity. Comparing specific heat and
thermal conductivity data led to the conclusion that the coupling
constant M must be similar in neutron-irradiated quartz to that in
vitreous silica within 35% (for some dose, an agreement within 5%
was even found). This is in agreement with previous acoustic
experiments quoted in Section III.A. by Golding et al. [26]. From the

thermal conductivity data, they could determine a density of states of TLS for three neutron doses; the results are given in Table III.

TABLE III [28]

Sample	dose* n/cm^2 $(E \gtrsim 0.3 \text{ MeV})$	$\dfrac{n_o}{n_o}$, vit.sil.
27 h	2.7×10^{18}	0.024
270 h	2.8×10^{19}	0.53
2700 h	2.1×10^{20}	0.40

*In our dose units for comparison

There is a good agreement between our sample N_2 and their 27 h sample. The result for the 2700 h sample is rather surprising and not understood. We are planning a more systematic study of the higher doses for which the samples are now being irradiated.

Specific heat and thermal conductivity measurements for one neutron dose have also been reported by Saint-Paul et al. in 1981 and 1982 [29,30]. The used dose was 5.3×10^{18} n/cm^2 $(E \gtrsim 1 \text{ MeV})$ which corresponds to a dose of about 7×10^{18} n/cm^2 in our dose units. They find a TLS density compared to vitreous silica of 0.018 taking M the same as in vitreous silica, which is smaller than our value and that of Gardner et al. for a similar dose. It must mean that the doses cannot be compared for reasons already given above (see note in Section III.A.).

C. Discussion

When discussing our ultrasonic attenuation measurements previously, quantitative values for any parameter (n_o, M or D) could not be deduced yet. Now, after having given the results for the thermal conductivity, a value for D can be found. Two samples irradiated to the same dose at the same time, and being in each others vicinity in the reactor, were used, one for thermal conductivity measurements (N_3) and the other one for ultrasonic attenuation measurements as a function of temperature (see Fig. 10, 6.6×10^{18} n/cm^2). For thermal conductivity we find the ratio:

$$\frac{(n_o M^2)_{N_3}}{(n_o M^2)_{vitr. sil.}} = \frac{n_{o,N_3}}{n_{o,vitr.sil.}} = 0.048$$

From ultrasonic attenuation we find the ratio:

$$\frac{(n_o M^2 D^2)_{N_3}}{(n_o M^2 D^2)_{vitr.sil.}} = \frac{(n_o D^2)_{N_3}}{(n_o D^2)_{vitr.sil.}} = 0.005$$

It is seen that the ultrasonic attenuation is a factor of ten too small to be consistent with a value of D similar in neutron irradiated quartz than in vitreous silica. We have to conclude that D is not the same as in vitreous silica. This is surprising since it was found that M is similar as in vitreous silica. Taking into account small differences in sound velocities and densities, we find:

$$D_{N_3} \simeq 0.4\ D_{vitr.silica}$$

Using this value for D, and M of vitreous silica, we can now calculate the TLS density compared to vitreous silica, from the ultrasonic attenuation curves as a function of temperature, and list them together with the numbers we found from our thermal conductivity experiments. The results are given in Table IV.

In Table V we give an overview of the density of states obtained in different studies, relative to that in vitreous silica.

TABLE IV

Sample	dose (n/cm^2)	TLS density compared to vitreous silica	
N_2	1×10^{18}	0.023	from K, Table I
N_3	6.6×10^{18}	0.048	
N_1	8×10^{18}	0.065	from ultrasonic velocity [26]
N_4	1.2×10^{19}	0.11	from ultrasonic attenuation as a function of temperature,
N_5	2.6×10^{19}	0.30	Fig. 11

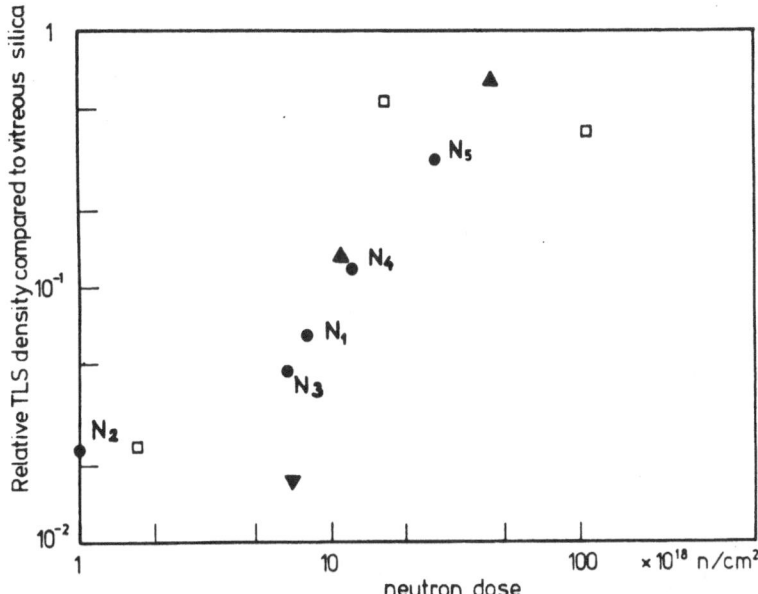

Fig. 14 Relative TLS density compared to vitreous silica as a
          function of neutron dose. For more details, see Table V.
          □ 27 h, 270 h and 2700 h; ▲ Pony I and III; ▼ sample SP.

Figure 14 is a logarithmic plot of these relative density of states
as a function of dose. As may be expected, the data do not lie
on a smooth curve; we already stated above that comparison between
data on samples irradiated at different reactor sites has to be
done with caution. (See note in Section III.A.). It would be
interesting to have more data for the high doses to know whether the
bending back of the curve is genuine. It is remarkable that the dots
for N_4 and N_5 seem to fall on a continuous curve formed by the
other data, since they are calculated with a D value which is only
0.4 of that in vitreous silica.

    Until now only the data giving evidence for TLS were discussed.
It is interesting also to consider the so-called "plateau" region
of the thermal conductivity. In noncrystalline solids at temperatures of about 10K, the thermal conductivity always shows a plateau.
This plateau is until now not yet understood and it is not known
whether or not it is related to the TLS. For discussion about the
plateau, see Refs. 33, 34, 35, and 36.

    It is interesting to remark, however, that upon neutron
irradiation the thermal conductivity of the quartz crystal gradually
evolves to the plateau. This is seen in Fig. 12 and Fig. 13, as
well as in some very old data of Berman [32]. We will see in the
next section that this is not the case for electron-irradiated

TABLE V

Sample	dose (n/cm^2) $E \gtrsim 0.1$ MeV	dose* (n/cm^2) (fission)	relative density of TLS compared to vit. silica.	Irradiated at	See
N$_2$		1x10^{18}	0.023	Mol	Table I
27 h	2.7x10^{18}	(1.7x10^{18})	0.024	Argonne	Table III
N$_3$	1.06x10^{19}	6.6x10^{18}	0.048	Mol	Table I
SP		(6.9x10^{18})**	0.017	Grenoble	[29]
N$_1$		8x10^{18}	0.065	Mol	Table I
Pony I	1.76x10^{19}	(1.1x10^{19})	0.13	Grenoble	Table II
N$_4$		1.2x10^{19}	0.12	Mol	Table I
270 h	2.6x10^{19}	(1.6x10^{19})	0.53	Argonne	Table III
N$_5$		2.6x10^{19}	0.30	Mol	Table I
Pony III	5.45x10^{19}	(3.4x10^{19})	0.65	Grenoble	Table II
2700 h	2.1x10^{20}	1.3x10^{20}	0.40	Argonne	Table III

* doses between brackets are converted to our dose units. The other doses are as given by the reactor staff.

** 5.3x10^{18} n/cm^3, $E \gtrsim 1$ MeV, the conversion factor was found from the doses of Pony I and III which are also in those dose units.

quartz where the thermal conductivity, in that region, can be explained by the presence of point defects. Another observation which is not yet understood, but which is of interest to quote in a review on neutron irradiated quartz, is that the density of states of the TLS in vitreous silica is influenced by neutron irradiation. Upon neutron-irradiation, the dynamical anomalies are reduced in vitreous silica; this has been observed in thermal conductivity and specific heat measurements [35] and in ultrasonic experiments [13].

IV.   ELECTRON-IRRADIATED QUARTZ: DYNAMICAL ANOMALIES

A.   Experiment

          Fast neutrons cause displacement cascades in the crystal, giving rise to extended damaged regions. The TLS may therefore be in the disordered regions, a topic to be discussed more in detail in Section VI. Electrons with energy of 2 MeV irradiated at low fluxes in a Van de Graaf accelerator, are generally accepted not to cause displacement cascades and therefore the damage has to be more limited than in case of the neutrons. As a matter of fact, it is known that the energy of the electron is high enough to create a primary knock on, but the energy of this primary (and this unlike for neutron irradiation) is too small to cause secondary displacements [36]. So at first sight, only point defects can be created. Very recently, however, we got evidence for more damage than only point defects.

          Because of the expected limited damage, it was thought that dynamical studies of electron-irradiated quartz would be an easier, even better approach to the investigation of the microscopic origin of the TLS, than the neutron-irradiated quartz, provided again similar low temperature anomalies are found. Indeed, they were found, but only for relatively high doses of 1.8×10^{20} e/cm^2 (E = 2 MeV) [37]. Figure 15 shows the evidence in the thermal conductivity data for two doses of electron-irradiated quartz. Figure 16 shows the departures of T^3. As in Fig. 13 for neutron irradiated quartz, an intermediate thermal conductivity is found for the electron-irradiated crystal, between the value for the pure crystal and that of the glass. The curve for E_2, at least for T < 1K the region in which we are mainly interested, is very similar to that of N_3 in Fig. 13. The similarity between E_2 and N_3, and the fact that it is generally accepted by now that in neutron-irradiated quartz TLS are present, leads us to the conclusion that also in electron-irradiated quartz these excitations cause anomalies. Below T < 1K a similar fit was done as for N_2 and N_3 in the previous section, calculating first K for the virgin crystal, and then introducing one parameter to account for the reduction of K, namely the relaxation time for the interaction with TLS (see Eq. (1)). This fitting again gives us a value for $n_0 M^2$. These values are listed in Table 6.

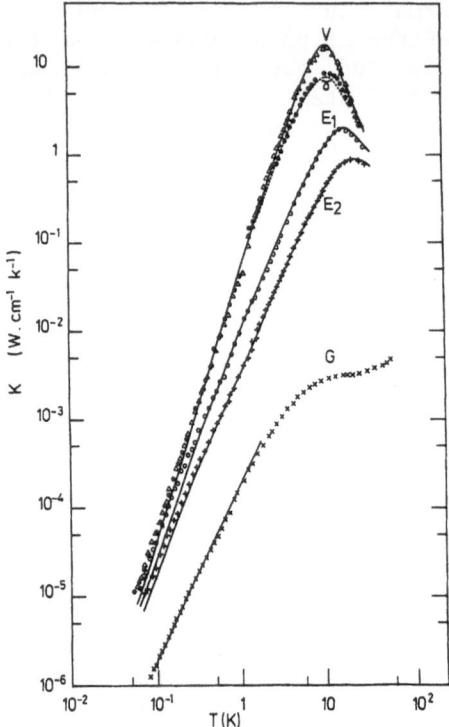

Fig. 15   Thermal conductivity versus temperature.   V: unirradiated
          crystal; E_1 and E_2 respectively 3×10^{19} e/cm² and 1.8×10^{20}
          e/cm² electron irradiated quartz; G: (soda) silicate
          glass; γ-irradiated quartz; full lines are calculated
          curves (from /37/).   Electron energy 2 MeV.

TABLE VI

Sample	dose e/cm² (E=2 MeV)	$n_o \overline{M}^2$ (erg/cm)	$\dfrac{(n_o \overline{M}^2)}{(n_o \overline{M}^2)^*_{vit.sil.}}$
E_1	3×10^{19}	2.6×10^6	0.022
E_2	1.8×10^{20}	8.8×10^6	0.073

$*$ $n_o \overline{M}^2 = 1.2 \times 10^8$ erg/cm is used as a calculated mean value.

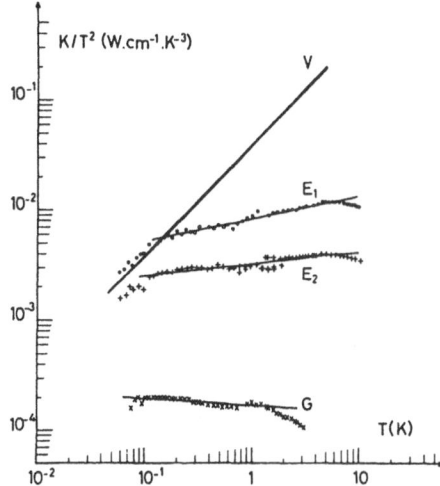

Fig. 16   Same data as Fig. 15 but plotted as K/T^3 versus temperature. Full lines are the calculated curves as in Fig. 15.

The temperature region above 1 K, the so-called plateau region, behaves very differently in electron-irradiated quartz than in neutron irradiated quartz. A dramatic change in thermal conductivity which evolves to the plateau is not found here. A fitting with a point defect relaxation time and umklapp processes can explain the data. Point defect scattering is known to influence the thermal conductivity above 1 K and umklapp processes are the dominant contribution to the thermal resistance above about 10 K [22]. As can be seen in the figure, a good fit could be obtained. Only at the very lowest temperatures there is a departure from the fit, but this is also the case for the virgin sample and is therefore unrelated to the irradiation. It is quite remarkable that while for T < 1 K neutron and electron-irradiated quartz show a similar behavior, that is not the case for the T ≃ 10 K region. This may be an indication that the plateau is unrelated to the TLS.

Thermal conductivity measurements on electron-irradiated quartz as well as specific heat were also done by other authors [38]. Unfortunately, their highest dose used was too low to be able to have conclusive evidence about the presence of TLS. The data are given in Fig. 17. As can be seen, their highest dose is smaller than our lowest dose. But the effects in the thermal conductivity and the specific heat can be entirely explained in terms of the creation of the aluminum-hole center, an electronic center, consisting of a hole trapped at an aluminum impurity [38].

This center is known from studies in γ-irradiated quartz and couples only weakly to phonons [66]. Indeed, the observed low

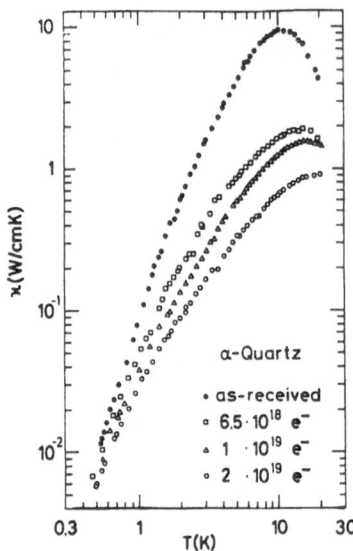

Fig. 17   Thermal conductivity versus temperature for electron irra-
          diated quartz (from /38/) for lower electron doses.

temperature specific heat in the electron-irradiated sample is very
similar to that for certain doses of γ-irradiation. This center
is believed to play only a minor role in our measurements since
upon saturating γ-dose the thermal conductivity is not very much
influenced (Fig. 15). This is a consequence of the low Al content,
12 ppm which is very low for natural quartz (normally at least
50 ppm).

B. Discussion

        The only convincing data about the glassy behavior until
now are the data for the highest dose in Fig. 15 (E_2: 1.8×10^{20} e/cm^2).
For T < 1K a strong effect is seen from the electron-irradiation on
the thermal conductivity. A $T^{2.1}$ behavior is found. Unlike the
data above T > 1K the effect cannot be explained by point defect
scattering since this mechanism is known to influence the thermal
conductivity only above 1K. The centers causing the scattering
at T < 1K should be strong phonon scatterers, like TLS in glasses.
A fitting with the relaxation time of the TLS using one fitting
parameter (n_oM^2) gave good results indicating that the TLS are a
possible explanation. It is difficult to think of another mechanism
giving such a large radiation-induced effect at T < 1K. Only
dislocations could give a similar behavior but an unreasonable
number (10^{10} cm^{-3}, Ref. 37) is necessary to explain the data. Also,
the aluminum hole center cannot explain the data in our samples as
already discussed in the previous section. At first sight, therefore, the radiation-induced TLS should be related to the point

defects created by the electrons. However, very recent ESR
spectra, taken in our laboratory, on sample E_2, probing the E'
center, show evidence for more disorder than just point defects
in E_2. The ESR spectra of the E' center which is believed to be
an oxygen vacancy (see also, Section VI.C.) in electron and neutron
irradiated quartz are in fact very similar. The results are not
published yet and more investigations are being done. Also,
more dynamical studies are being done.

V. WHAT WAS KNOWN ABOUT THE DISORDER

Irradiation studies of quartz date back to 1950. Therefore,
before discussing in detail recent damage studies, which were done
in relation to the dynamical studies, we will first give a short
overview of the general knowledge of radiation damage in quartz.
Since during neutron irradiation in a reactor, also γ's are
produced and since electrons cause also ionization damage, some
attention will be paid to ionization damage, and to γ and X-ray
irradiation. For references, see references in review articles
and books [36,39-43].

A. γ-rays and X-rays

The interaction of ionizing radiation, such as X-rays
and γ-rays, with matter is quite well understood. The basic loss
mechanism of the X- and γ-quanta is due to ionization and excitation of bound electrons in target atoms, a mechanism usually denoted
by ionization. Because of the rigid character of the binding forces
in SiO_2, it is understandable that electro-magnetic radiation is
not capable of causing a significant number of structural defects.
The covalent bond is highly directional with a well-known bond
length and causes the rigid structure which prevents diffusion of
lattice atoms, exchange of lattice positions, and of interstitial-
vacancy pairs.

Ionization effects are therefore limited to changes in the
electronic distribution. The radiation frees electrons, primarily
by Compton processes, which can move through the lattice losing
energy by interacting with and freeing less tightly bound electrons.
A large number of electrons along electron deficient sites, commonly
called holes, is produced this way. The vast majority of the freed
electrons recombine with holes within a very short time. Some
electrons and holes become trapped permanently, especially at lower
temperatures. Lattice defects or impurities act as trapping sites.
In quartz (and silica) the defects are generally present in the
material before irradiation and only in a few cases, an indication
exists that the defect is produced by the irradiation; and even in
these cases, the network probably had been disturbed before.

The calculated displacement energies, however, appear sufficient to displace atoms, at least in the case of γ-irradiation. This irradiation is usually caused by a Co^{60} source and also reactor irradiation causes γ quanta. In both cases, the γ-quanta have an energy centered around 1 MeV. For a nucleus as a target particle, the energy transfer is of the order of some tens eV, which is just on the brink of creating structural damage in solids. Hence, interaction between γ particles and atomic nuclei is not very effective; in addition, the corresponding cross-section is so small that collision between γ-particles and nuclei can be neglected. If, however, the target particle is an atomic electron, then the maximum energy transfer becomes of the order of the energy of the γ-particle. This implies that, although gammas cannot displace atoms in solids efficiently by direct energy transfer, they can, indirectly, by creating energetic electron recoils, the Compton electrons which in turn interact with the atomic nuclei. The scattering cross-section for energy transfer in such a "Compton-scattering" event is about 1 barn in most cases. Semenov et al. [42] calculated the concentration of Frenkel pairs which can be created in such a process. They found that for the usual irradiation doses (< 10^{10} rad) no appreciable concentration of Frenkel pairs is formed. For similar irradiation doses, they found a concentration of Frenkel defects 10^4 times smaller for γ quanta than for electron irradiation.

Therefore X- and γ-irradiation is "ionizing" irradiation. There we may distinguish two processes in quartz: 1) the formation of elementary excitations (electron-hole pairs, excitons, plasmons), 2) the capture of electrons, holes and various mobile ions (N_a^+, L_i^+, H^+) released by the radiation at impurity and intrinsic primary trapping defects under competitive conditions, with the formation of centers which reveal their presence, by the creation of optical maxima, maxima of the mechanical and dielectric losses and by the appearance of ESR signals. These centers are extensively studied; reviews and references are found in Refs. 39-43.

B. Fast Electrons

When a fast electron (energy ∿ 1 MeV) traverses the lattice of a crystalline solid, the basic energy-loss mechanism is due to ionization, which brings us back to the interaction of ionizing radiation with matter described earlier. Ionizing radiation was found to cause no appreciable structural damage in quartz.

Although the electron loses most of its energy in ionization, the little fractional loss which is spent in elastic collisions with atomic nuclei, accounts for the production of lattice defects. If in an elastic process the lattice atom receives an energy in excess of the energy necessary to displace it, the atom leaves its site and a vacancy is created. If this "primary-knock-on" atom comes to rest in a non-equilibrium or interstitial site, an interstitial

is also created, hence a Frenkel pair. Because of the small mass
of the electron (compared to other charged particles), the energy
of the primary after the collision of the electron is mostly too
small to cause secondary displacements as opposed to other particles which create displacement cascades. Therefore, complex localized
regions of damage that result from a cascade are excluded and the
Frenkel defects are randomly distributed throughout the affected
region of the specimen. Seitz and Koehler have determined the
displacement cross-section as a function of electron energy for a
number of target materials: σ_d increases rapidly from zero at the
so-called threshold (~ 0.5 MeV) to a nearly constant value for high
electron energies. Depending on the atomic mass, σ_d at saturation
takes values of 10^{-23} to 10^{-22} cm^{-2} or 10 to 100 barn. The Frenkel
defect concentration in quartz caused by electrons of 1 MeV was
calculated by Semenov and Fothenkov to be 1.1×10^{14}/cm^3 for a dose
of 10^{14}/cm^2.

   Although it was generally accepted that _ionization_ during fast
electron irradiation causes no structural damage, more recently
some authors attribute their observations in electron microscopy
to the ionization damage. At a prolonged electron exposure, in situ
electron microscopy studies, reveal crystalline to amorphous transition [44]. It is attributed to a process which is called radiolysis (= ionization and subsequent displacement). It has a displacement cross-section inversely proportional to the electron energy and
is therefore mainly important at low energies (kV range). Also,
Revesz [45] attributes his observations rather to changes of bond
angles, as a consequence of ionization, than to the creation of
Frenkel pairs. This change of bond angles would be caused by
changes in electron distribution due to the ionizing irradiation.
Extrapolation of the data of Das and Mitchell [44] give a dose
necessary to render quartz amorphous, by this process, of about 10^{22}
electrons/cm^2 for 2 MeV electrons. This dose is much higher than
doses usually used and an irradiation time of many months would be
necessary in a Van de Graaf accelerator. In addition for such a
dose, almost every particle of the quartz will have been displaced
by the head-on collision process described earlier. Since in an
in situ observation of electron irradiation damage in an electron
microscope, the electron flux is about 10^6 times higher as in a
Van de Graaf accelerator and the specimens are very thin, the above
described processes were not expected to play a role in our
irradiations. The very recent ESR spectra showing evidence for
more disorder than only point defects, however, may indicate that
indeed, also at low electron fluxes and high energies, radiolysis
takes place. This will now be studied in detail in the future.

C. Neutron Irradiation

   In contrast to γ and electron irradiation, fast neutrons
are capable of doing severe damage to the intrinsic structure of the

quartz by displacement of lattice atoms which can even cause
secondary displacements and result in displacement cascades causing
disordered regions. Already in 1954, Wittels and Sherril [46]
reported that from a dose (mostly authors do not give specification
about the used dose units, if they do, we quote them) of 3×10^{19}
n/cm^2 on long range order is gradually diminished until the quartz
has suffered a bombardment of about 1.2×10^{20} n/cm^2. For lower
doses than 3×10^{19} n/cm^2, structural defects are according to
Wittels [47] point defects and slightly disordered regions. In the
dose range 3×10^{19} n/cm^2-2×10^{20} n/cm^2, many structure sensitive
properties have been investigated and they all indicate that the
quartz structure becomes increasingly glass-like. For instance,
the density decreases by more than 14% to a value which is close
to, but not the same as in vitreous silica; as a matter of fact,
the density evolves to a value which is the same as in heavily
neutron-irradiated vitreous silica, which is compacted by neutron
irradiation. This can be seen in Fig. 18 [39]. In fact, all forms
of silicon dioxide are transformed in optically isotropic, glass-
like material with virtually identical density, thermal expansion,
elastic properties and absence of X-ray diffraction pattern [39].

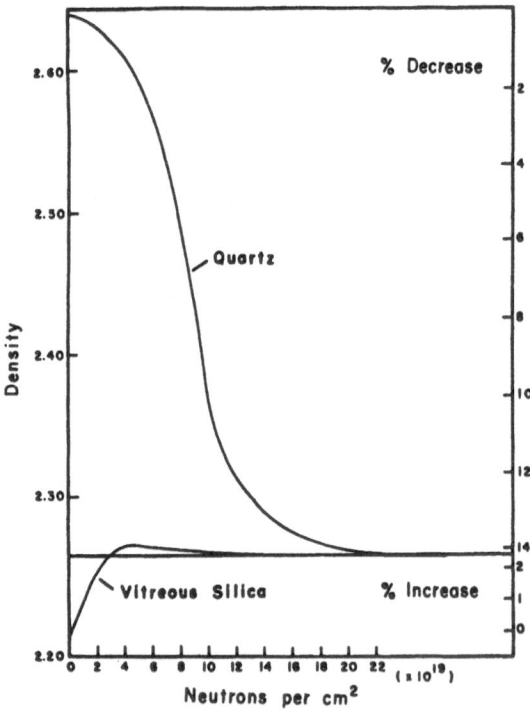

Fig. 18   Density change of quartz and vitreous silica as a function
          of fast neutron exposure (from /39/).

This led to the conclusion that neutron irradiation produces a
new phase of silicon dioxide, sometimes named the metamict phase,
adapting the earlier designation for minerals disordered by
radiation in the geological past. The fact that the metamict
phase is different from vitreous silica indicates that the disordered regions introduced at doses even lower than $3x10^{19}$ n/cm^2 may
be glass-like, but are probably different from vitreous silica. An
indication for that is the high paramagnetic defect concentration
in neutron irradiated quartz, even for doses lower than $3x10^{19}$
n/cm^2 as opposed to vitreous silica.

Most of the studies, until recently, involve doses higher than
10^{19} n/cm^2. The most detailed studies are by Wittels [47],
Comes, Lambert and Guinier [48] and Bates, Hendriks and Shaffer [49].
Wittels showed that at doses smaller than $3x10^{19}$ n/cm^2, the crystal
expands anisotropically ($\Delta a/a > \Delta c/c$) as with thermal expansion.
Also, taking into account other studies, a model of the basic of
point defects and slightly disordered regions is developed and
the point defects are thought to be oxygen vacancies. The presence
of point defects was revealed by ESR studies [50], a method extensively used to study the radiation induced defects in quartz. For
doses higher than $3x10^{19}$ n/cm^2, the observations are compatible
with the formation of vitreous regions. At $7x10^{19}$ n/cm^2, a diffuse
X-ray scattering halo appears in the diffraction pattern although
the crystal is still long-range ordered. The peak position of the
diffuse halo is close to that for unirradiated vitreous silica,
but the shape and the position of this diffuse scattering is progressively altered with increasing dose until the rate of density
change becomes zero. Therefore, it is evident that the configuration of the atoms within the amorphous regions changes upon
increased irradiation: the shift of the peak is indicative of
an alteration of the average interatomic distances within the
disordered regions. In addition, it is found that the interatomic
distances within the disordered regions, is smaller than those
found in unirradiated silica.

Comes et al. did a very detailed diffuse X-ray study of the
transition of crystalline quartz into the amorphous phase upon
neutron irradiation. For doses higher than $2x10^{19}$ n/cm^2, they
also find islands of highly perturbed matter in a still crystalline
matrix. Before becoming completely amorphous, irradiated quartz
consists of two phases, the crystal phase and the amorphous phase,
the proportion of which grows with the irradiation dose. However,
the crystal has no more a perfect lattice and the amorphous phase
is not completely disordered. The disordered phase for intermediate
stages is found to be anisotropic: it keeps some relation with the
parent crystal structure. This is a apparent in the diffraction
pattern which is not yet the simple ring pattern of the amorphous
silica. On the other hand, the crystalline matrix is somewhat
disturbed. A characteristic diffuse scattering is observed which

seems to be similar to the scattering observed in unirradiated
quartz near the α-β transition. What is permanent and nearly
unchanged during the whole process of radiation damage is the
building unit of silica, the tetrahedron SiO_4 and what is modified
is the mutual arrangement of these tetrahedra.

Bates et al. used Raman and IR spectroscopy and small angle
X-ray scattering to study the irradiation of crystalline and vitreous silica. Their results are consistent with the investigations
above. Recently, very detailed diffuse X-ray studies were reported
which also show the presence of disordered clusters in neutron
irradiated quartz for doses as low as 3×10^{18} n/cm^2. These experiments will be discussed in the next section.

VI. RELATED DAMAGE STUDIES

A. Diffuse X-ray Scattering Studies

Grasse et al. [51,52] did careful diffuse X-ray scattering
studies of neutron-irradiated quartz and also of the electron-irradiated sample E_2 of Section IV [53]. They were able to investigate
lightly irradiated quartz in a dosage regime which was until now
uncharacterized and for which a lot of "anomalous" dynamical data
are available. They did measurements of the diffuse X-ray scattering
close to the Bragg reflections, of lattice parameter changes and
of glass-like, or amorphous, scattering from quartz crystals in the
dose range 3 to 49×10^{18} n/cm^2 (E $\gtrsim$ 0.1 MeV) corresponding to
$1.9-30 \times 10^{18}$ n/cm^2 is our dose units. They find evidence for the
creation of defective regions even for their lowest dose. With
increasing dose, the diameter of the defective regions (about 20 Å)
does not change very much but their number increases. An amorphous
scattering, very similar to that in a fully amorphized crystal,
and in heavily irradiated vitreous silica, is observed and in the
more lightly irradiated samples this amorphous scattering seems to
grow with increasing irradiation in proportion to the number of
defective regions leading to the conclusion that the defective
clusters are most likely "amorphous".

"Amorphous" here means similar as heavily irradiated vitreous
silica and not similar to unirradiated vitreous silica: as already
seen in Section V.C, Wittels (however, for higher doses) concludes
that upon increasing irradiation dose, the configuration of the
atoms in the disordered regions changes. The results are summarized
in Tables VII and VIII, taken from Ref. 52.

Comparing the TLS density in neutron-irradiated quartz and the
volume fraction of amorphous SiO_2 led Grasse et al. [51] to the
conclusion that the locations of the TLS are most likely the amorphous clusters and therefore glass-like in nature. This conclusion
was however not followed by other authors [54] who found that the

TABLE VII [52]

Influence of irradiation and effective number of atoms associated
with defect cluster, $\Delta V/\Omega$, concentration of clusters, c, and
approximate cluster radius r_{cl}.

Dose $E \gtrsim 0.1$ MeV	Dose (n/cm^2) $E \gtrsim 0.3$ MeV	$\Delta V/\Omega$	c x 10-5	r_{cl}
$3x10^{18}$	$1.9x10^{18}$	29	13.5	10 Å
$8.2x10^{18}$	$5.1x10^{18}$	54	22.5	12 Å
$16x10^{18}$	$10x10^{19}$	54	71	12 Å

TABLE VIII [52]

Estimate of the volume fractions of amorphous SiO_2

Dose $E \gtrsim 0.1$ MeV	Dose (n/cm^2) $E \gtrsim 0.3$ MeV	From amorphous scattering	From Huang scattering
$3x10^{18}$	$1.9x10^{18}$	2.7%	4.5%
$8.2x10^{18}$	$5.1x10^{18}$	10 %	13 %
$16x10^{18}$	$10x10^{18}$	25 %	41 %

TLS density does not scale with the volume fraction of the amorphous
regions. A reason for the fact that both conclusions do not agree
may be that in the first, samples were compared which in fact may
have suffered very different damage. Since they were irradiated
in different reactors, this would not be surprising.

The conclusion that the TLS do not necessarily reside in such
large amorphous clusters is supported by the studies on electron

irradiated quartz. As already seen in Section IV, the thermal
conductivity in sample E_2 (1.8×10^{20} e/cm^2) is very similar as that
in the neutron-irradiated sample N_3 (dose 6.6×10^{18} n/cm^2). According to the diffuse X-ray study discussed above, N_2 should have a
volume fraction of about 6% (see Table VIII). If TLS would have to
be in amorphous clusters, also E_2 should have a similar volume
fraction of amorphous clusters. This is not the case: similar
diffuse X-ray scattering measurements [53], as in the neutron
irradiated samples of Table VIII, were done in the electron-irradiated sample E_2 of Table VI. No evidence was found for amorphous
clusters. The absence of amorphous scattering indicates that the
volume fraction of amorphized regions must be smaller than 3%.
In addition, no diffuse scattering intensity could be detected.
If the defects with the same strain as in neutron-irradiated quartz
are created, it can be deduced that the defect concentration has to
be 30 times smaller for the e$^-$ irradiated samples as for a
1.9×10^{18} n/cm^2 neutron-irradiated sample. From that, it was
concluded that in the electron-irradiated sample no such large
defective regions, as in neutron-irradiated quartz, are present.

B. Positron Annihilation Lifetime Investigations

Positron annihilation is now a well-established method to
study electronic and structural properties of solids. Information
about the material under investigation can be obtained by measuring
the angular or energy distribution of the annihilation radiation
or by positron lifetime measurement. Recent publications have
discussed various aspects of positron annihilation in quartz, silicates, and neutron-irradiated quartz using lifetime measurements
[53,56,57,58]. Amorphous and crystalline quartz turn out to have
different annihilation characteristics and neutron-irradiated quartz
behaves as a mixed phase of amorphous and crystalline quartz in
agreement with diffuse X-ray scattering investigations (see Section
V.C. and VI.A.). While the lifetime spectra in pure quartz can be
fitted by two lifetimes (120 ps and 285 ps), the data for vitreous
silica can only be explained with the use of three lifetime components. Also, neutron-irradiated quartz shows a third lifetime
component, showing the existence of disordered regions. Figure 19
shows a typical spectrum for pure quartz, neutron-irradiated quartz
and vitreous silica. Table IX gives the lifetimes and their intensities for a γ-irradiated quartz sample which behaves as an
unirradiated sample [56], the neutron-irradiated samples N_2 and N_3
of Section III.A., the electron irradiated sample E_2 of Section IV
and a vitreous silica sample. It is clearly seen that neutron
irradiation induces a long lifetime component as in vitreous silica,
consistent with the presence of amorphous clusters. The intensity
of the long lifetime increases with increasing dose (for higher
doses, see Ref. 56). For saturating doses (> 1×10^{20} n/cm^2), this
third lifetime even becomes higher than in vitreous silica [56] and
reaches a value similar as in irradiated vitreous silica [58],

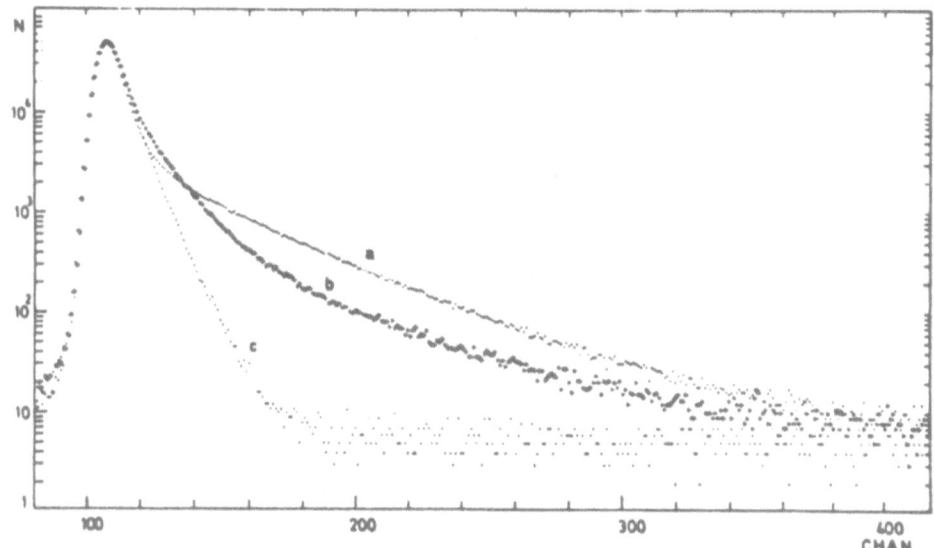

Fig. 19   Positron annihilation lifetime spectrum of vitreous silica
(a), neutron irradiated quartz, 12×10^{19} n/cm^2 (b),
α-quartz after low γ-dose (c). From (/56/).

TABLE IX

Results of the positron lifetime measurements [57].

Sample	Received Doses	Lifetimes (with t_1 fixed at 120 ps)		Intensities (in %)		
		t_2(ps)	t_3(ps	I_1	I_2	I_3
γ	10^8 rad gamma	285 ± 7		60.0 ± 6	40.0 ± 6	
E_2	1.8×10^{20} e/cm^2	287 ± 5		58.1 ± 3.	41.9 ± 3	
N_2	1×10^{18} n/cm^2	330 ± 6	980 ± 80	61.8 ± 1.8	36.7 ± 1.3	1.5 ± 0.5
N_3	6.6×10^{18} n/cm^2	343 ± 6	950 ± 90	63.3 ± 3.8	34.8 ± 3.7	1.9 ± 0.4
G'	---	460 ± 20	1540 ± 20	63.8 ± 1.5	10.7 ± 0.3	25.2 ± 1.5

G' = vitreous silica sample 471 of Ref. 56.
γ: unirradiated quartz or γ-irradiated quartz for such a low γ-dose
give the same result [56].

supporting the earlier observations (see Section V.C.) that neutron
irradiated quartz and vitreous silica evolve to a common phase,
different from vitreous silica. The fact that the third lifetime
is different from that in vitreous silica and changes with dose is
consistent with the findings of Wittels (see Section V.C.) that the
configuration of the atoms within the amorphous regions is not the
same as in vitreous silica and changes with dose. The behavior of
the second lifetime is comprehensible, considering the difference
between the positron lifetime t_2 in unirradiated samples (t_2 is
about 280 ps) and in quartz irradiated to saturation where t_2 is
much longer [56]. The lifetime t_2 in neutron-irradiated quartz
can be considered as a weighted mean value taking into account the
crystalline and amorphous part of the crystal. Both the behavior
of the second and the third lifetime show that amorphous regions
are present for doses as low as 1×10^{18} n/cm^2. Previous diffuse
X-ray scattering studies (see Section VI.A.) gave evidence for the
presence of amorphous clusters for doses as low as 1.9×10^{18} n/cm^2
in our dose units.

Contrary to neutron irradiation, the positron annihilation
behavior in α-quartz is found not to be influenced by electron
irradiation. As can be seen in Table IX, there is no third lifetime
present and the second lifetime is, within experimental accuracy,
the same as that in the unirradiated material (or γ-irradiated).
As discussed in Section V.B., this is expected since electrons,
because of their small mass, do not easily induce cascades. It is
also in agreement with the diffuse X-ray scattering experiments,
which were done in the same sample. In view of the thermal conductivity results for $T < 1K$, also obtained in the same sample E_2,
which are very similar to those obtained in the neutron-irradiated
sample N_3, this must mean than the thermal conductivity anomaly is
not necessarily related to such 20 Å amorphous regions as found in
neutron-irradiated quartz. This means that TLS are not necessary
in such amorphous regions as already discussed above.

C. Paramagnetic Susceptibility and ESR

The structural damage of solids by irradiation involves
a variety of processes, the simplest being the production of atomic
displacements and in covalently bonded crystals (what quartz partly
is) of broken bonds, or free radials. There will, in general, be
several ionization states available to such defects and some of
these may have an odd number of electrons giving a paramagnetic
center. It should be possible then to study some aspects of the
defects produced upon irradiation by magnetic techniques. Static
paramagnetic susceptibility measurements were done and ESR spectra
taken of some of the samples on which dynamical studies (see
Section IV and V) or other damage studies (see Section VI) were done.
In order to obtain evidence about damage versus low temperature
anomalies the studies were done on samples irradiated, placed very

close together, during the same reactor cycle, to rule out the influence of the difference in (reactor) spectrum and flux calibration.

Paramagnetic susceptibility measurements in neutron-irradiated quartz have been done before and Fig. 20 shows the number of paramagnetic centers as a function of dose [59]. Our own data are also given [60]. Generally it is accepted that the paramagnetic center production is equal to 3.7 per incident fast neutron [39] up to a dose of about 3×10^{19} n/cm^2. For higher doses, the center production goes down again and finally saturates at a dose of 2×10^{20} n/cm^2 [61]. Figure 20 also shows the paramagnetic centers in electron-irradiated quartz (E_1 and E_2). For neutron and electron doses which give similar low temperature thermal conductivity anomalies (E_2 and N_3) the number of paramagnetic centers produced is very different. According to these results, there is no relation between the paramagnetic center production and the low temperature properties. This is also not expected, since in (unirradiated) vitreous silica the number of paramagnetic centers is very small (10^{15}/cm^3).

There are numerous ESR studies in irradiated quartz trying to identify the defects produced by the irradiation. The most studied center is E' center [50,39,62, and references therein]. It is the only intrinsic point defect which has been firmly established in crystalline quartz, as opposed to vitreous SiO$_2$ where more intrinsic point defects are well known [63]. Apart from the E' center, which is identified as an oxygen-vacancy center, some impurity related defects give rise to ESR signals and have been studied extensively, as the so-called aluminum hole defect which is an Al. substituted for a Si+trapped hole. There has been some question

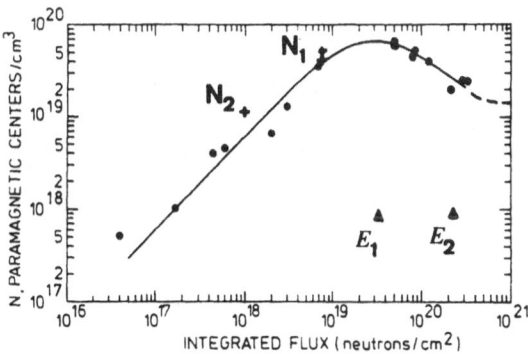

Fig. 20 Paramagnetic center production upon neutron irradiation.
• Ref. 59; + our data. The ▲ are the values for electron-irradiated quartz E_1, and E_2, the dose is than in e/cm^2.
For E_1 it is an upper limit (from /60/).

as to whether SiO_2 defects are of the separated vacancy-interstitial type, or whether they occur as "valence alternation pairs", or VAPS.

Present experimental evidence favors the vacancy-interstitial type of defects, albeit with large atomic relaxations [63]. The E_1' center in α-quartz, is thought to be an O^- vacancy, with the remaining electron localized on one silicon via, large asymmetric relaxations of the silicons [64]. Other oxygen vacancy, or E' centers, also exist. Although it is unlikely that the E' centers are directly related with the TLS (very low E' density in vitreous silica) they may turn out to be a tool to probe their environment. A remarkable observation was done very recently in sample E_2 by R. Weeks and A. Stesmans [65]: the ESR line showed a background which only can be attributed to the presence of more disorder than just a point defect. The lines are still under investigation and the results will be published later. The damage may be a consequence of disordering due to radiolysis, a process discussed in Section V.B. In view of our low temperature thermal conductivity anomalies and of the microscopic origin of the TLS, these observations are very interesting. The TLS may be related to such small defect centers, which are not just point defects, but also are not such extended amorphous regions as observed in neutron-irradiated quartz (see Sections VI.A. and VI.B.).

ACKNOWLEDGMENTS

The author is grateful to S. Hunklinger, M. von Schickfus, B. Golding, R. Orbach, K. Dransfeld, R. Weeks and A. C. Anderson for stimulating discussions on the physical principles underlying this work. Particular thanks go to A. M. de Goër, A. Van den Bosch, D. Segers for the extended collaborations on the irradiated quartz problem. Also M. Locatelli, Mbungu-Tsumbu, M. Dorikens, L. Dorikens-Vanpraet, D. Grasse, H. Peisl, D. Müller, V. Esteves and J. Cornelis have contributed to that study.

Financial support by the Belguim I.I.K.W. is gratefully acknowledged.

REFERENCES

1. W. A. Phillips, ed., _Amorphous Solids: Low Temperature Properties_, Springer Verlag (1981).
2. P. W. Anderson, B. I. Halperin and C. M. Varma, Philos. Mag. 25:1 (1972); W. A. Phillips, J. Low Temp. Phys. 7:351 (1972).
3. C. Laermans, Phys. Rev. Lett. 42:250 (1979).

4. R. Comes, M. Lambert, A. Guinier, Interaction of Radiation with Solids, A. Bishay, ed., Plenum Press, New York, p. 319 (1967).
5. D. Grasse, O. Kocar, J. Peisl, S. C. Moss and B. Golding, Phys. Rev. Lett. 46:261 (1981).
6. R. C. Zeller and R. O. Pohl, Phys. Rev. B4:2029 (1971).
7. R. B. Stephens, Phys. Rev. B8:2896 (1973); J. C. Lasjaunias, A. Ravex, M. Vandorpe and S. Hunklinger, Solid State Comm. 17:1045 (1975).
8. A. M. de Goer, M. Locatelli and C. Laermans, J. Physique 42:C6-78 (1981).
9. M. P. Zaitlin, A. C. Anderson, Phys. Rev. B12:4475 (1975).
10. S. Hunklinger, W. Arnold, S. Stein, R. Nava, K. Dransfeld, Phys. Lett. 42A:253 (1972).
11. B. Golding, J. E. Graebner, R. J. Schutz, Phys. Rev. Lett. 30:233 (1973).
12. C. Laermans, W. Arnold, S. Hunklinger, unpublished results.
13. C. Laermans, L. Piche, W. Arnold, S. Hunklinger, in Non-Crystalline Solids, G. H. Frischat, ed., Trans. Tech. Aedermanns-dorf, p. 562 (1977).
14. L. Piché, R. Maynard, S. Hunklinger, J. Jäckle, Phys. Rev. Lett. 32:1426 (1974).
15. B. Golding and J. E. Graebner, Phys. Rev. Lett. 37:852 (1976).
16. B. Golding, Proc. Ultrasonic Symp., IEEE, 692 (1976).
17. M. von Schickfus, S. Hunklinger, Phys. Lett. 64A:144 (1977).
18. B. Golding, M. von Schickfus, S. Hunklinger, and K. Dransfeld, Phys. Rev. Lett. 43:1817 (1979).
19. C. Laermans, W. Arnold, S. Hunklinger, J. Phys. C10:L161 (1977).
20. P. Doussineau, A. Levelut, Ta Thu-Thuy, J. Phys. Lett. 38:L37 (1977).
21. S. Hunklinger and W. Arnold, Physical Acoustics, Vol. 12, W. P. Mason and R. N. Thurston, eds., Academic Press, New York, 155 (1976).
22. R. Berman, Thermal Conduction in Solids, Clarendon Press, Oxford (1976).
23. G. Federle and S. Hunklinger, J. de Physique, 43:C9-505 (1982).
24. C. Laermans, V. Esteves, Phonon Scattering in Condensed Matter, W. Eisenmenger, K. Lassman, S. Döttinger, eds. 407 (1983).
25. J. Jäckle, L. Piché, W. Arnold, S. Hunklinger, J. Non-Cryst. Sol. 20:365 (1976).
26. B. Golding, J. E. Graebner, W. H. Haemmerle and C. Laermans, Bull. Am. Phys. Soc. 24:495 (1979); B. Golding and J. E. Graebner, Phonon Scattering in Condensed Matter, H. J. Maris, ed., Plenum Press, 11 (1980).
27. A. M. de Goër, N. Devismes, B. Salce, C. Laermans, Int. Conf. on Defects in Insulating Crystals, Salt Lake City, 8:20-24 (1984).
28. J. W. Gardner and A. C. Anderson, Phys. Rev. B23:474 (1981).
29. M. Saint-Paul, J. C. Lasjaunias, J. Phys. C.:Sol. St. Phys. 14:L365 (1981).

30. M. Saint-Paul, J. C. Lasjaunias, M. Locatelli, J. Phys. C.:Sol.
    St. Phys. 15:2375 (1982).
31. R. Nava, private communication.
32. R. Berman, Proc. Roy. Soc. London, A208:90 (1951).
33. The "plateau" is duscussed in a paper by A. C. Anderson in
    Amorphous Solids, Low Temperatures Properties, W. A. Phillips,
    ed., Springer Verlag (1981). This paper gives references
    to other papers.
34. A possible origin of the "plateau" is also given in: S.
    Alexander, C. Laermans, R. Orbach, H. M. Rosenberg, Phys.
    Rev. B28:4615 (1983).
35. T. L. Smith, P. J. Anthony, A. C. Anderson, Phys. Rev. B17:4997
    (1978).
36. D. S. Billington, J. H. Crawford, Radiation Damage in Solids,
    Princeton U.P., NJ (1961).
37. C. Laermans, A. M. de Goër and M. Locatelli, Phys. Lett. 80A:331
    (1980).
38. M. Hofacker, H. von Löhneysen, Z. Physik B:42:291 (1981).
39. E. Lell, N. J. Kreidl, J. R. Hensler, Radiation Effects in Quartz,
    Silica and Glasses, Progress in Ceramic Science, Vol. 4,
    J. E. Burke, ed., Pergamon Press, 3 (1966).
40. C. Lehman, Interaction of Radiation with Solids and Elementary
    Defect Production, North Holland (1977).
41.
42. K. P. Seminov and A. A. Fotchenkov, Soc. Phys. Crystallography
    22:571 (1977).
43. W. Primak, The Compacted States of Vitreous Silica, Gordon
    and Breach, New York (1975).
44. Gopal Das, T. E. Mitchell, Rad. Eff. 23:49 (1974); L. W. Hobbs,
    M. R. Pascussi, J. de Physique 41:C6-237 (1980).
45. A. G. Revesz, Sol. St. Comm. 10:127 (1972).
46. M. C. Wittels, F. A. Sherril, Phys. Rev. 93:1117 (1954).
47. M. C. Wittels, Phil. Mag. 2:1445 (1957).
48. R. Comes, M. Lambert, H. Guinier, in Interaction of Radiation
    in Solids, A. Bishay, ed., Plenum Press, New York 319 (1967).
49. J. B. Bates, R. W. Hendrickx, L. B. Shaffer, J. Chem. Phys.
    61:4163 (1974).
50. R. A. Weeks, J. Appl. Phys. 27:1376 (1956).
51. D. Grasse, O. Kocar, H. Peisl, S. C. Moss and B. Golding, Phys.
    Rev. Lett. 46:261 (1981).
52. D. Grasse, O. Kocar, H. Peisl and S. C. Moss, Rad. Eff. 66:61
    (1982).
53. D. Grasse, M. Müller, H. Peisl and C. Laermans, J. de Physique,
    43:C9-119 (1982).
54. A. C. Anderson, J. A. McMillan and F. J. Walker, Phys. Rev.
    B24:1124 (1981).
55. G. Brauer, G. Boden, A. Balogh, A. Andreeff, Appl. Phys. 16:231
    (1978).

56. Mbungu Tsumbu, D. Segers, F. Van Brabander, L. Dorikens-Vanpraet,
        M. Dorikens, A. Van den Bosch, Rad. Eff. 61:22 (1982).
57. C. Laermans, Mbungu-Tsumbu, D. Segers, M. Dorikens, L. Dorikens-
        Vanpraet, A. Van den Bosch, J. Phys. C.:Sol. State Phys.
        17:763 (1984).
58. Mbungo-Tsumbu, D. Segers, M. Dorikens, L. Dorikens-Vanpraet,
        C. Laermans, A. Van den Bosch, J. Non-Cryst. Solids 65:131
        (1984).
59. D. K. Stevens, W. J. Sturm and R. H. Silsbee, J. Appl. Phys.
        29:66 (1958).
60. C. Laermans, A. Van den Bosch, J. Cornelis and J. Vansummeren,
        J. de Physique 43:C9-119 (1982).
61. A. Van den Bosch, private communication.
62. L. E. Halliburton, B. D. Perlson, R. A. Weeks, J. A. Weil and
        M. C. Wintersgill, Sol. St. Comm. 30:575 (1979); and
        references herein.
63. W. Beal Fowler in Semiconductors and Insulators, Vol. 5,
        Gordon and Breach, 583 (1983).
64. F. J. Feigl, W. B. Fowler, and K. L. Yip, Sol. State Comm.
        14:225 (1974); K. L. Yip and W. B. Fowler, Phys. Rev. B11:2327
        (1975).
65. R. A. Weeks and A. Stesmans, private communication.
66. M. Rodriguez, F. Garcia-Golding and R. Nava, Phys. Lett. 79A:241
        (1980); J. Chaussy, J. Le G. Gilchrist, J. C. Lasjaunias,
        M. Saint-Paul, and R. Nava, J. Phys. Chem. Solids 40:1073
        (1979).

NOTE ADDED IN PROOF:

After this article has been completed ultrasonic experiments
in neutron-irradiated quartz have shown the presence of a shoulder
around 10 K giving evidence for the relaxation ultrasonic attenuation due to TLS in the regime $\omega\tau \ll 1$ (C. Laermans, V. Esteves,
A. Vanelstraete, Rad. Effects, May 1986 to be published). Also
evidence for a thermally activated relaxation process giving rise
to a peak around 50 K has been observed (V. Esteves, A. Vanelstraete,
C. Laermans, in "Phonon Physics" eds. J. Kollar, N. Kroo, N. Menyhard,
T. Siklos, World Scientific 1985, p. 45). In electron-irradiated
quartz additional evidence was found for the presence of TLS in the
low temperature ultrasonic attenuation which shows a T^3 behavior
(C. Laermans, A. Vanelstraete, in "Phonon Scattering in Condensed
Matter," eds. A. C. Anderson, J. P. Wolfe, Springer-Verlag 1986,
to be published).

THERMAL BLEACHING OF γ-RAY-INDUCED DEFECT CENTERS IN HIGH PURITY

FUSED SILICA BY DIFFUSION OF RADIOLYTIC MOLECULAR WATER

David L. Griscom

Naval Research Laboratory
Washington, DC 20375-5000

ABSTRACT

Isochronal and isothermal anneal curves are presented for paramagnetic defect centers in high purity Type III (1200 ppm OH) fused silicas which had been γ irradiated. Successful mathematical fits to the data were achieved by means of a modified second order kinetic theory formulation under the operating hypothesis that E' centers are annealed in the temperature range 500-700K by reaction with radiolytically-formed interstitial water molecules. Thus, H_2O becomes the third small molecule (after O_2 and H_2) recently demonstrated to account for the thermal bleaching of radiation-induced defect centers in amorphous silica by diffusion-limited processes. Significant dependences of the annealing behavior on γ-ray dose and time after irradiation are reported and discussed in the context of the H_2O diffusional model. Finally, it is emphasized that removal of the paramagnetic centers by this mechanism would leave behind an ensemble of charged, diamagnetic defects which could be reactivated by subsequent irradiations unless care is taken to remove them by higher temperature anneals.

I. INTRODUCTION

Amorphous silicon dioxide ($a-SiO_2$) is a critical material component in many modern technologies, including conventional optics, fiber optics, and metal-oxide-semiconductor (MOS) devices. For this reason, considerable current interest centers on the influences of nuclear and space radiations on the optical and electronic properties of $a-SiO_2$. Radiation-induced defect centers can be responsible for stress build-ups [1] in the surfaces of glass lenses or mirrors, optical losses in fiber waveguides [2-4],

369

or unwanted electrical property shifts in MOS structures [5-10].
Not only is it desirable therefore to understand the natures of
these defect centers, but it is also important as well to determine
the mechanisms by which they thermally bleach. For example, the
well-documented room temperature fading of radiation-induced
attenuation in silica-based optical fibers [2-4] is not yet understood in detail. Similarly, it is not known for certain that the
anneal schedules employed in the manufacture of MOS devices are
sufficient to totally heal the damage created in the oxide layer
by processing steps such as ion implantation [6-8] or x-ray
lithography [9].

The natures of radiation-induced defect centers in pure and
doped a-SiO$_2$ are becoming increasingly well understood as a
consequence of intensive spectroscopic studies. From electron
spin resonance (ESR), three generic defect types have been
elucidated in high-purity silicas: the E' center ($\equiv$Si$\cdot$) [11-14],
the nonbridging oxygen hole center ($\equiv$Si-O$\cdot$) [15-16] and the
superoxide radical ($\equiv$Si-O-O$\cdot$) [16-17]. (In the present notation,
three parallel lines are used to denote chemical bonds between a
silicon atom and three oxygen neighbors in the glass network; a
dot designates an unpaired electron.) However, until recently
little progress had been made toward understanding the mechanisms
by which these centers thermally bleach.

Edwards and Fowler [18] were the first to propose a mechanism
based on the diffusion of small molecules. Specifically, the
decay of E' centers observed to occur near 200°C in γ-irradiated
low-OH a-SiO2 accompanied by a one-for-one growth in superoxide
radical concentration [15] was proposed to be due to the diffusion-
limited reaction [18]

$$\equiv Si\cdot + O_2 \xrightarrow{\text{T>400K}} \equiv Si\text{-}O\text{-}O\cdot \tag{1}$$

Shortly thereafter, Griscom [19,20] presented an entirely separate
case for the annealing of nonbridging oxygen hole centers in
x-irradiated high-OH silicas by the diffusion of radiolytic molecular hydrogen according to the sequence:

$$\equiv Si\text{-}OH \xrightarrow{h\nu} \equiv Si\text{-}O\cdot + H^0 \tag{2}$$

$$2H^0 \xrightarrow{\text{T>130K}} H_2 \tag{3}$$

and

$$\equiv Si-O^{\cdot} + H_2 \xrightarrow[\rightarrow]{T>200K} \equiv Si-OH + H^0 \qquad (4)$$

But, by contrast with both of the aforementioned cases, in high-OH silicas which are γ-irradiated, defect anneal temperatures as high as 600K are commonly reported [15-21].

It has been proposed [14,22] that when high-OH silicas are exposed to radiations capable of producing knock-on damage (e.g., neutrons, γ rays, or electrons of $\geq$ 150 keV) the displaced oxygens will become decorated with hydrogen to form interstitial water molecules. In such an event, the thermal bleaching process would be limited by the diffusion of these radiolytic H_2O molecules, thus accounting for a characteristic anneal temperature different from the temperatures which characterize H_2 or O_2 diffusion-limited processes [14,22]. Brown et al. [10] have already provided some support for this hypothesis deriving from isochronal anneal data for MOS devices. The present paper explores in some detail the applicability of the H_2O diffusional model to the anneal kinetics of γ-irradiated high purity bulk fused silicas of high-OH contents.

II. EXPERIMENTAL

Samples for this experiment comprised 4-mm dia. rods of manufacturers' stock Type III fused silicas (containing 1200 ppm by weight of water in the form of hydroxyl groups) [23]. Most data presented here pertain to Spectrosil, although similar results were obtained for Suprasil 1 and Corning 7940. These samples were exposed to ^{60}Co γ rays at ambient temperature for accumulated doses ranging from $3x10^6$ to $6x10^8$ rads (Si). They were investigated at either room temperature or $\sim$ 100K on a Varian E-9 spectrometer operating at 9.15 GHz and utilizing 100 kHz field modulation. Spin concentrations were determined by double numerical integration and comparison with a Varian strong pitch standard. In the case of the E' center, the latter measurement was carried out at room temperature and a microwave power level of 0.2 μW to avert microwave saturation [13].

Most isochronal and isothermal anneal experiments were performed by removing the sample to a Hotpack electric furnace regulated by a Wheelco temperature controller. Measurements with a digital thermometer indicated that the target temperature could be set with an accuracy of $\pm$ 2.5°C. These measurements also showed the temperature to re-equilibriate to within 1°C of target 90 sec following the operation of inserting the sample. The total time at temperature was thus taken to be the total time in the furnace less 90 sec for each insertion. The samples were recooled to the measurement temperature by allowing them to rest 3 min in a temperature regulated stream of N_2 gas. One isochronal anneal experiment was carried out <u>in situ</u> in the N_2-flow-through

temperature regulator associated with the spectrometer. In the
latter case, samples exposed to a range of radiation doses were
observed versus time at a common fixed temperature $\sim 200^{\circ}$C.

III.   RESULTS AND DISCUSSION

        Figure 1 displays a set of isochronal anneal curves (350 sec
at temperature) for a sample of Spectrosil examined 78 days

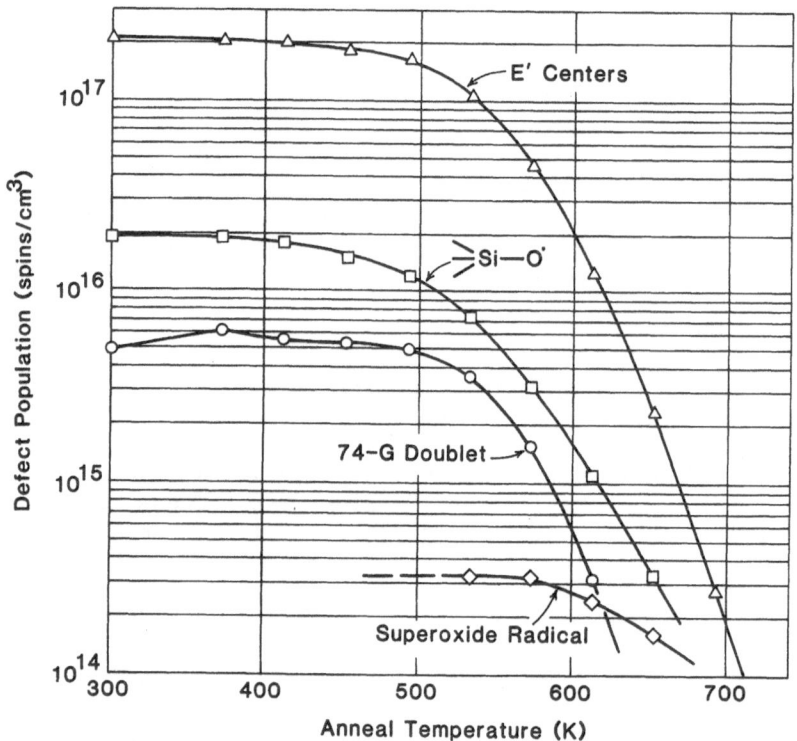

Fig. 1 Paramagnetic defect concentrations versus anneal temperature
        for a 4-mm dia. Spectrosil rod observed 78 days after a
        γ-ray dose of 1.26×10^8 rads. ESR spectra were obtained at
        room temperature following furnace anneals of ~ 350 sec
        duration at the indicated temperatures.

following a γ ray exposure of 1.26×10^8 rads. All distinguishable
radiation-induced paramagnetic defects are quantitatively represented in this figure (the well known 74-G doublet arises from a
proton-associated defect center of still uncertain origin) [24,25].
It can be seen that all defects, save the weakly abundant superoxide radical, display similar anneal kinetics and, moreover, that
the E' center is the most abundant species by approximately an
order of magnitude. These two facts strongly suggest that the

anneal kinetics must be dominated by a single mechanism wherein
the E' center plays a central role. Consistent with the thesis
set out in the introduction, we shall proceed to test the proposition that the anneal mechanism is primarily limited by H_2O
diffusion. Although perhaps an oversimplification, the data of
Fig. 1 can formally be interpreted in terms of the direct reaction

$$\equiv Si \cdot + H_2O \xrightarrow{T>500K} \equiv Si-OH + H^0 \tag{5}$$

where the (highly mobile) H^0 on the righthand side of Eq. (5)
would be swiftly consumed in processes such as the reverse
reaction of Eq. (2).

[A digression is necessary at this point to consider the
subsidiary reactions which must be invoked to balance the chemical
equations. First, it is to be expected under the central operating
hypothesis that there should be at least one water molecule for
each E' center, since

$$\equiv Si-0-Si \equiv \xrightarrow{\text{knock-on process}} \equiv Si-Si \equiv + 0_{\text{interstitial}} \tag{6}$$

and

$$0_{\text{interstitial}} + 2H^0 \rightarrow H_2 0_{\text{interstitial}} \text{,} \tag{7}$$

while trapping of a hole on the silicon-silicon bond on the righthand side of Eq. (6) yields an E' center according to the Feigl-
Fowler-Yip theory [26]:

$$\equiv Si-Si \equiv + h^+ \rightarrow \equiv Si \cdot {}^+ Si \equiv . \tag{8}$$

The hydrogen for reaction (7) is of course provided by reaction
(2). Thus, two nonbridging oxygen hole centers ($\equiv Si-0 \cdot$) would
be formed for each E' center, were it not for the necessity that
one electron (e^-) be trapped for each hole (h^+) trapped in reaction
(8). These electrons could trap on nonbridging oxygen hole
centers [27]:

$$\equiv Si-0 \cdot + e^- \rightarrow \equiv Si-0^- . \tag{9}$$

Equations (2), (6), (7), (8), and (9) could then be simultaneously
balanced by producing $\equiv Si \cdot$, $\equiv Si-0 \cdot$, $\equiv Si-0^-$, and H_2O in the ratio
of 1:1:1:1. However, the experimental data of Fig. 1 indicate a
substantially lower concentration of $\equiv Si-0 \cdot$ centers than predicted

in this scenario. The experimental outcome might result from the dimerization of paired nonbridging oxygen hole centers to form peroxy linkages [27]:

$$\equiv Si-O^{\cdot} \quad {}^{\cdot}O-Si\equiv \rightarrow \equiv Si-O-O-Si\equiv . \tag{10}$$

In such an event, much of the hydrogen yielded on the righthand side of Eq. (5) would have to be consumed in the reaction

$$\equiv Si-O-O-Si\equiv \quad + H^0 \rightarrow \equiv Si-O^{\cdot} \quad HO-Si\equiv . \tag{11}$$

In Fig. 1, the slower decay of the $\equiv Si-O^{\cdot}$ species vis-a-vis the E' center above $\sim$ 570K provides some tantalizing support for the last mentioned process, since Eq. (11) should tend to restore the $\equiv Si-O^{\cdot}$ population back toward a 1:1 ratio with the E' center.]

Regardless of the details of the necessary subsidiary reactions, if Eq. (5) is assumed to govern the E' center anneal process, it is appropriate to compare the E' center anneal data with the predictions of standard second-order rate theory. The result due to Waite [28], as applied to the reaction of Eq. (5), is given by

$$\frac{[E']}{[E']_0} = \frac{[E']_0 - [H_2O]_0}{[E']_0-[H_2O]_0 \; \exp\left\{- 4\pi r^{*}D([E']_0-[H_2O]_0) \left[1+ \frac{2r^{*}}{(\pi Dt)^{1/2}}\right]t\right\}} , \tag{12}$$

where $[E']_0$ and $[H_2O]_0$ correspond to the initial concentrations of the E' center and the mobile water molecules, respectively, r^{*} is the capture radius (generally taken to be $\sim$ 5Å) [18,20], and D is the diffusion coefficient expressible as an Arrhenius relation:

$$D = D_0 \exp [-E/kT] . \tag{13}$$

The pre-exponential factor D_0 and activation energy E for H_2O diffusion in silica glass were measured by Moulson and Roberts [29] to be 1×10^{-6} cm^2/sec and 0.79 eV, respectively, using a method which assumed the diffusivity to be concentration independent. However, Drury and Roberts [30], using more sophisticated methods, found concentration dependent D values and $\log D$-versus-T^{-1} curves exhibiting complicated behaviors across the transition range ($\sim$ 900–1100°C). Subsequently, Roberts and Roberts [31] demonstrated a distinct dependence of water diffusivity on fictive temperature. These data suggest that sizeable deviations ($\sim$ 50%) from Moulson and Roberts' parameters might be expected in practice, particularly at temperatures < 700°C where direct experimental data have yet to be obtained.

In Figure 2, the predictions of Eq. (12) are compared with two sets of isochronal anneal data for γ-irradiated Spectrosil. The

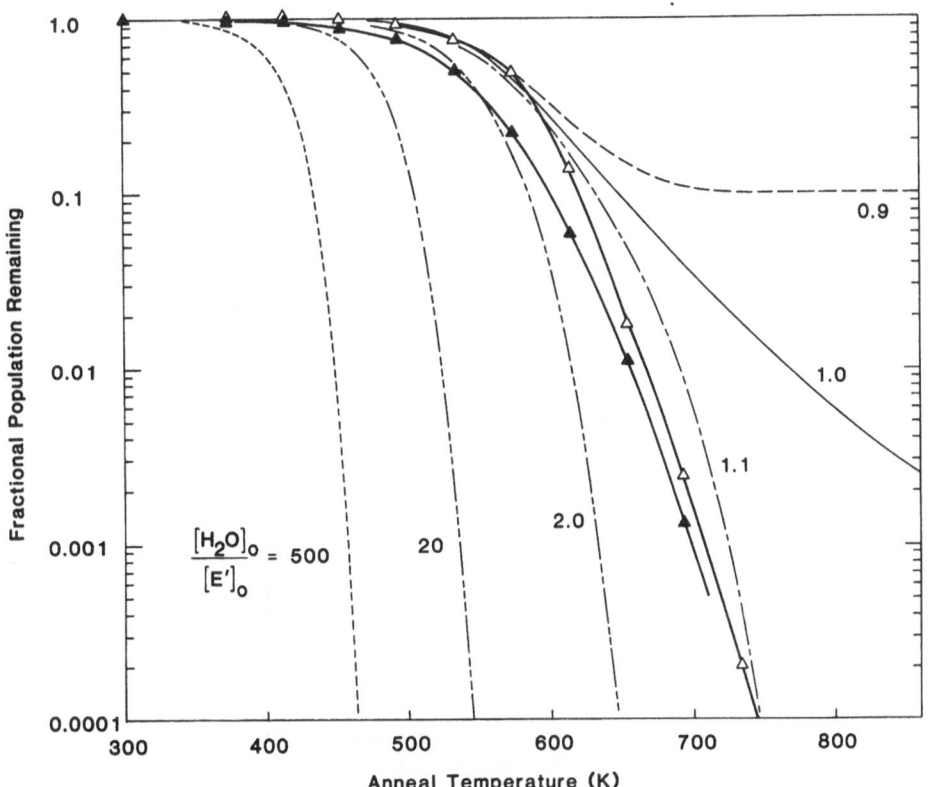

Fig. 2   E' center isochronal anneal data for γ-irradiated Spectro-sil rods are compared with theoretical predictions based on an H_2O diffusional model. Experimental parameters were: (open symbols) 1.97×10^8 rads, $[E']_0 = 2.5 \times 10^{17}$ cm^{-3}, 18 mo. after irradiation; (filled symbols) 1.26×10^8 rads, $[E']_0 = 2.0 \times 10^{17}$ cm^{-3}, 78 days after irradiation. Theoretical curves (lightly drawn) were calculated by means of Eq. (12) assuming $[E']_0 = 2 \times 10^{17}$ cm^{-3}, $r^* = 5$ Å, $D_0 = 1 \times 10^{-6}$ cm^2/sec, $E = 0.86$ eV, $t = 350$ sec, and values of $[H_2O]_0/[E']_0$ as indicated. Bold curves are passed through the data as an aid to the eye and represent no theory.

calculated curves employ values of $[E']_0 = 2 \times 10^{17}$ cm^{-3} (as measured),
$D_0 = 1 \times 10^{-6}$ cm^2/sec, and $E = 0.86$ eV. Curves corresponding to
different relative concentrations of diffusing water are shown,
since further diffusible water molecules in addition to those
produced radiolytically could become available at higher temperatures due to the equilibrium

$$\equiv Si-OH \; HO-Si\equiv \; \rightleftharpoons \; \equiv Si-O-Si\equiv \; + \; H_2O_{interstitial} \tag{14}$$

The case $[H_2O]_0/[E']_0 = 500$ would correspond to the equilibrium of
Eq. (14) shifting entirely to the right. Comparison of theory and
experiment in Fig. 2 shows clearly that $1.1 \lesssim [H_2O]_0/[E']_0 \lesssim 2$ over
the entire ranges of temperature and defect concentration spanned
by the data.

[A second digression is appropriate here in order to consider
the differences in shape between the two experimental curves of
Fig. 2. Both data sets correspond to the same material and
approximately the same radiation doses and initial E' center
concentrations. The sole experimental parameter which differed
significantly was the time between the removal of the sample from
the source and the performance of the experiment: 78 days for
the filled symbols vis-à-vis 540 days for the open symbols. A
reasonable hypothesis to account for the discrepancy between the
two curves is the existence of a distribution in activation
energies (c.f. Ref. 20). The sample remaining the longest at room
temperature prior to the experiment would then suffer a pre-
annealing of the low energy tail of the activation energy
distribution; hence the remaining centers, being characterized by
the higher activation energies, would be less prone to thermal
bleaching upon heating to temperatures $\lesssim$ 600K. But above $\lesssim$ 600K
the annealing behaviors of the two samples would be expected to
converge, as observed, since at that stage only the higher
activation energy sites would remain in either sample. As an
illustration of the plausibility of this argument, Figure 3
displays the room temperature isothermal anneal curves predicted
by Eq. (12) for a range of activation energies. Clearly, after a
room temperature aging period of 540 days, virtually all of the
activation energy components $\lesssim$ 0.7 eV would have been removed
from any sample which initially contained them. Note that the
three data sets shown in this figure suggest that the peak in the
effective activation energy distribution function may in fact be
lower for samples receiving lower γ-ray doses (additional data
supporting this conclusion will be presented.)]

An interesting property of the data of Fig. 2 is the asymptotic
decay law above 600K which approaches an $\exp[-\lambda T]$ behavior. It
seems significant that this type of behavior cannot be reproduced
by any parameterization of the standard second order kinetic theory,

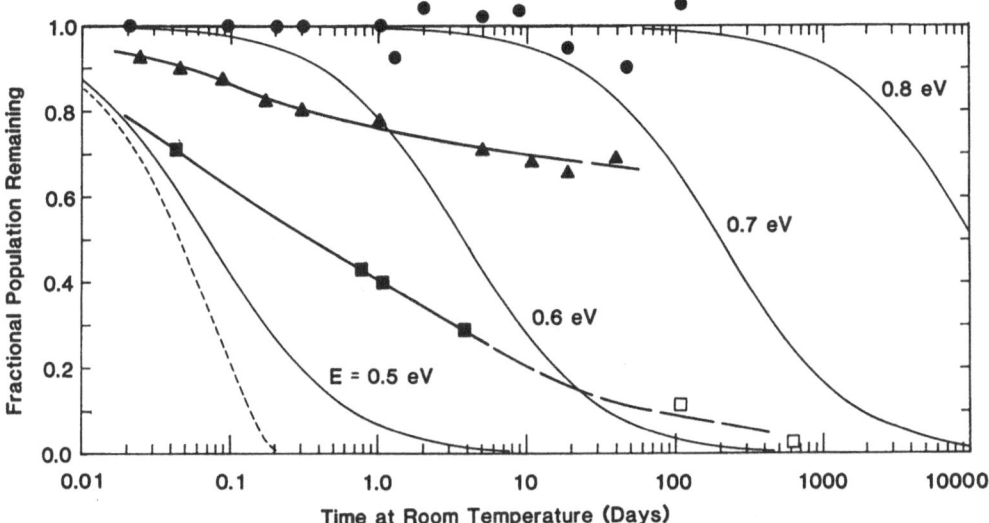

Fig. 3 Room temperature (293K) isothermal decay data for E'
        centers in γ-irradiated Spectrosil (symbols) are compared
        with the predictions of Eq. (12) for $[E']_0 = 10^{17}$ cm^{-1},
        $r^* = 5$ Å, $D_0 = 1 \times 10^{-6}$ cm^2/sec, and various values of E
        (light, fully drawn curves). Also illustrated for compari-
        son is the behavior expected under a first-order decay law
        (light dashed curve). Filled symbols pertain to a single
        sample which was successively administered ^{60}Co γ ray
        doses of 2.7×10^6 (squares), 2.6×10^7 (triangles), and
        4.4×10^8 rads (circles). Open symbols pertain to less
        well controlled experiments on other samples. Bold curves
        are passed through the data as an aid to the eye.

Eq. (12). For $[H_2O]/[E']_0 = 1$ the calculated curve is slightly
concave upward and does not descend steeply enough with increasing
temperature, while for $[H_2O]_0/[E']_0 > 1$ the calculated curves are
constantly concave downward and eventually descend too steeply.
Moreover, additional discrepancies of a similar nature were noted
between the predictions of the Waite theory and a set of isothermal
anneal curves (vide infra). In order to circumvent these difficulties it appeared useful to attempt an a posteriori modification
of the standard rate theory.

The starting point for the modified theory is the standard
expression for nth order kinetics ($n \geq 1$):

$$[E']/[E']_0 = [1 + (n-1)\nu(T)[E']_0^{(n-1)}t]^{1/(1-n)} \qquad (15)$$

For n=2, Eq. (15) approaches the Waite solution for the case

$[E']_0 = [H_2O]_0$ if one takes $\nu(T) = 4\pi r^*D$ and provided $2r^*/(\pi Dt)^{1/2} \ll 1$. In fact, for the temperatures and times involved in the experiment and using Moulson and Roberts' values [29] for D_0 and E, one calculates $2r^*/(\pi Dt)^{1/2} \lesssim 0.005$, and hence this term can indeed be neglected.

From Fig. 2 it should be apparent that second order rate theory of Eq. (12) <u>approximately</u> describes the data if it is assumed that $[H_2O]_0$ slightly exceeds $[E']_0$. We note that for $[H_2O]_0 \gg [E']_0$ the anneal kinetics would be effectively first order (the curve for $[H_2O]_0/[E']_0 = 500$ is indistinguishable from the standard first order solution). But since the standard first and second order theories fail to fit the experimental anneal curves in detail, it seems worth considering the formal predictions of a <u>fractional-order</u> kinetic scheme where $1 < n < 2$. With some ad hoc changes to preserve nondimentionality, Eq. (15) can be adapted to the case of non-integer n values as follows:

$$[E']/[E']_0 = [1 + (n-1)D(4\pi r^*)^{(3n-5)}[E']_0^{(n-1)}t]^{1/(1-n)} \qquad (16)$$

Figure 4 compares the predictions of Eq. (16) with two sets of isochronal anneal data for E' centers in γ-irradiated Type-III silicas. It can be seen that the data are fit rather well by the modified theory using values of $n \sim 1.4$-1.6. But more importantly, use of Eq. (16) now makes it possible to fit the <u>isothermal</u> anneal data presented in Fig. 5.

With respect to the isothermal anneal experiments, the standard second order kinetic solution of Eq. (15) (as well as the Waite solution for $[E']_0 = [H_2O]_0$) yields curves which approach t^{-1} dependence in the long-time limit, while the Waite solution for $[H_2O]_0 > [E']_0$, Eq. (12), approaches an $\exp[-\nu t]$ dependence at long times. By contrast, the data of Fig. 5 tend to exhibit t^{-x} behaviors in the long-time limit, where $x \gtrsim 1$. Such a behavior cannot be reproduced by introducing a distribution of activation energies into Eq. (12), since this artifice has the effect of flattening the isothermal decay curves rather than steepening them [20]. Thus, the experimental isothermal anneal data cannot be accounted for in terms of the standard rate theory but, as shown in Fig. 5, can be fit to Eq. (16) for $1 < n < 2$.

It will be noted in Fig. 5 that the value of n required to fit the data tends to decrease with increasing temperature (from $n=2$ for $T=573K$ to $n=1.78$ for $T=633K$). Several points should be emphasized in this connection. First, n is not an additional free variable relative to the standard theory. Rather, the variable $[H_2O]_0/[E]_0$ which must be assigned in Eq. (12) is, in effect, replaced by the variable n in Eq. (16). That n should be a function of temperature is therefore easily understood on the

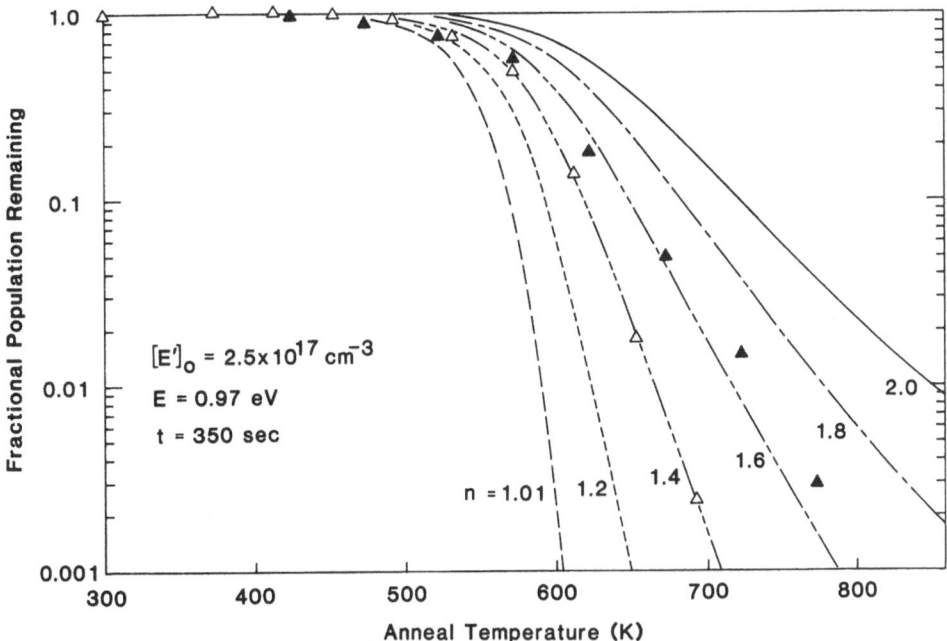

Fig. 4   E' center isochronal anneal data for γ irradiated Type-III
fused silicas are compared with calculated curves based
upon a non-integral-order kinetic theory. Experimental
data were for (open symbols) Spectrosil receiving 1.97x10[8]
rads and (filled symbols) Suprasil 1 receiving 6x10[8] rads;
both samples were observed over a year following irradiation. Calculated curves assume $[E']_0 = 2.5 \times 10^{17}$ cm^{-3},
r* = 5 Å, $D_0 = 1 \times 10^{-6}$ cm^2·sec, E = 0.97 eV, t = 350 sec,
and various values for the parameter n in Eq. (16).

same basis that $[H_2O]_0/[E']_0$ should be a function of temperature:
namely, the equilibrium of Eq. (14) will shift to the right at
higher temperatures. In fact, the necessity of invoking fractional
order kinetics here may be cryptically related to the fact that the
concentration of water is dynamically adjusted by its tendency to
seek equilibrium with the OH groups in the sample and hence cannot
be simply specified in terms of a fixed value of $[H_2O]_0$. Thus,
reality is probably best represented by a reaction order that
decreases with increasing temperature (Fig. 5). Consequently,
the fits of the isochronal anneal data of Fig. 4 involving a single,
temperature-independent value of n should be considered as instruc-
tive but perhaps somewhat less realistic (particularly as regards
the values derived for the activation energy E).

Figure 6 presents some preliminary isochronal anneal data
for a series of samples receiving γ ray doses ranging from 3x10[6] to
1.65x10[8] rads. The initial E' center concentrations in these

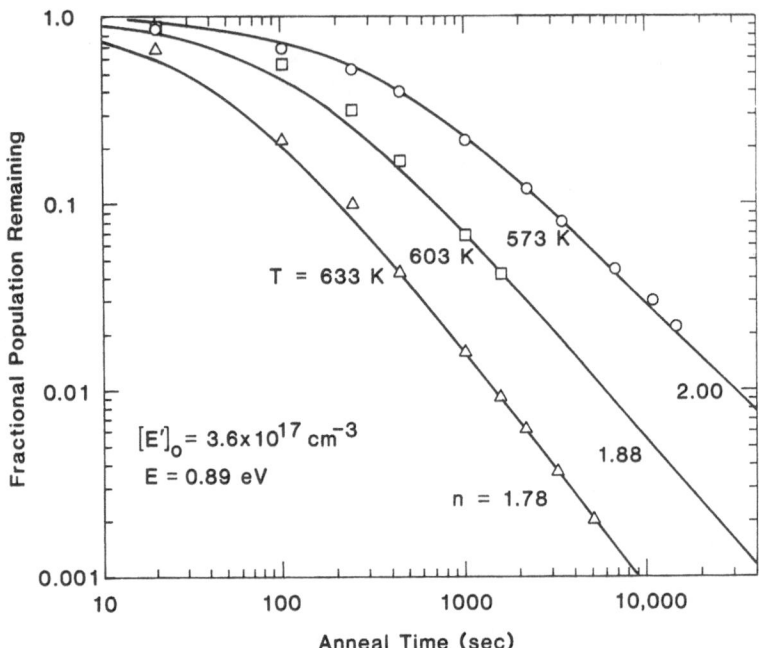

Fig. 5   Isothermal anneal data for E' centers in a suite of 4-mm-dia.
Spectrosil rods observed 3-5 weeks after a γ ray dose of
1.65×10^8 rads. ESR data were obtained at room temperature
following the indicated anneal schedules. Calculated curves
are based on Eq. (16) assuming $[E']_0 = 3.6 \times 10^{17}$ cm^{-3} (as
measured), $r^* = 5$ Å, $D_0 = 1 \times 10^{-6}$ cm^2/sec, $E = 0.89$ eV, and
the indicated values for the parameter n.

samples also varied widely from ∿ 1×10^{15} at the lowest dose to
3.6×10^{17} cm^{-3} at the highest dose. (It should be borne in mind
that time after irradiation is also an important parameter influen-
cing the value of $[E']_0$ (see Fig. 3).) The curves shown passing
through the data were calculated by means of Eq. (16). While the
fits shown certainly do not represent unique simultaneous determinations of both n and E, it became clear in generating these curves
that both n and E must decrease in order to fit the data for samples
receiving progressively lower γ-ray doses. That n should decrease
with decreasing $[E']_0$ is a corollary of the reasoning developed
in the preceding paragraph: as the initial defect concentration
becomes less than the equilibrium molecular water content, the
effective order of the kinetics must decrease from second toward
first. The physical origin of the dependence of the activation
energy on radiation dose is less obvious and will be addressed
briefly in the concluding section.

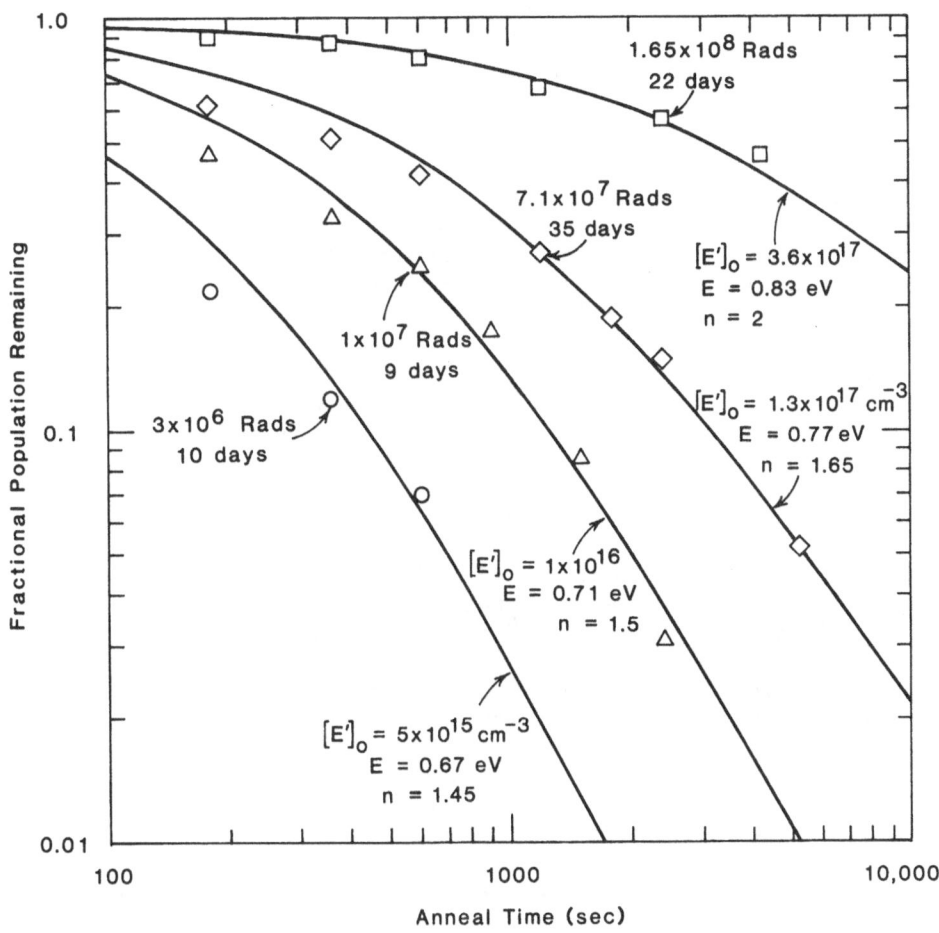

Fig. 6   Isothermal anneal data for a suite of Spectrosil rods
receiving various γ ray doses. The anneals were performed
in situ at ∿ 200°C. Other experimental conditions are
noted on the figure, as are the parameters used in Eq. (16)
to generate the smooth curves.

IV. SUMMARY AND CONCLUSIONS

Isochronal and isothermal anneal data have been presented for
γ-ray-induced paramagnetic defect centers in high-OH Type III
fused silicas. After showing the E' center ($\equiv$Si·) to be the most
prevalent defect, arguments were presented that this oxygen-
vacancy-associated defect should thermally bleach by a diffusion-
limited process wherein radiolytic molecular water is the diffusing species. These arguments were substantially buttressed
by demonstrating that the experimental anneal curves can be
reasonably well reproduced using standard second order kinetic

theory and published diffusivity data for molecular water and
assuming $[H_2O] \approx [E']$. Some minor discrepancies between data and
theory were shown to be ascribable to the existence of
statistical distributions in the activation energy for H_2O
diffusion in a-SiO_2, analogous to distributions previously
demonstrated for H_2 diffusion in these materials [20]. Other
discrepancies of a separate nature were noted to occur for anneal
temperatures $\gtrsim$ 570K. In order to fit the data in this higher
temperature range, an a posteriori modification of the standard
kinetic theory was proposed which involved the formal admission
of fractional order kinetics, 1<n<2. Excellent fits to sets of
isochronal and isothermal anneal data demonstrated the usefulness
of the new theory. Physically, it was argued that the defect
anneal kinetics should become less than second order as the
number of (nonradiolytic) water molecules in equilibrium with the
hydroxyl groups in the sample becomes comparable to the number of
H_2O's produced radiolytically. Clearly, such could occur at high
enough temperatures for any given reference defect concentration
and at lower temperatures for lower defect concentrations. Both
effects were demonstrated. Based on the results presented, the
equilibrium molecular water content of Type-III fused silicas can
be crudely deduced to be $\sim 10^{16}$ cm^{-3} at 600K.

Activation energies for water diffusion determined by curve
fitting the experimental data ranged to both sides of the value
E = 0.79 eV given by Moulson and Roberts [29]; the pre-exponential
factor $D_0 = 1 \times 10^{-6}$ cm^2/sec determined by these authors was used
for all of the calculated curves. Some variation in the values
of E are to be expected on the basis of later work [30,31]. However, the highest activation energies were invariably deduced for
samples receiving the highest γ ray doses and for measurements
performed at the highest temperatures. As discussed in some
detail elsewhere [20], these systematic effects can be ascribed at
least in part to the existence of a variety of percolation
pathways characterized by a distribution in activation energies.
Thus, at low enough temperatures and large enough distances
between defect centers, the more numerous high energy pathways
would be "frozen out" and a lower effective activation energy
would be measured [20]. Nevertheless, at a fixed low temperature
this description does not predict that lower defect concentrations should bleach more rapidly than higher defect concentrations--
in fact, the contrary is predicted. Yet the results presented in
Figs. 3 and 6 seem to indicate more rapid bleaching of E' centers
in the samples receiving lower γ ray doses. Although it has been
shown that the presumed excess of diffusing water molecules over
the number of defects must contribute to this effect, the curve
fitting procedures also point toward lower effective activation
energies for the samples receiving the lower doses.

Several possible mechanisms could account for the observation of lower activation energies at lower γ ray doses. For example, at low doses the displaced oxygens could be incompletely decorated by hydrogens, resulting in interstitial OH groups which would be both more diffusible and more reactive than the water molecules which seem to be the dominant product at high doses (OH-diffusion limited processes in irradiated high-water alkali silicate glasses have recently been described) [32]. But whatever the physical mechanism giving rise to the low activation energy component(s), account must eventually be taken of the fact that the lowest energy components will anneal out during the period of the irradiation (ranging from $\sim$ 7 h for 3×10^6 rads to 30 d for 3×10^8 rads). Thus, the "instantaneous" activation energy spectrum could perhaps be the same for all doses, in which case the higher average activation energies measured at higher doses would be an artifact of the long irradiation times. This question could be resolved by irradiations performed at higher dose rates or lower temperatures.

The present paper has demonstrated that for γ ray doses in excess of $\sim 10^8$ rads, administered at a rate of $\sim$ 0.5 Mrad/h, the subsequent thermal bleaching of the induced defect centers can be consistently interpreted in terms of the diffusion-limited recombination of E' centers with radiolytic molecular water. In the context of MOS device physics, it is necessary to point out however that the anneal mechanism of Eq. (5) removes E' centers at temperatures $\sim 350^{\circ}$C -- but not the species $^{+}Si\equiv$ and $\equiv Si\text{-}O^{-}$ found on the righthand sides of Eq. (8) and (9), respectively. As seems to be supported by the recent work of Devine [8], the latter species must be available for reactivation by subsequent irradiations unless they are removed by anneals to still higher temperatures.

REFERENCES

1. W. Primak, J. Appl. Phys. 53:7331 (1982); ibid. 55:852 (1984); ibid. 55:3315 (1984).
2. E. J. Friebele, Opt. Engineering, 18:552 (1979).
3. E. J. Friebele, M. E. Gingerich, K. J. Long, P. S. Levin, and D. A. Pinnow, J. Lightwave Tech. LT1:462 (1983).
4. E. J. Friebele, C. G. Askins, M. E. Gingerich, and K. J. Long, Nucl. Inst. & Methods B1:355 (1984).
5. E. P. EerNisse and C. B. Norris, J. Appl. Phys. 45:5196 (1974).
6. R. A. B. Devine and A. Golanski, J. Appl. Phys. 54:3833 (1983).
7. A. Stesmans, J. Braet, J. Witters, and R. F. DeKeersmaecker, J. Appl. Phys. 55:1551 (1984).
8. R. A. B. Devine, J. Appl. Phys., 56:563 (1984).
9. C. M. Dozier and D. B. Brown, IEEE Trans. Nuc. Sci. NS-28:4137 (1981).

10. D. B. Brown, D. I. Ma, C. M. Dozier, and M. C. Peckerar, IEEE
    Trans. Nucl. Sci. NS-30:4059 (1983).
11. R. A. Weeks, J. Appl. Phys. 27:1376 (1956); C. M. Nelson and
    R. A. Weeks, J. Am. Ceram. Soc. 43:396 (1960); R. A. Weeks
    and C. M. Nelson, ibid. 399 (1960).
12. D. L. Griscom, E. J. Friebele, and G. H. Sigel, Jr., Sol.
    Stat. Commun. 15:479 (1974).
13. D. L. Griscom, Phys. Rev. B 20:1823 (1979); ibid. B 22:4192
    (1980).
14. D. L. Griscom, Nucl. Inst. & Methods B 1:481 (1984).
15. M. Stapelbroek, D. L. Griscom, E. J. Friebele, and G. H. Sigel,
    Jr., J. Non-Cryst. Solids 32:313 (1979).
16. D. L. Griscom and E. J. Friebele, Phys. Rev. B 24:4896 (1981).
17. E. J. Friebele, D. L. Griscom, M. Stapelbroek, and R. A. Weeks,
    Phys. Rev. Lett. 42:1346 (1979).
18. A. H. Edwards and W. B. Fowler, Phys. Rev. B 26:6649 (1982).
19. D. L. Griscom, M. Stapelbroek, and E. J. Friebele, J. Chem.
    Phys. 78:1638 (1983).
20. D. L. Griscom, J. Non-Cryst. Solids, 68:301 (1984).
21. D. L. Griscom, G. H. Sigel, Jr., and E. J. Friebele, Proc.
    11th Int. Cong. on Glass, Prague (Dum Techniky, Praha (1977)
    p. 3.
22. D. L. Griscom, J. Non-Cryst. Solids 73:51 (1985).
23. G. Hetherington and K. H. Jack, Phys. Chem. Glasses 3:129
    (1962).
24. N. G. Cherenda, A. V. Shendrik, and D. M. Yudin, Phys. Stat.
    Solidi B 69:687 (1975).
25. J. Vitko, Jr., J. Appl. Phys. 49:5530 (1978).
26. F. J. Feigl, W. B. Fowler, and K. L. Yip, Sol. Stat. Commun.
    14:225 (1974); K. L. Yip and W. B. Fowler, Phys. Rev. B
    11:2327 (1975).
27. W. B. Fowler, private communication.
28. T. R. Waite, Phys. Rev. 107:463 (1957).
29. A. J. Moulson and J. P. Roberts, Trans. Brit. Ceram. Soc.
    59:388 (1960).
30. T. Drury and J. P. Roberts, Phys. Chem. Glasses 4:79 (1963).
31. G. T. Roberts and J. P. Roberts, Phys. Chem. Glasses 5:26 (1964).
32. A. A. Wolf, E. J. Friebele, D. L. Griscom, J. Acocella, and
    M. Tomozawa, J. Non-Cryst. Solids 56:349 (1983).

LOCAL ATOMIC STRUCTURE IN TRANSITION METAL/METALLOID GLASSES: Ni-P

L. H. Bennett, G. G. Long, M. Kuriyama and A. I. Goldman*

Center for Materials Science, National Bureau of Standards
Gaithersburg, MD 20899; *Department of Physics, Brookhaven
National Laboratory, Upton, NY 11973

INTRODUCTION

Metallic alloy glasses are often obtained in systems characterized by strong chemical bonding between atoms of different species
within the material. A prominent example of this class of solids
is the transition metal/metalloid (TM/M) alloy glass, which is
easily produced either by electrodeposition or by liquid quenching
and is very stable at room temperature. The compositional range
of the TM/M glasses usually lies near deep eutectics in the equilibrium phase diagram of the alloy system where the liquid phase is
most stable compared to the solid phases.

Metallic alloy glasses, in common with the semiconductor and
the oxide glasses, are often called "amorphous" in order to emphasize
that these noncrystalline materials do not display the long range
order characteristic of crystalline solids. Nevertheless, all of
these glasses display a high degree of short range order. Indeed,
it is this short range order that accounts for their reproducibility
and for those physical properties (such as the electronic, magnetic,
superconducting, mechanical, and optical properties) which have
attracted wide ranging technological interest.

The term "glass" has traditionally been reserved for those
noncrystalline solids formed by quenching from the liquid state.
Bagley, Chen and Turnbull [1] have suggested broadening the definition to include all solids displaying the following qualities
of a metastable liquid: (i) x-ray diffraction displays a few
broad halos; (ii) electron microscopy reveals no diffraction contrast; and (iii) the annealing response to a continuous increase
in temperature is precipitous over a very narrow temperature range

(i.e., there is heterogeneous nucleation). Qualities labelled i
and ii do not generally distinguish between a glass that may be
heterogeneously composed of very small (∿ 1 nm) crystallites or a
homogeneously disordered arrangement of topologically stable cells.

The glass transition is generally considered to be a kinetic
phenomenon and not a thermodynamic phase transition. This emphasizes
the fact that the glass temperature (T_g) is not fixed but is dependent on the rate of cooling from the liquid. This would also imply
that the liquid and the glass belong to the same structural and
thermodynamic phase. However, it is also true that there is evidence
[2] of a higher degree of short range ordering in the glass than
in the metallic liquid. In particular, the contribution of configurational entropy to the free energy of the liquid system is
high and would favor random mixing. In that case, M-M contact
would be allowed in the TM/M liquid.

Many of the gross features of the structure of TM/M glasses
can be understood within the dense random packing of hard spheres
[3-5] (DRPHS) model. Chemical (bond) ordering was introduced into
this otherwise random model by Polk [6] who suggested that DRPHS
could be applied to TM/M structure by expanding the TM structure
and accommodating the M atoms into the larger Bernal holes. His
prediction [6] of M-M avoidance in the glass was later shown [7]
to be most likely correct, and is in contrast to the possiblity of
M-M contact in the TM/M liquid.

The metallic alloy glass structure has, in certain cases, been
shown to possess a local symmetry similar to that of an appropriate
crystalline counterpart. Since the density of the glasses is
universally only a few percent less than the value for the crystalline material, it would not be unexpected to find similar short
range packing of the atoms and similar near neighbor interactions
in the glassy state and in the crystalline solid. This assumption --
i.e., that the short range order in the glass and in the crystalline
solid are the same -- is at the basis of the "microcrystalline" [8],
the "random network" [9], and other structural models of glasses.
As Warren [9] has stated concerning vitreous silica: "If a man
sitting on a silicon atom could look no further than the nearest and
next nearest atoms, he would not know whether he was in a piece
of silica glass or in a cristobalite crystal." For those systems
in which many stable and metastable crystalline polymorphs are
known to exist, it is not clear which form of short range order
will be preferred in the glass, or if the glass will be a mixture
of two or more forms of order. There is some experimental evidence
for more than one glass-like state dispersed locally in a specimen.
Walter, et al., [10] found that there were two distinct glassy
states in melt-spun ribbons of Fe-B alloy. Their interpretation
of this result is that there were clusters of differing compositions in the ribbons. This can be compared to the presence of

two liquids of differing compositions in a liquid miscibility
gap. More will be said about this in Section IIC.

Bennett, Lashmore et al., [11-13] have, in contrast, found the
glassy analog of polymorphism in Ni-P glasses prepared by different
methods. Polymorphism differs from phase separation in that the
two (or more) structurally distinct phases occur at the same composition, and usually only one of these is prevalent in a sample
at a time. Furthermore, a phase transformation between polymorphs
differs from structural relaxation in that the latter is a continuous
process which occurs during low temperature annealing, whereas the
former is an abrupt structural transformation between two distinct
phases. Goldman et al. [14] have proposed an identification of
the character of the local environment for two of the glasses, and
demonstrated in an extended x-ray absorption fine structure experiment that the two glasses are not related through structural relaxation.

The main preoccupation of this review will be with the Ni-P
metallic glass system. This alloy system has a long history,
having been the first [15] material in which the special metallic
glass state was observed to exist. Ni-P glasses have been prepared
by a wide variety of techniques, including chemical deposition,
electrodeposition, and splat quenching. This fact also serves as
an indication of its ease of preparation and its relative stability.
Finally, as a binary system, Ni-P offers a simplification in the
interpretation of experimental results that may be masked in
ternary, or higher systems.

In the section which follows, the fundamental issues of equilibrium and relaxation are discussed, and we consider the separate role
of the glassy state among other condensed states. The next section
deals with the Ni-P alloy system in particular, while the following
two sections go into some detail concerning the nuclear magnetic resonance (NMR) and extended x-ray absorption fine structure (EXAFS)
results which form the basis for many of our conclusions. Lastly,
a discussion and summary are offered.

ALLOY PHASES: EQUILIBRIUM AND RELAXATION

A. Stable and Metastable Equilibrium Phases

The representation of equilibrium phases in heterogeneous
systems is historically the phase diagram introduced by Gibbs.
The phase diagram displays the changes which take place in a system
as a function of the thermodynamic state variables, such as
temperature and composition. The most stable phase or the most
stable coexisting phases are, of course, the ones with the lowest
free energy. If the system is in equilibrium, the chemical
potential is the same in all the phases present, and thus there

is no tendency for any constituent to move from one phase to another.
If the system is displaced from equilibrium by a small amount, it
will return to equilibrium in a reversible manner.

All of the fundamental properties described above for equilibrium can be applied, with some limitations, to metastable equilibrium. If a phase in a particular system cannot be formed for
some reason, an appropriate metastable phase diagram can be constructed by searching for the lowest free energy state among the
remaining phases. This is an example [16] of "constrained equilibrium". If, at a given temperature and composition, two stable
phases would coexist at equilibrium, but a constraint prevents one
of these phases from nucleating, then it is possible to form a
metastable phase where the stable phases actually dissolve into
the metastable phase. An interesting example of such constrained
behavior was recently given by Schwarz and Johnson [17] using the
Au-La system. They prepared thin layers of crystalline Au and La
in intimate contact with one another and then gently annealed the
sample at a temperature low enough so that stable crystalline AuLa
alloy could not nucleate. Since Au diffuses readily into La, and
the diffusion distances are short, interdiffusion took place and
an amorphous metastable structure was formed. In this situation,
two phases which are stable when separated dissolve into one
another to form the metastable rather than the stable phase.

B. Nonequilibrium, Quenching and Relaxation

When a crystalline solid such as a pure metal, for example,
is quenched from a high to a low temperature such that it remains
in the original phase, then the quenched solid will, strictly
speaking, not be in equilibrium. At the higher temperature, there
will be more vacancies in equilibrium with the crystalline lattice
than are allowed at the lower temperature. Many of these vacancies
may remain after the quenching process, leading to a nonequilibrium
vacancy concentration in the metal at the lower temperature. With
time, or upon heating, these vacancies can be annealed out. This
process, which is continuous, has been called "structural relaxation". There will be measurable changes in some of the physical
properties of the material during relaxation. Notably, the density
increases and the electrical resistivity generally decreases as the
number of vacancies is reduced.

Structural relaxation is theoretically distinguishable from a
first order phase transformation in that relaxation involves
kinetics and is a continuous process. However, if two neighboring
phases have similar physical properties, and there are also excess
vacancies present, it may be difficult to distinguish experimentally
between a phase transformation and structural relaxation. This is
especially so because structural relaxation may be reversible [18]
(e.g., elastic) or irreversible (e.g., annealing out of excess free
volume).

C. Miscibility Gaps, Phase Separation and Phase Transforma-
    tions

        In this discussion, it will be useful to first consider a
binary system in which the enthalpy of solution is positive. At
low temperatures, this system will decompose into two stable phases.
An alloy with average composition inside this "miscibility gap" will
naturally be made up of a combination of two phases rather than
a homogeneous solution. These two phases differ from one another
in composition and density.

        Since metallic alloy glasses are often formed from supercooled
liquids, some of these glasses could similarly be expected to be
composed of two states, with these states differing from one
another in composition and density. There are a number of examples
of such separation of states or "phases" in metallic glass systems.
Chou and Turnbull [19] interpreted their small angle x-ray scattering results on Pd-Au-Si glasses to suggest that, while the
as-quenched samples of this material appeared homogeneous, samples
which were annealed at temperatures just above the glass transition
temperature included large scale ($\sim$20 nm) inhomogeneities. This
result suggests a separation of glassy states upon annealing.
Kim and Johnson [20] used x-ray diffraction and measurements of the
superconducting temperature and the critical field in Co-Nb-B
alloys to identify a separation of glassy states in this material.

        There are also a number of examples in the literature of
structural transformations between glassy states where only one
state predominates in the sample. Mak, et al. [21], in measuring
the density and resistivity of $(Zr_{0.67}Ni_{0.33})_{1-x}B_x$ alloys, detected
abrupt changes in bulk properties near $x = 0.05$. Corb, et al. [22]
presented evidence of a reversible first-order transformation
between two local structural configurations in a glassy Co-Nb-B
using measurements of magnetic parameters.

D. The Metallic Glass as a Structural State

        In the random microcrystalline model of glasses [8], the
glass structure is envisioned to consist of small regions of
crystalline material, presumably separated by regions of disordered
liquid-like structure which are not otherwise specified. In this
model, the glass is heterogeneous, with the disordered regions
filling the volume between the randomly ordered crystallites. This
microcrystalline model is now regarded [1] as unphysical because
the volume fraction of the liquid-like regions is too large. A
variation on this model, which decreases the disordered volume
fraction while increasing the heterogeneity, requires [23] that two
or more microcrystalline structures are present. While there is
yet no direct experimental evidence supporting this model, it has
been used successfully [23] to account for some of the observed
properties of glasses.

An alternate approach is the quasicrystalline model as put forward for example by Gaskell [24]. Despite the unfortunate similarity of this name to that of the microcrystalline model, this is, in contrast, a homogeneous view of the glass in which the short range order is nevertheless similar to that of an appropriate crystalline counterpart. The glass structure is derived by introducing small distortions in the bond angles and the bond distances of the crystalline counterpart, such that long range order is lost. Since the quasicrystalline model of glass is thermodynamically homogeneous [25], it may be appropriate to think of it as a distinct metastable phase.

An early attempt to predict glass formation was put forward by Zachariasen [26] in 1932, and has since become known as "Zachariasen's rules" for the formation of oxide glasses. These requirements were derived by considering the possible atomic packings in a nonperiodic network for the oxide to form a glass with energy comparable to that of a crystalline form. More recently, the method to discover metallic alloy glasses has been to search for structures near a eutectic point in equilibrium phase diagrams. Boettinger et al. have shown [27] that the easiest glass formation can be shifted from the eutectic composition due to thermodynamic factors and growth kinetics. Watson and Bennett [28] have proposed a classification of crystal structures into "orderly", "disorderly", and "intermediate". Under this scheme, three structures in the Ni-P system, Ni_3P, $Ni_{12}P_5$, and Ni_2P, are disorderly. This disorderliness is reflected in the multiple Ni/Ni and Ni/P distances in their radial distribution functions. (The distribution of P and Ni neighbors about an average Ni atom is shown below).

The extent to which the local environment in a glass resembles that of a crystalline structure remains a matter of conjecture and controversy. Cargill [29] suggested that DRPHS is a more successful representation of the structure than a microcrystalline model because of the poor fit of calculations for even very small crystallites to the x-ray scattering data. Fujiwara and Ishii [30], in contrast, favor a quasicrystalline model in which the symmetry of the environment around the P atoms in glassy Fe-P, for example, is "very similar" to that in crystalline Fe_3P. Sadoc et al. [31] found that their model for Ni-P glasses was in good agreement with experimental results predicting nine-fold coordination of P by Ni, as in crystalline Ni_3P. This model explicitly requires that the metalloid atoms have only TM nearest neighbors, thereby requiring chemical short range order. Gaskell [24] proposed a model for glasses based on the random packing of trigonal prismatic units resembling the cementite (Fe_3C) structure. And, as a final example also favoring quasicrystalline picture, NMR electric-quadrupole studies [32] are said to have given "conclusive evidence that the amorphous structure retains to a significant extent the local symmetry of the crystalline counterpart."

It can surely be said that all of these models (microcrystalline, random network, DRPHS and quasicrystalline) of the structure of glasses are, to zeroth order, not mutually exclusive. Small changes or relaxation of one structural type can be used to derive another. The problem in trying to choose among the foregoing models is that they begin to differ only in a first order approximation. Unfortunately, there are as yet no clearly defined criteria to distinguish the subtle differences among them.

Ni-P ALLOYS

A. The Ni-P Alloy System

The phase diagram for Ni-P, shown in Fig. 1, is based [33] primarily on a study made in 1908. The solid solubility of P in Ni is about 0.3 atomic percent. The crystal structures of various stable or metastable intermetallic compounds, up to 50 atomic percent P, are given in Table I. Atomic positions are well known in only three of the crystal structures with P concentration less than 40 atomic percent: Ni_3P, $Ni_{12}P_5$, and Ni_2P.

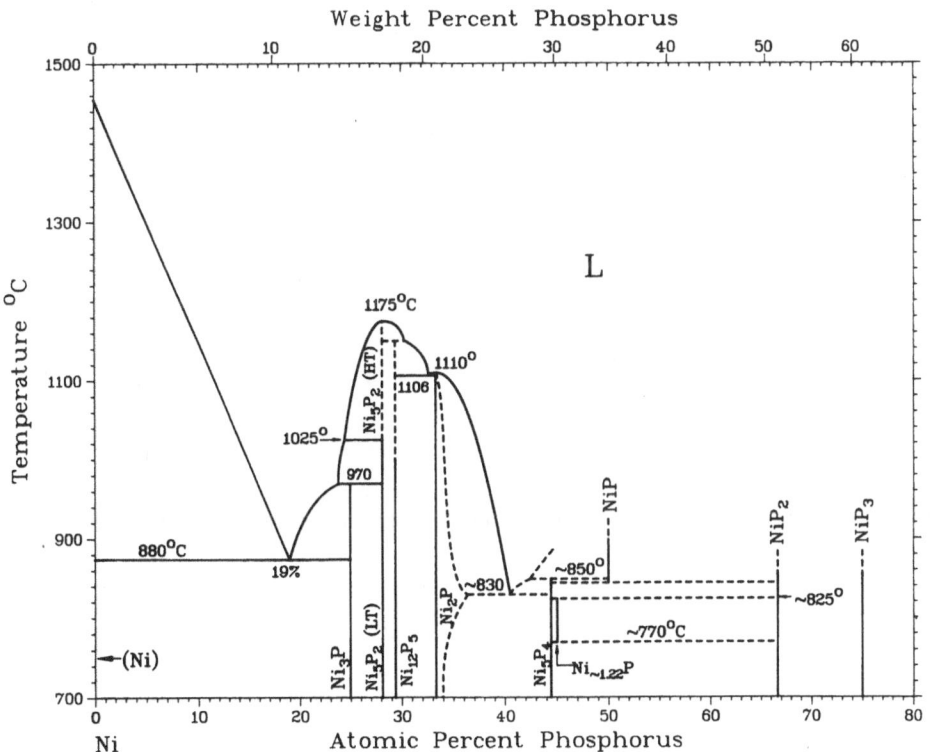

Fig. 1 The Ni-P phase diagram.

Table I. Compositions, structures and lattice constants of
compounds in the Ni-rich Ni-P alloy system.

Phase	At.% P	Wt.%P	Structure*	Space group	Lattice constants, nm		
					a	b	c
Ni_3P	25	14.96	tI32	$I\bar{4}$	0.8954		0.4386
Ni_5P_2	28.2	17.43	hP168		1.3220		2.4632
$Ni_{12}P_5$	29.41	18.02	tI34	I4/m	0.8646		0.5070
Ni_7P_3	30	18.44	cI60		0.8630		
Ni_2P	33.33	20.88	hP9	$P\bar{6}2m$	0.5864		0.3385
Ni_5P_4	44.44	29.68	hP36	$P6_3mc$	0.6789		1.0986
Ni_6P_5	45.45	30.54	?				
NiP	50	34.54	oP16		0.605	0.488	0.689

*Pearson symbol: t = tetragonal   I = body centered
                  h = hexagonal   P = primitive
                  c = cubic
                  o = orthorhombic

The Pearson symbol also shows the number of atoms per unit cell.

In Ni_3P, there are 32 atoms in the unit cell with three inequivalent Ni sites (8 Ni(1), 8 Ni(2), 8 Ni(3) atoms per unit cell) and one P site (8 P atoms per cell). The number of neighbors at distances less than 0.3 nm are:

Ni_3P structure

Atom	P neighbors	Ni neighbors
Ni(1)	2	12
Ni(2)	4	10
Ni(3)	3	10
P	0	9

In $Ni_{12}P_5$, there are 34 atoms in the unit cell, with two inequivalent Ni sites (16 Ni(1) and 8 Ni(2)) atoms per cell) and two inequivalent P sites (8 P(1) and 2 P(2) atoms per cell). The number of neighbors at distances less than 0.3 nm are (with additional neighbors slightly further away shown in parentheses):

$Ni_{12}P_5$ structure

Atom	P neighbors	Ni neighbors
Ni(1)	4	8
Ni(2)	4	7(+4)
P(1)	0	10
P(2)	0	8

In Ni_2P, there are 9 atoms in the unit cell, with two inequivalent Ni sites (3 Ni(1) and 3 Ni(3) atoms per cell) and two P (2 P(1) and 1 P(2) atoms per cell). The number of neighbors at distances less than 0.3 nm are:

Ni_2P structure

Atom	P neighbors	Ni neighbors
Ni(1)	4	8
Ni(2)	5	6(+4)
P(1)	0	9
P(2)	0	9

The distribution of P and Ni neighbors around an average Ni atom is shown in Fig. 2. The spread of Ni-Ni distances is broadest in Ni_3P and narrowest in Ni_2P.

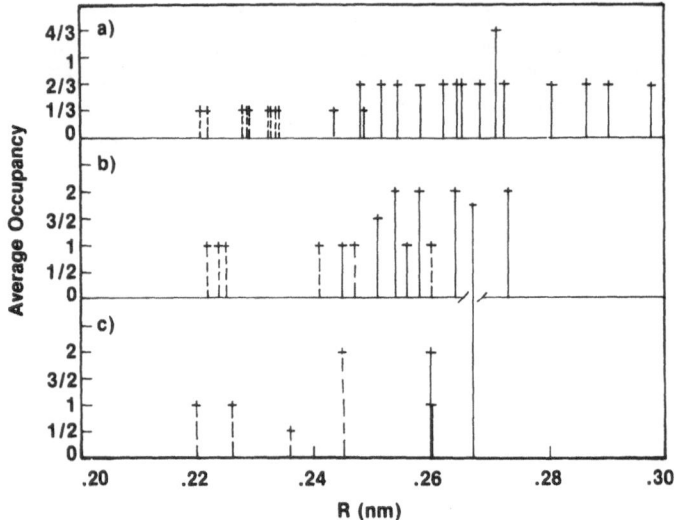

Fig. 2   Calculated radial distribution functions averaged from the
Ni atoms at the origin for (a) Ni_3P, (b) $Ni_{12}P_5$ and (c) Ni_2P.
The solid lines represent Ni neighbors, the dashed lines P
neighbors.

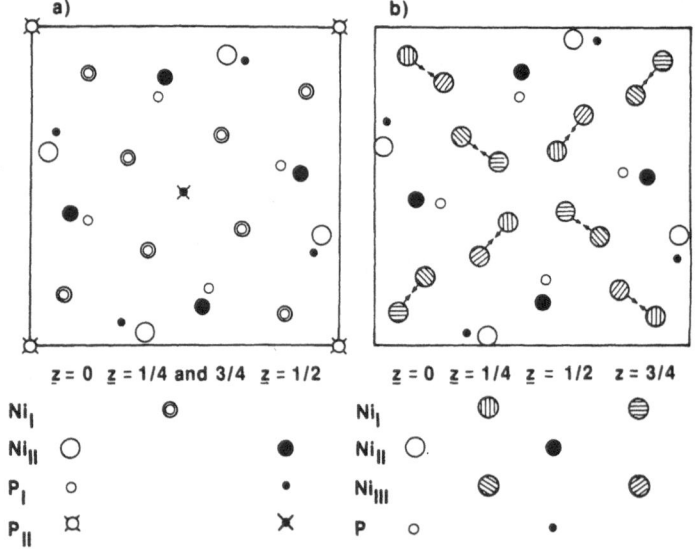

Fig. 3   The unit cell of (a) $Ni_{12}P_5$ and (b) Ni_3P, projected on the
basal plane.   In (b), the arrows on the Ni(1) and Ni(2)
indicate imagined displacements towards the Ni atom arrangement in $Ni_{12}P_5$.   (From Ref. 34.)

Although they are distinct crystal structures, the $Ni_{12}P_5$ and Ni_3P atomic arrangements are similar to each other, as noted by Rundquist and Larsson [34]. We mentioned above the zeroth-order similarities between models of glasses; this is an example of the same kind of similarities among crystalline structures. The basal plane projections shown in Fig. 3 emphasize this point. The Landau rules permit two structures with these space groups to be related via a second-order phase transformation.

Another indication of the similarity between Ni_3P and $Ni_{12}P_5$ is given in the measured heats of formation [35] shown in Fig. 4. There is a sizeable heat to form any of the Ni-P compounds from the elements, i.e., there is strong chemical bonding between the Ni and the P atoms. However, the differences in the heats of formation among the compounds are quite small.

B.   Ni-P Glassy Alloys

Electroless nickel (which, it turns out, is actually a Ni-P alloy) was the first metallic glass reported [15] in the literature. These glasses were prepared using an autocatalytic chemical deposition process resembling but not requiring electrodeposition. Differential thermal analysis showed [36] that the heat evolved in converting electroless Ni to Ni_3P was 11 to 17% of the heat of formation of Ni_3P. Since the metallic alloy glass has about 86% of the heat of formation of the stable cyrstalline form, Randin et al. [36] concluded that the phosphorous atoms were most likely chemically bonded ("alloyed") with the Ni, despite the fact that no sharp lines were observed in the x-ray diffraction.

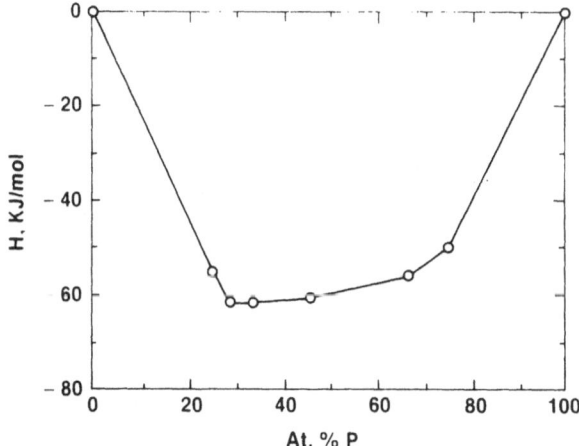

Fig. 4   The heats of formation, in kJ/mol of atoms, of $Ni_{1-x}P_x$ crystalline alloys [35]. The reference states are white P and fcc Ni at $25^{\circ}C$.

The Ni-P metallic glasses can also be prepared by conventional electrodeposition from solution. A compilation of density measurements from the literature [37-40] are shown in Fig. 5. The densities of electroplated glassy Ni-P alloys in the composition range from ∿5 to 25 at. % P are ∿1% less than the densities of the concentration weighted averages of crystalline Ni and Ni₃P. The densities of electroless Ni-P alloys are ∿5% less than electroplated alloys over the same range of compositions.

When these alloys are annealed at sufficiently high temperatures, or for sufficiently long times, a variety of crystallization phenomena occur. An alloy with composition in the eutectic region will

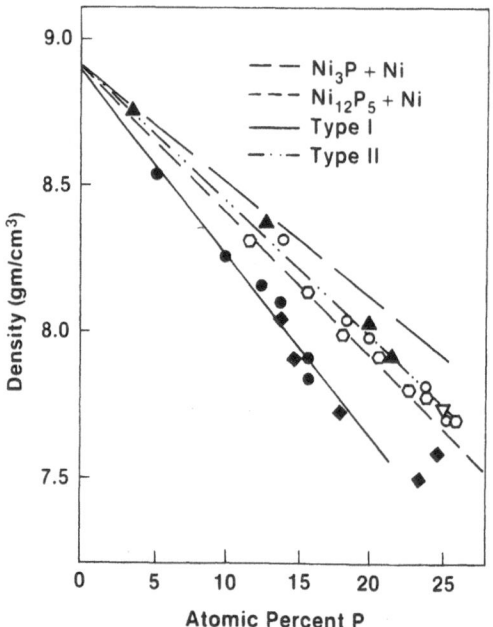

Fig. 5 Density measurements reported for glassy Ni-P alloys prepared by different techniques. Dashed lines show the average concentration weighted density for a mixture of Ni and Ni₃P, and for a similar mixture with Ni₁₂P₅. The dotted line is a fit to the type I data: the dashed-dotted line is a fit to the type II. (Type I and II Ni-P glasses are defined in the NMR section below.

Electroplated: ▽ Ref. 37   Electroless: ● Ref. 38
                 ▲ Ref. 38   ◆ Ref. 39
                 ○ Ref. 39
                 ⬡ Ref. 40

ultimately crystallize to Ni + Ni_3P, as would be expected from
Fig. 1. Before this occurs, Bagley and Turnbull [41] noted that
glassy Ni-P alloys may crystallize to some "complex" intermediate
phase or phases. A hexagonal intermediate phase has been reported
[42] upon annealing. In two other studies [39,43], mixtures of
Ni_7P_3 and Ni_5P_2 or $Ni_{12}P_5$ were found. Crystallization studies
[1,44] have also been demonstrated to be very sensitive to composition. At compositions close to Ni_3P, the crystallization was
heterogeneously nucleated with faceted inclusions. At slightly
higher Ni compositions, the crystallization was dendritic.

The atomic structure of the Ni-P glasses has been a subject of
investigation almost since their discovery. Brenner [15] originally
refused to consider what their structure might be since he believed
that they were truly amorphous, i.e., lacking both long and short
range order. During the 1960's the structure of Ni-P glasses was
studied by Dixmier, Doi and Guinier [45] and by Bagley and Turnbull
[37]. On the basis of x-ray diffraction work, the former group
concluded that these glasses are neither liquid metal-like nor
microcrystalline, but rather consist of clusters with the character
of the crystalline bonding and containing an adequate number of
distortions to remove the possibility of long range order. The
latter study found that electrodeposited samples were "more disordered" than chemically deposited samples. They reported that the
atomic distributions in the electrodeposited samples are rather
similar to that in a liquid metal.

The literature comes full circle with the x-ray work of
Waseda [46] who found that liquid $Ni_{0.75}P_{0.25}$ has a short range
structure similar to Ni_3P as well as to DRPHS plus chemical ordering. Earlier Cargill [40] carefully ruled out fcc, hcp and the
Ni_3P microcrystalline models on the basis of a poor fit to x-ray
diffraction data. He later suggested [28] that DRPHS yields a
much better fit. While x-ray diffraction results are dominated by
the Ni/Ni pair distributions, recent EXAFS work [47,48] using the
Ni K-edge structure, claims that the backscattering is dominated
by the P scatterers, partially because of the sharpness [48] of
the Ni/P distribution. More will be said about EXAFS measurements
in Section V.

NUCLEAR MAGNETIC RESONANCE

A. Structure

Measurements of nuclear magnetic resonance (NMR) Knight
shifts, which sample the local electronic structure through the
hyperfine interaction, are useful for probing local atomic structure. These experiments are a very sensitive measure of differences
among structures. Knight shift measurements on Ni-P glassy alloys
[11-13] have established that there are at least two, and possibly

three, distinct glassy states. For compositions between 14 and 25
at. % P, a constant Knight shift was measured for the conventional
(i.e., direct-current) electroplated samples. These were called
type II in Ref. 13. Over a similar composition range, type I
samples, which were prepared by chemical deposition, splat quenching, and other methods showed an increasing Knight shift with decreasing P concentration.   Of particular significance is the bimodal
nature of the results; i.e., as Fig. 6 shows, no Knight shifts were
recorded in the region between the type I and the type II. The
absence of data in this region is in contrast to what would be
expected if these materials were not structurally distinct, but
were related by a relaxation process. If the structures are
related via a first order transition, then it is possible that
a two phase region may be found in the future. Since the crystalline phases Ni_3P and $Ni_{12}P_5$ can be related by a second order phase
transition, and if the two glass structures bear a similarity to
these, it is possible that the two types of glass transform similarly. Then an average Knight shift in this region would not be
expected. Coexistence of two glassy structures could then only be
found as a result of improper processing.

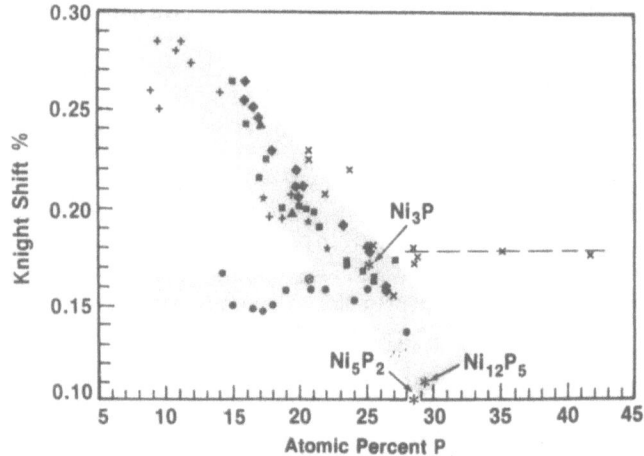

Fig. 6   NMR Knight shifts in Ni-P amorphous alloys prepared by
         different techniques:

         o   dc electroplated (NBS
         ■   dc electroplated (European)
         ⊠   pulse plated (99.9% on time)
         x   pulse plated (91% on time)
         ▲   splat quenched
         ▮   electroless (NBS)
         +   electroless (European)

     Also included are data (*) for three crystalline alloys:
     Ni_3P, Ni_5P and $Ni_{12}P_5$.

In order to identify the lower energy state, annealing experiments were performed [11-13]. Annealing of type I samples resulted in new samples with Knight shifts in the type II range. These results identify II as the lower energy state in this composition range, consistent with the fact that the measured densities for the type II samples were greater than those for the type I samples. Furthermore, a compilation from the literature of density measurements as a function of composition similarly results in a bimodal distribution, i.e., the densities fall along two linear curves (Fig. 5). The scatter in the data for the densities, and for the Knight shifts shown in Fig. 6, may be due to experimental errors or to undetected structural differences, but it is most likely that they are primarily due to measurements on samples which have undergone varying amounts of structural relaxation.

Zero-field NMR linewidth measurements have established [12] that the phosphorous has only Ni nearest neighbors in either structure. The linewidth in either type I or II glass is proportional to the applied magnetic field, implying an inhomogeneous Knight shift mechanism. This is consistent with the random nature of the glasses. The zero-field width, which results from nuclear dipolar interactions between the ^{31}P nuclei (the contribution from ^{61}Ni is negligible), is ~ 3 kHz, the same as the (field independent) linewidth in crystalline Ni_3P. If the nickel and phosphorous were randomly distributed, the calculated Van Vleck second moment would give an expected linewidth of ~ 6 kHz.

The local environment about the P atom was determined from the field dependent NMR linewidths to be more homogeneous in the type II samples. Noting that there is only one P environment in Ni_3P, whereas in $Ni_{12}P_5$ there are two inequivalent P environments, this NMR result is consistent with the suggestion in the next section that the dc electroplated samples (type II) more nearly resemble Ni_3P, whereas the glasses made by other techniques (type I) more nearly resemble $Ni_{12}P_5$.

In Fig. 7, a schematic free energy diagram for the two Ni-P glass phases is proposed, corresponding to the above ideas. In the glasses with low P concentration, the phase labelled II is more stable. For intermediate P concentrations, Phase I would be more stable. The NMR data for high P concentration (> 30 at .% P) appear to fit on a third line, not collinear with either the type I or II. This suggests the possibility of still another phase which is not shown here since it has not yet been studied in detail.

B. Bonding

The NMR Knight shift [11-13] and relaxation time [12] results lead to the conclusion that the chemical bonding in the

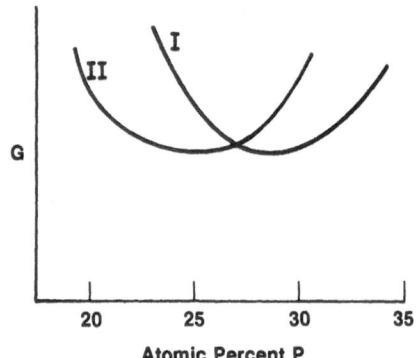

Fig. 7   Schematic free energy, G, diagram for the two glassy
         structures. Structure type I has its minimum near the
         composition $Ni_{12}P_5$; type II near Ni_3P. The relative mag-
         nitudes of G at the minima have been arbitrarily placed.

Ni-P glasses is similar to the bonding in stoichiometric Ni-P
crystals. Compound formation between transition metals, such as
Ni, and polyvalent metalloids, such as P, involves strong chemical
hybridization between the transition metal d and the metalloid p
valence electrons. An NMR relaxation time measurement [12] for
an electroless glass displayed a Korringa enhancement of 3.0 (where
1.0 corresponds to pure s electrons), consistent with such hybridization. The smaller Knight shifts in the type II samples are
indicative of smaller s-electron hyperfine fields due to increased
p-electron occupancy arising from increased hybridization with the
Ni d bands. This, in turn, implies increased chemical bonding.
This result is supported by the greater densities of the type II
samples and by the annealing experiments.

X-RAY ABSORPTION

A.   The Structure of the Absorption Edge

        X-ray absorption spectra were recorded [14] above the Ni
K-edge of type I and type II Ni-P glasses. (All measurements were
performed at the Cornell High Energy Synchrotron Source.) The
near edge regions of the type I, type II and nickel foil samples
are shown in Fig. 8, where the experimental uncertainty in the
energy is 0.5 eV. The edge of the type I specimen is shifted ∿1 eV
relative to that of type II, and the detailed edge shapes are quite
different for the two glass samples, suggestive of differences in
the chemical bonding in the two structures. Since this region of
the absorption spectrum, extending to approximately 50 eV above
the absorption edge, is sensitive to the chemical bonding, it is
excluded from the data set used for the EXAFS analysis below.

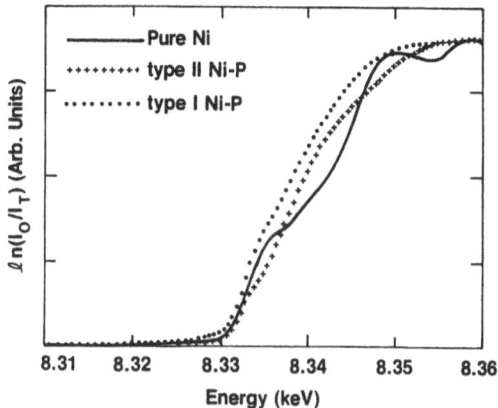

Fig. 8   A comparison of the Ni K-near edge region of the x-ray
         absorption spectra of Ni metal (solid line), a type II
         sample (++++), and a type I sample of the same composi-
         tion (....).

B.   Extended X-ray Absorption Fine Structure

        The extended x-ray absorption fine structure (EXAFS) refers
to the oscillations in the linear abosrption coefficient observed
on the high energy side of the x-ray absorption edges of atoms for
molecules and solids. This fine structure reflects the modulation
in the absorption coefficient due to the interference between the
outgoing photoelectron wave, and waves backscattered by neighboring
atoms.   The period of the oscillations is related to the distances
between the absorbing atom and its nearest neighbors, and the amplitude of the oscillations is related to the number of nearest neighbors, their elemental type and the degree of disorder in the
nearest neighbor bonds.   (It is important to note that a wide
distribution of bond lengths appears in the EXAFS analysis as
"static" disorder.)   Therefore, a study of the fine structure can
provide a variety of information about the local environment of
a particular element in a complex system.

        Both x-ray scattering and the EXAFS yield a one dimensional
averaged view as a function of the radial distance from the probe
atoms of the short range order in noncrystalline samples.   The
data range of a typical scattering experiment extends from about
$|\vec{q}| = 5$ to 150 nm^{-1}, where $\vec{q}$ is the photon momentum transfer.   For
EXAFS, the useable data range extends from approximately $|2\vec{k}| = 60$
to more than 250 nm^{-1}, where $\vec{k}$ is the photoelectron wavevector
and $2\vec{k}$ is equivalent to $\vec{q}$.   Thus, the EXAFS experiment is more
sensitive to short range correlations than x-ray scattering.
However, the loss of low $\vec{k}$ data has been shown [49] to introduce
ambiguities in the coordination number and distance information
that can be extracted.   Essentially, the EXAFS are most sensitive

to the sharp features in the pair distribution function; broad features (e.g., asymmetric tails) in the distribution may be difficult to detect. Even within these restrictions, much information regarding the character of the local environment surrounding particular elements in disordered materials may be extracted.

A primary advantage of the EXAFS technique in the study of TM/M glasses lies in its elemental specificity and sensitivity to low Z metalloid backscatterers. Since the K-absorption edge of most elements are well separated in energy, the local environment around each constituent of the system may be studied separately. For multicomponent glassy structures, this is an important simplification in the analysis of short-range order. For a system containing N elemental species, a conventional x-ray scattering experiment is a measure of a summation involving $N(N+1)/2$ independent pair correlation functions. The contribution of each pair is weighted by the concentration of each species, and by the magnitude of the atomic scattering factors of each element. Consequently, the scattering experiment is primarily sensitive to the more numerous, and higher Z, TM/TM pairs in these glasses. The EXAFS measures only the N subsets of the $N(N+1)/2$ pair correlations. Often, each contribution (e.g., Ni/Ni or Ni/P) may be distinguished since the energy dependence of their photoelectron backscattering amplitude varies with atomic number. In addition, the peak amplitude of the backscattering from light elements is comparable, and in fact, sometimes greater than that of higher Z backscatterers.

C.   EXAFS Results [14]

After removing the smooth atomic absorption contribution, and normalizing the spectra to a per-atom basis using standard procedures [50], the EXAFS oscillations, X(k), measured from the type I and type II samples were extracted (Fig. 9). Striking differences between the spectra of each sample are evident. Also shown in Fig. 9 are the photoelectron backscattering amplitude functions [51] for P and Ni. For the type II sample, the envelope of the oscillations peaks at a photoelectron momentum (defined as $|\vec{k}| = \sqrt{[2m(E-E_o)]}/\hbar$, where E_o is the threshold energy of the free photoelectron) characteristic of primarily phosphorous backscattering. As previously noted by Cargill [48] this is indicative of the presence of a sharply peaked Ni/P pair distribution, and a rather broader Ni/Ni pair distribution. Since broad features in the pair distribution are smeared out in the useable range of EXAFS data, the Ni/P contribution dominates. In contrast, the EXAFS of the type I sample has a more pronounced Ni backscattering character, consistent with a broader Ni/P contribution, a narrower Ni/Ni contribution, or perhaps both.

The data shown in Fig. 9 were Fourier transformed using k^3 weighting on the k range from 30 to 140 nm^{-1}, and a Hanning window

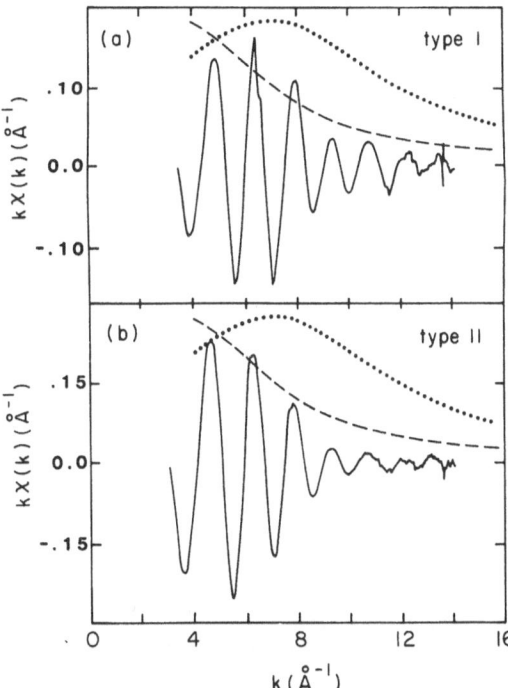

Fig. 9 EXAFS Signals measured from the (a) type I and the (b)
        type II Ni-P glasses, compared to backscattering by Ni (...)
        and P (---).

function of 2.1. The magnitudes of these Fourier transforms (Fig.
10) of the EXAFS spectra provide a more intuitive picture of the
local environment around the Ni atoms in these glasses. The
transform is related to the radial distribution function of the
Ni/Ni and Ni/P pairs, although it is shifted to lower radial distance R and distorted due to the photoelectron scattering phase
shifts and finite k-space data length. For both types of glass,
the first major peak of the transform contains the bond length
information for the nearest Ni/Ni and the Ni/P pairs. There is
a significant difference in the widths of the first peak of the
transforms. It is difficult to measure absolute widths from the
EXAFS because $X(k)$ cannot be measured below 3 A^{-1} (or above 16 A^{-1}),
and thus the window function and the damping factors are important.
However, using the same bandwidth in k-space and the same damping
factors, very precise <u>comparisons</u> of features can be made between
materials having the same composition. The first peak in the
Fourier transform of the data for the type II is significantly
broader than for the type I. Modelling calculations were performed
in which the number of nearest neighbors was arbitrarily chosen
and the amplitude reduction factor, S_0^2, and the relative

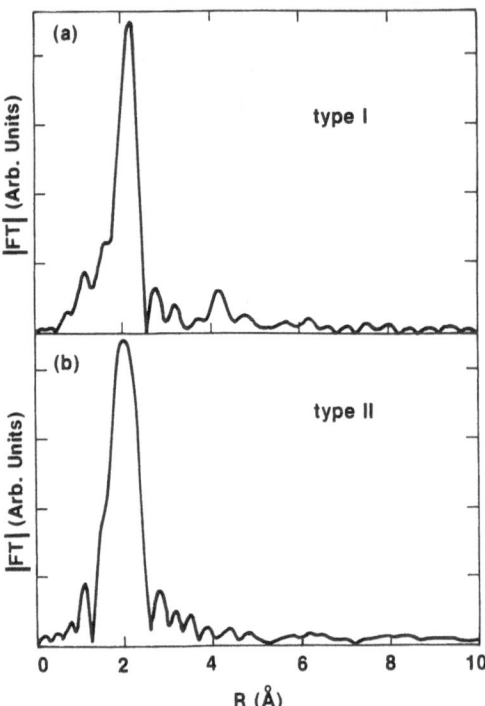

Fig. 10 Magnitude of the Fourier transform $|FT|$, for the (a)
type I and the (b) type II Ni-P glasses. R is uncorrected
for phase.

Debye-Waller factor, σ^2, were determined. The function $D = S_0^2 \exp(-2k^2\sigma^2)$ is then an inverse measure of the actual number of different bond lengths contributing to the peak. These results show that the Ni/Ni bond distribution in type I is much narrower than in type II, and the Ni/P bond distribution in type I is significantly broader than in type II:

Ni/Ni $D(I)/D(II) = 1.5$

Ni/P $D(I)/D(II) = 0.67$

The transform of the type I glass exhibits a second (analysis-independent) peak above the noise at higher R, which is not evident in the result for the type II glass. This suggests that the Ni/Ni distribution in the type I glass may involve more intermediate range correlations than that in the type II glass.

DISCUSSION

The detailed atomic rearrangements accompanying structural relaxation are not well understood. What is known is that the density increases in the "fully" relaxed" Ni-P system. Since the nearest neighbor distance does not decrease in the radial distribution function, it is assumed that the average coordination number or the "structural order" increases in the relaxed state. Egami [18] has used a local fluctuation theory to predict that the distribution of atomic distances becomes narrower in the relaxed state as defects of different types are annealed out. Our data, in contrast, suggest that the lower energy state (II) shows a greater distribution of Ni/Ni atomic distances around the central Ni atoms. A second neighbor peak is also seen in the result for the higher energy (I) state which is not evident for the lower energy glass. This indicates there is greater "structural order" in the higher energy state than in the lower energy state for this system. The better structural definition (i.e., narrower peaks and more peaks in the Fourier transform) is seen in the higher energy glass; the greater distribution of atomic distances (a single broad peak) is seen in the lower energy glass.

These results, which cannot be reconciled using a continuous structural relaxation mechanism, are understandable when viewed in terms of two distinct local structural environments. A comparison of the local structure of type II and I glasses to the Ni_3P and $Ni_{12}P_5$ structures, respectively, is illuminating (Fig. 2). Both crystal structures are very complex, and many different Ni/Ni and Ni/P pairs are summed into the pair distribution functions. The Ni/Ni pair distribution in $Ni_{12}P_5$ is narrower than that in Ni_3P, and the Ni/P pair distribution is four times broader. Thus the distribution of atomic distances in the type II structure is similar to that of Ni_3P, whereas that of type I is similar to that of $Ni_{12}P_5$. This proposed identification is consistent with the NMR data (Fig. 6). It is also consistent with the density measurements (Fig. 5), in which the lower energy, higher density, glass curve falls to just below the value for an average of crystalline Ni and Ni_3P, and the higher energy, lower density, glass curve falls just below the value for Ni and $Ni_{12}P_5$. The linear curves in the NMR and density data throughout this compositional range suggest that the two structural types persist, even in the phosphorous deficient range far from perfect stoichiometry.

The combination of evidence from density, NMR, and x-ray absorption measurements outlined above suggests that the local environments in the two types of Ni-P glasses studied here are similar to the local environments in the two stoichiometric crystals,

Ni_3P and $Ni_{12}P_5$. Using this proposed association, it is possible to speculate on the reason the Knight shift in regime II is independent of composition, while there is a composition dependence in regime I. Since there is only one P environment in Ni_3P, glassy alloys in regime II would form at compositions below 25 at.% P by decreasing the number of occupied P sites. However, the local effective coordination of the remaining P atoms would continue to be 9 Ni atoms. In regime I, the glasses do not extend to the 29.4 at.% P composition of $Ni_{12}P_5$, presumably because a third glassy structure intervenes. In $Ni_{12}P_5$, some P atoms have 10 Ni neighbors whereas other P atoms have 8 Ni neighbors (average = 9.6). Assuming that the Knight shift is mainly determined by the number of near Ni neighbors, and assuming additivity, the shift in regime I may be the result of preferential depletion of the P(1) site with its 10 Ni neighbors.

In equation form, the coefficients obtained from the crystalline Ni_3P and $Ni_{12}P_5$ Knight shift values,

$$1.07 - 0.10 \ N = K$$

Here N is the average number of Ni neightbors for the P sites, and K is the Knight shift in percent. Then, when all the P(1) sites in the regime I glass are depleted, K = 0.26%, and this should occur at 2/26 or $\sim$ 8 at.% P. The agreement between this oversimplified theory and the experimental results is embarrassingly good and may be coincidental since we cannot be sure that the P(1) sites are preferentially depleted.

In the EXAFS results, the type II alloys are characterized by a sharp Ni/P distribution, and a significantly broadened Ni/Ni distribution, compared to type I. The better Ni definition found in the EXAFS (i.e., narrower Ni peaks and more peaks in the Fourier transform) for the higher energy (type I) glass provides strong evidence that the two glasses are not related by a continuous structural relaxation in the traditional sense [14]. Instead this is an example of polymorphism in the Ni-P glassy alloys. There appear to be at least two, and perhaps three, distinct forms of short range order in the glassy Ni-P system with the processing and thermal history of the samples determining the local configurations.

The fact that a bimodal distribution was obtained in the NMR and density measurements, and that x-ray absorption measurements could distinguish between the two types of glasses, emphasizes that only one type of glass was generally present in any single sample. It is possible, however, that under certain conditions the two types of glasses could be found mixed in a single sample, in the same way that two crystalline polymorphs can be found in the same sample. An investigation of this would require detailed kinetic studies together with effective characterization tools.

SUMMARY

The picture of the metallic glass which emerges from this work resembles most closely the quasicrystalline model (Section II.D). The underlying qualities which define this model are: (1) that the structure is homogeneous; (2) that the local packing and bonding takes on a character similar to a specific crystalline type; and (3) that the structure is minimally disordered with a loss of long range order.

The evidence for (at least) two structural types in Ni-P glasses is found in several kinds of experimental data. Perhaps among the most convincing is the density data (shown in Fig. 5) which strongly suggests two different local structures for P concentrations between 5 and 25 atomic percent. Equally convincing are the NMR Knight shift results (Fig. 6), which are really a more sophisticated way of looking at the same phenomenon, in this case observing the local coordination with good sensitivity through the hyperfine interaction. The EXAFS results go one step further by providing a proposed identification of the character of the local ordering present in the different glass samples, and by demonstrating that the different glasses are not related through structural relaxation. Other peripheral data similarly support the underlying idea of different polymorphic forms in Ni-P glasses. The mechanical properties of the types I and II appear to be quite different: the type I samples have a tendency to be brittle while the type II samples are ductile. Furthermore, the modes of crystallization are very sensitive to composition. While the mechanical and cyrstallization properties may reflect different structural states, it is also possible that they are a function of the defect concentration and structure and it has not been shown that they are necessarily evidence of polymorphism.

This paper has dealt with a model glass system based on Ni and P. It is expected that many of the features of this system are present in related metal-metalloid glasses, and perhaps more generally.

REFERENCES

1. B. G. Bagley, H. S. Chen, and D. Turnbull, Mater. Res. Bull., 3:159 (1968).
2. H. S. Chen, Rep. Prog. Phys., 43:353 (1980).
3. J. D. Bernal, Nature, 185:68 (1960).
4. J. D. Bernal, Proc. Roy. Soc., A280:299 (1964).
5. J. L. Finney, Proc. Roy. Soc., A319:479 (1970).
6. D. E. Polk, Acta Metall., 20:485 (1972).
7. T. M. Hayes, J. W. Allen, J. Tauc, B. C. Giessen, and J. J. Hauser, Phys. Rev. Letters, 40:1282 (1978).
8. N. Valenkov, and E. Poray-Koshitz, Zeits. f. Krist., 95:195 (1936).

9. B. E. Warren, J. Appl. Phys., 8:645 (1937).

10. J. L. Walter, S. F. Bartram and I. Miller, Matls. Sci. and Eng'g., 36:193 (1978).

11. L. H. Bennett, Bull. Am. Phys. Soc., 20:644 (1975).

12. L. H. Bennett, H. E. Schone and P. S. Gustafson, Phys. Rev. B., 18:2027 (1978).

13. D. S. Lashmore, L. H. Bennett, H. E. Schone, P. Gustafson, and R. E. Watson, Phys. Rev. Letters, 48:1760 (1982).

14. A. I. Goldman, G. G. Long, L. H. Bennett and M. Kuriyama, Submitted; A. I. Goldman, Ph.D. Thesis, S.U.N.Y. at Stony Brook, NY (1983).

15. A. Brenner and G. E. Riddel, J. Research NBS, 37:31 (1946); 39:385 (1947).

16. J. W. Cahn, Bull. Alloy Phase Diag. 1, 2:27 (1982).

17. R. B. Schwarz and W. L. Johnson, Phys. Rev. Letters, 51:415 (1983).

18. T. Egami, Annals N.Y. Acad. Sci., 371:238 (1981).

19. C. P. Chou and D. Turnbull, J. NonCryst. Solids, 17:168 (1975).

20. C. O. Kim and W. L. Johnson, Phys. Rev., B 23:143 (1981).

21. A. Mak, K. Samwer and W. L. Johnson, Phys. Lett. 98A:353 (1983).

22. B. W. Corb, R. C. O'Handley, J. Megusar and N. J. Grant, Phys. Rev. Letters, 51:1386 (1983).

23. C. H. L. Goodman, Nature 257:370 (1975); C. H. L. Goodman, in Structure of Monocrystalline Materials, P. H. Gaskell, Ed., Taylor & Francis, London, p. 197 (1977).

24. P. H. Gaskell, J. NonCryst. Solids 32:207 (1979); P. H. Gaskell, J. Phys. C 12:4337 (1979).

25. N. Rivier, Revista Brasileria de Fisica, Special Issue on Amorphous Systems (Proc. of March 1984 Symposium) in press.

26. W. H. Zachariasen, J. Am. Chem. Soc., 54:3841 (1932).

27. W. J. Boettinger, F. S. Biancaniello, G. M. Kalongi and J. W. Cahn, in Rapid Solidification Processing: Principles and Technologies, II, R. Mehrabian, B. H. Kear and M. Cohen, Eds., Claitor's Baton Rouge, LA, p. 50 (1980); W. J. Boettinger, Proc. 4th Intl. Conf. on Rapidly Quenched Metals, Sendai, p. 99, (1981); W. J. Boettinger, in Rapidly Solidified Amorphous and Crystalline Alloys, by B. H. Kear, B. C. Giessen and M. Cohen, Eds., Elsevier, NY, p. 15, (1982).

28. R. E. Watson and L. H. Bennett, Scripta Metall., 17:827 (1983).

29. G. S. Cargill, Solid St. Phys., 30:227 (1975).

30. T. Fujiwara and Y. Ishii, J. Phys. F: Metal Phys., 10:1901 (1980).

31. J. F. Sadoc, J. Dixmier and A. Guinier, J. NonCryst. Solids, 12:46 (1973).

32. P. Panissod, D. Aliaga Guerra, A. Amamou, J. Durand, W. L. Johnson, W. L. Carter and S. J. Poon, Phys. Rev. Letters, 44:1465 (1980).

33. M. Hansen and K. Anderko, Constitution of Binary Alloys, McGraw-Hill, (1958); R. P. Elliott, Constitution of Binary Alloys, First Supplement, McGraw-Hill, (1965).

34. S. Rundquist and E. Larsson, Acta Chem. Scand., 13:551 (1959).

35. O. Kubaschewski and J. A. Catterall, Thermochemical Data of Alloys, Pergamon Press, London, p. 134 (1956).

36. J. -P. Randin, P. A. Maire, E. Sauer and H. E. Hintermann, J. Electrochem. Soc., 114:442 (1967).
37. B. G. Bagley and D. Turnbull, J. Appl. Phys., 39:5681 (1968).
38. W. H. Safranek, The Properties of Electroplated Metals and Alloys, Elsevier, Amsterdam, p. 465 (1974).
39. P. S. Gustafson, Ph.D. Thesis, College of William and Mary, Williamsburg, VA, (1981).
40. G. S. Cargill, III, J. Appl. Phys., 41:12 (1970).
41. B. G. Bagley and D. Turnbull, Acta Metall. 18:875 (1970).
42. E. Vafaei-Makhsoos, E. L. Thomas and L. E. Toth, Metall. Trans., 9A:1449 (1978).
43. A. Czriraky, B. Fograssy, I. Bakonyi, K. Tompa, T. Bagi and Z. Hegedus, J. de Phys. Coll. C8, 41:C8 (1980).
44. B. G. Bagley, Ph.D. Thesis, Harvard University, Cambridge, MA (1967).
45. J. Dixmier, K. Doi and A. Guinier, in Physics of NonCrystalline Solids, North Holland, Amsterdam, p. 67, (1965).
46. Y. Waseda, in Liquid Metals, 1976, R. Evans and D. A. Greenwood, Eds. Inst. Phys. Conf. Ser. #30, Bristol, Chap. 2, p. 230, (1977).
47. P. Lagarde, J. Rivory and G. Vlaic, J. NonCryst. Solids, 57:275 (1983).
48. G. S. Cargill III, in Proc. of 4th Int. Conf. on Rapidly Quenched Metals, T. Masumato and K. Suzuki, Ed. JIM, Sendai, p. 389, (1982).
49. P. Eisenberger and G. S. Brown, Solid State Commun., 29:481 (1979).
50. P. A. Lee, P. H. Citrin, P. Eisenberger and B. M. Kincaid, Rev. Mod. Phys. 53:769 (1981).
51. B. K. Teo and P. A. Lee, J. Amer. Chem. Soc., 101:2815 (1979).

CRYSTAL-STRUCTURE DATA AS A TOOL FOR INTERPRETING LAXS (LARGE ANGLE

X-RAY SCATTERING) STUDIES OF AMORPHOUS INORGANIC COMPOUNDS

A. Mosset and J. Galy

Laboratoire de Chimie de Coordination du CNRS associé
à l'Université Paul Sabatier, 205 route de Narbonne
31400 Toulouse, France

ABSTRACT

The building of a structural model for amorphous solids requires the extensive knowledge of the geometry of the compound
studied. The collection of numerous structural data from various
sources (x-ray diffraction, spectroscopic methods, conformational
calculations...) may be a promising method to obtain this preliminary knowledge. This is illustrated by the Large Angle X-ray
Scattering study of several amorphous compounds chosen in different
fields of chemistry: V_2O_5, (adenosine-triphosphate) $(H_2O)_4$ Mn(II)Na$_2$,
(cytidine) (glycyl-L-leucine)Cu(II)$(H_2O)_3$, Cu(dithiooxamide)(H_2O),
Ni$_4$(dithiooxamide)$_5$ $(H_2O)_{0.25}$, MM'(EDTA)$(H_2O)_4$. $2H_2O$ with [MM'] =
[CoNi] or [NiNi].

INTRODUCTION

A literature survey of structural studies devoted to amorphous
or poorly-organized materials surprisingly leads to the statement
that the Chemistry of Complexes is hardly represented. One main
reason could be the seemingly intricate problem of short- and
long-range order of these materials: the structural parameters
requested to master the problem are numerous.

(i) Coordination polyhedron of the metal. Structural studies
carried out on single crystals show the large variety of their
geometries and principally the major distortions affecting
coordination polyhedra.

(ii) Coordination scheme of the organic ligand. The possibilities for such schemes rapidly increase with the number of potential

coordination sites: ten sites for the EDTA ligand, thirteen sites for adenosine triphosphate...

(iii) <u>Conformation of the organic ligand</u>. The situation becomes really perplexing when several dihedral angles vary correlatively.

(iv) <u>Interactions between complex molecules</u>. Only elementary geometries, e.g., chains or planar molecules (porphyrins), may generate stackings whose modeling still remains easy.

These difficulties oppose to the building of a structural model for amorphous solids and make it a critical exercise if no preliminary studies are carried out.

The recent development of the EXAFS technique allowed to solve some structural problems related to very short distances.

We believe that the only promising method consists of the preliminary collection of as many structural data as possible. Such structural data subsequently allow to impose constraints when building a structural model. But for very few propitious circumstances, it is generally impossible to limit to the transposition of a model achieved by x-ray diffraction on a single crystal. Consequently, it is necessary to vary data sources, to allow a better knowledge of the geometry of the molecule studied. A possible nonrestrictive list for these sources could be:

- structural studies on single crystals of analogous complexes producing data on partial geometries;

- checking of the results achieved through spectroscopic techniques for designing potential coordination schemes;

- exhaustive bibliography for determining the limits within which the conformation of one class of ligands varies;

- calculation of the lowest energy conformation by theoretical methods.

This is the investigation process the present article endeavours to enlighten. The amorphous materials studied are divided into two groups.

1. Materials which could not be obtained as single crystals while presenting fundamental interest or physical properties which justifies the study in the amorphous phase:

(i) The metal-nucleic acid constituent complex (adenosine triphosphate)$(H_2O)_4Mn(II)Na_2$;

(ii) The ternary peptide–metal–nucleic acid constituent complex, (cytidine)(glycyl-L-leucine)Cu(II)(H$_2$O)$_3$;

(iii) The dithiooxamide metallic complexes, CuL(H$_2$O) and Ni$_4$L$_5$(H$_2$O)$_{0.25}$.

2. Materials exhibiting important or unusual physical properties in the amorphous state differing from that observed in the crystalline phase:

(i) vanadium pentaoxide, V$_2$O$_5$;

(ii) EDTA complexes; MM'(EDTA)(H$_2$O)$_4$,2H2O, with [MM'] = [CoNi] or [NiNi].

EXPERIMENTAL SECTION

1. Sampling

The synthesis of the compounds has been conducted in aqueous medium. The complexes are obtained either as an amorphous powder (MM'-EDTA; M-dithiooxamide) by precipitation or as a vitreous lac subsequent to evaporation (Mn-ATP; Cu-cytidine-glycyl-L-leucine).

Subsequent to washing and drying, the powders are finely ground and pelletized under a pressure of 280 kg.cm^{-2}. The pellets are 0.2-0.3 mm thick. The vitreous lacs are used as obtained for data collection.

NiNi-EDTA was obtained under both forms. Two data collections have consequently been performed: the experimental radial distribution curves are comparable by direct superposition.

Amorphous V$_2$O$_5$ was prepared by ultrafast quenching from the melt (splat-cooling), at rates of about 10^6 °C s^{-1}. The oxide was melt at 800 °C in an induction-heated platinum crucible which has a hole 1 mm in diameter at the bottom. High-pressure argon is suddenly admitted, above the crucible, forcing the liquid oxide through the hole and spreading between two rollers. This ultrafast quenching yields dark blue platelets a few micrometers thick.

2. Collection of scattered intensities

Two instruments have been successively used for data collection: (i) an automatic diffractometer for single crystals, (ii) a two-circle goniometer equipped with a position-sensitive diffractometer especially designed and built in the laboratory [1].

A number of characteristics are common to both collections:

(i) the sample is mounted on a goniometric head, just as a
single crystal, and oriented perpendicularly to the incident beam
when the goniometer angles are at zero; (ii) the MoKα radiation
monochromatized by a graphite crystal was used for all data
collections; (iii) the scattered intensities are measured with the
scan mode θ-2θ.

With the CAD4 diffractometer, the data collection is carried
out step-by-step with a constant s value between each measured
Bragg angle (s=4 sinπ/θ; Δs=0.035363). Some 400 reflections were
recorded for $1.5°<θ<70°$. The periodicity was chosen equal to 15
min for a valuable counting statistics. Five reference θ angles
allowed to check the stability of the x-ray beam and of the sample
towards the scattering.

With the position-sensitive diffractometer, the conditions for
data collection changed. The Bragg angle covered for such a
position sensitive detector is $2θ=12°$. The integration of the
nonlinearity errors due to the detector wire requires that the detector
sweeps across the 2θ area under investigation. Each point of the
detector successively proceeds before each data point of the angle
θ, the full data collection in the range $1.5<θ<66°$ roughly requires
half-a-day.

3. Data processing - calculations - LASIP system

The LASIP (Liquid and Amorphous Structure Investigation
Package) system is a full software for a small microcomputer, i.e.,
Apple II Plus, which has been especially organized to undertake all
the calculations necessary to perform the structural analysis of
compounds in the amorphous state or in solution. This software
was specifically prepared for structural chemists [2].

Scattered intensities were corrected for polarization and absorption effects (namely $I_c(s)$), then normalized by comparison with the
sum of coherent and incoherent independent intensities in the
vicinity of high θ angles. The normalization factor K derived
from this data treatment is in good agreement with those determined
from Norman's [3] and Krogh-Moe [4] methods. The atomic scattering factors, $f_i(s)$ for all the atoms, were those proposed by
Cromer and Waber [5]; Compton diffusion factors were taken from
tables published by Cromer [6].

The processing of the scattered intensities gives a curve
$F(r)=D(r)-4\pi\rho_o r^2$, showing the distribution of the interatomic
distances in the complex molecule. The experimental curve results
from the processing of the scattered intensity using Zernicke and
Prins' [7] relation.

$$D(r) = 4\pi\rho(r)r^2 = 4\pi\rho_o r^2 + 2r/\pi \int_{s_{min}}^{s_{max}} s.i(s)_{exp}.M(s).\sin(rs).ds$$

(ρ_o: average electronic density in the sample; $s=4\pi\sin\theta/\lambda$ with 2θ: scattering angle and λ: wavelength of the beam utilized; $i(s)$: reduced intensity, $M(s)$: modification function

$$= [\,^2_{Mn}(0)/\,^2_{Mn}(s)]\ \exp(-0.01\ s^2)$$

The theoretical intensities have been calculated using Debye's formula:

$$i(s)_{th} = \Sigma \Sigma f_i(s)f_j(s) \frac{\sin(r_{ij}s)}{r_{ij}s}$$

The curve $F(r)_{th}$ is obtained by Fourier transform:

$$F(r)_{th} = 2r/\pi \int_{s_{min}}^{s_{max}} s.i(s)_{th}.M(s).\sin(rs).ds.$$

RESULTS AND DISCUSSION

1. Spat-cooled V_2O_5

        Vanadium pentoxide is a non-stoichiometric compound in which the loss of oxygen is compensated by vanadium ions in a lower oxidation state, namely V^{4+}, giving rise to semiconducting properties arising from electron transfer between V^{4+} and V^{5+} ions [8,9,10].

        It is therefore important, in a study of these properties of amorphous V_2O_5 to obtain detailed information about the local structure in the glass.

        Such information can be obtained from studies of ESR associated with the paramagnetic V^{4+} ions. Published results dealing with amorphous V_2O_5 obtained by splat-cooling show that short-range order around V^{4+} ions remains almost unchanged in the glass [11].

        ESR gives information on the paramagnetic V^{4+} ions only, which correspond to about 1% of the total amount of vanadium. The ESR determination of the local environment of V^{4+} does not

allow the investigation to extend to the whole material, due to
structural information dealing with only 1% vanadium. The literature shows that V^{5+} can accommodate various coordination polyhedra:
tetrahedra (CN4), square pyramids, and trigonal bipyramids with all
intermediate distortions (CN5) and octahedra more or less distorted
(CN6).

V_2O_5 crystallizes in the orthorhombic system with a layer
structure built up with square pyramids, VO_5, sharing edges and
corners (Fig. 1) [12]. Some interatomic distances are indicated
on the schematic drawing: it is to be noticed that the layers are
separated by 4.3 Å (c parameter). Such a disposition of the layer
locates an oxygen at the summit opposite to the apex of the oxygen
square pyramid surrounding the vanadium, shaped as a kind of
elongated octahedron.

The V_2O_5 network can be described as resulting from a "crystallographic shear" (CS) operation in a pseudo-ReO_3-type structure (see,
for example [13,14]) resulting in edge sharing between square pyramids.

The analysis of the experimental radial distribution curve
(Fig. 2), has to be carried out in the light of these structural
features.

The first peak in the curve $D(r)-4\pi r^2\rho_0$ is centered around the
value r=1.75 Å, which obviously corresponds to V-O distances. Such
a value is shorter than the average V-O distance in crystalline
V_2O_5 (1.83 Å). There is a shoulder at r=2.70 Å which can be
reasonably attributed to O-O interatomic distances between oxygen

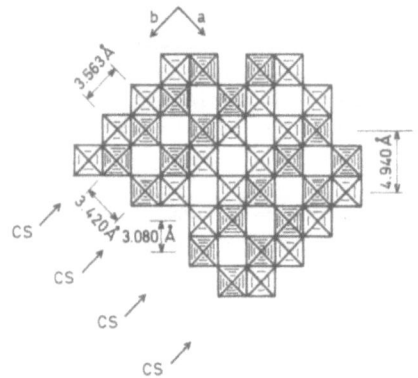

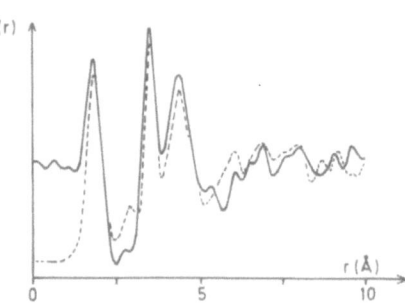

Fig. 1 Crystalline structure of
        V_2O_5. The arrows indi-
        cate the crystallographic
        shear (CS) present in the
        structure.

Fig. 2 Experimental (full line)
        and theoretical (dashed
        line) radial distribu-
        tion curves.

atoms surrounding the vanadium. The main peak, centered around
r=3.45 Å, corresponds to V-V interatomic distances; that has to be
compared with the V-V distances observed in the crystal (3.42 and
3.56 Å) when these atoms are situated in corner-sharing pyramids.
It is to be noticed that the shortest V-V distance, i.e., 3.08Å,
existing in the crystalline state along the crystallographic shear
corresponding to edge-sharing pyramids, does not appear in the
amorphous V_2O_5. The same observation can be made for V-V=4.94 Å
(Fig. 1). These comparisons suggest indeed that some drastic
changes have occurred in the vanadium coordination and polyhedra
association after melting and quenching.

The first step in developing a possible structure for "amorphous"
V_2O_5 is to settle the problem of vanadium coordination. The average
V-O distance of 1.75 Å has already been mentioned; such a distance
is typical of tetrahedrally coordinated vanadium. This possibility
is strengthened by a calculation of the number of oxygen atoms, n_o,
surrounding the vanadium. This is obtained by evaluating the peak
area at 1.75 Å and yields $n_o \approx 3.6$. Then the structural model can
be built as in silica glasses, using chains of VO_4 tetrahedra sharing corners characterized by V-O=1.75 Å and O-O=2.70 Å. Another
rule has to be followed in order to maintain the right chemical
formula V_2O_5: each tetrahedron shares three corners with the
neighboring VO_4 tetrahedra.

Obviously, the predominant peaks at 3.45 and 4.35 Å suggest
the presence of some kind of sub-unit. A satisfying model, in the
sense that a reasonable agreement is obtained between experimental
and calculated data, was built using three tetrahedra. The optimal
configuration of such a unit, drawn in Fig. 3 has been obtained
after refinement of the angles between edges ($O_1O_2O_5$ for example)
and dihedral angles between faces ($O_1O_2O_3/O_2O_6O_7$).

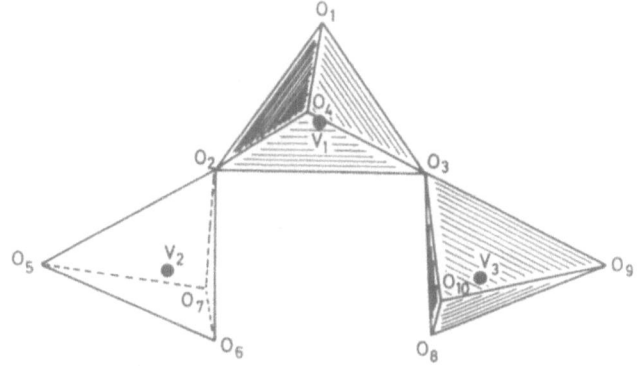

Fig. 3 Optimal configuration of the basic unit.

Such basic units are then connected as chains and their relative positions refined. A chain containing several sub-units has
been built, and the lateral chains connected in order to fill the space
(Fig. 4). The theoretical curve corresponding to this model is
given in Fig. 2.

Using this method it would be fallacious to try to build a
more precise and extended model concerning these VO_4 tetrahedra
chains and their relative orientation within the amorphous solid.
It cannot be asserted that the proposed model is unique; indeed its
correlation with the experimental radial distribution is not
unequivocal. Nevertheless, the architecture of the model must
follow reasonable structural and chemical features, such as are
present in the model described here.

Structural relationships between the amorphous and the crystalline form of V_2O_5 have been drawn up by Mosset, Lecante, Galy and
Livage [15] as well as a proposal to describe how the crystalline
network destroys itself before melting. The amorphous structure
described includes some features -- i.e., VO_4 tetrahedra -- already
mentioned to exist in liquid V_2O_5 [16].

Amorphous V_2O_5 contains some reduced vanadium atoms, i.e., V^{4+}.
Such a cation does not adopt a tetrahedral coordination [17]; it
is well known that the only plausible coordination polyhedra are:
CN5 with a square pyramid (SP) or a trigonal bipyramid (TBP) and
CN6 with an octahedron (O), all these polyhedra being more or less
distorted.

The small amounts of V^{4+} contained in the glass are probably
situated in square pyramids VO_5, which could easily be randomly
distributed and connected by corner sharing to the tetrahedra chains.
An illustration of such accommodation for the proposed structure of
amorphous V_2O_5 is indicated in Fig. 5.

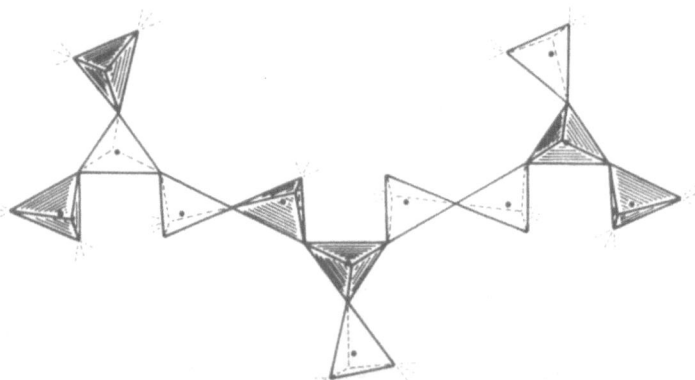

Fig. 4 Schematic drawing of the proposed model.

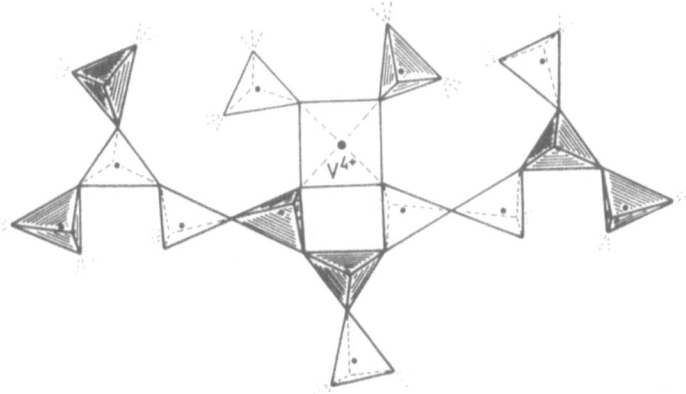

Fig. 5   The possibility to insert $V^{4+}O_5$ square pyramids within the
         VO_4 tetrahedra structure.

2.   The complex Manganese(II)-Adenosine Triphosphate

        Enzymatic reactions involving nucleotide triphosphate
require the use of divalent metallic cations as catalysts. In most
cases, magnesium (II) is used but manganese (II) may be substituted
as well.   Numerous solution studies of this system have led to the
formulation of mutually inconsistent hypotheses on the coordination
of adenosine triphosphate (ATP).

        Thus, the structural study of the complex ATP-manganese(II)
was of fundamental importance and fitted in the work developed at
the laboratory on the metallic complexes of nucleic acid constituents.

        Attempts to obtain single crystals suitable for an x-ray
diffraction study failed as only amorphous phases were obtained.
We then considered to use the LAXS technique for the study of this
amorphous compound [18].

        The definition of the initial geometry is more intricate than
in the preceding study.   The structural investigation of the
complex manganese-adenosine triphosphate by x-ray diffraction has
not been performed so far while structural studies of analogous
complexes are extremely rare.   Besides, the structure of ATP is
known with a low resolution.   Finally, due to the biochemical
importance of the system Metal-ATP, many studies in solution have
been carried out.   Their authors have proposed various coordination
schemes among which none could be sorted out.

        Thus, no sure basis is available either for the coordination
or for the conformation of the ATP.   The problem even becomes
more complicated if coordination involves both the phosphate
chain and the puric base.   The molecule "closes" and it is no

longer possible to have the dihedral angles vary independently for the refinement of the conformation (Fig. 6).

This is why we made use of a method differing from the method presented above. The models studied for this complex involve structural constraints and have been defined by the Cartesian coordinates of the atoms. These coordinates have been determined from two projections of a "Kendrew-type model". They result from an exhaustive investigation of the structural data related to the metallic complexes of nucleic acid constituents and of free ligands [19].

(i) The bond distances and angles adopted for the adenine base are the mean values calculated from 20 non-complexed puric bases and 24 bases involved in the coordination to one metallic center. The dispersion of the geometrical values averages 0.03 Å for bond distances and 2.5° for bond angles.

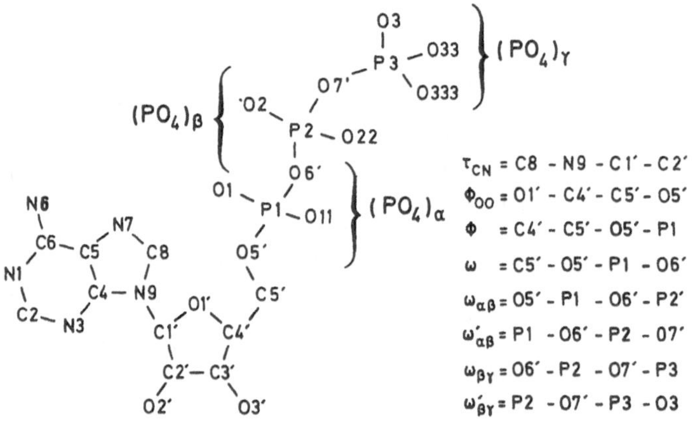

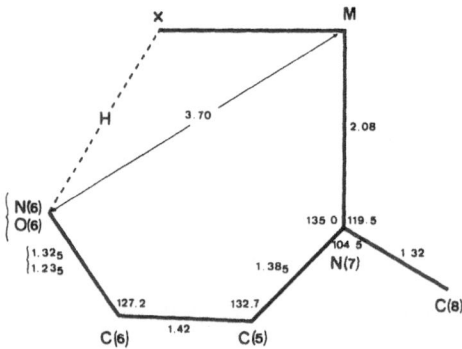

$(PO_4)_\gamma$
$(PO_4)_\beta$
$(PO_4)_\alpha$
τ_{CN} = C8 - N9 - C1' - C2'
Φ_{OO} = O1' - C4' - C5' - O5'
Φ = C4' - C5' - O5' - P1
ω = C5' - O5' - P1 - O6'
$\omega_{\alpha\beta}$ = O5' - P1 - O6' - P2'
$\omega'_{\alpha\beta}$ = P1 - O6' - P2 - O7'
$\omega_{\beta\gamma}$ = O6' - P2 - O7' - P3
$\omega'_{\beta\gamma}$ = P2 - O7' - P3 - O3

Fig. 6 Numbering and definition of dihedral angles in ATP.

X
M
H
3.70
2.08
N(6)
O(6)
135.0
119.5
104.5
1.32
N(7)
C(8)
1.385
127.2
132.7
C(6)
1.42
C(5)

Fig. 7 Purines: average geometry for the coordination in N(7).

(ii) Coordination of the adenine base may happen through the nitrogen atoms, N(1), N(3), N(6) and N(7). However, in 70% cases only N(7) is involved. We have proposed a mean geometry of the coordination on this site (Fig. 7). In addition, we have checked the other possibilities of coordination.

(iii) The adenine base has been assumed as planar, as the average dihedral angle between the two cycles forming the base is equal to 2° and 1.2° for the complexed and for the non-complexed bases, respectively.

(iv) The angle of torsion τ_{CN} (Fig. 6) allows to describe the conformation of the nucleoside with reference to the glycosidic bond N(9)–C(1'). If τ_{CN} ranges from 0° to -180°, the conformation is called anti while it is noted syn if τ_{CN} ranges from 0° to $+180^\circ$ (Fig. 8). Most of nucleic acid constituents (either complexed or not) have an anti conformation. Furthermore τ_{CN} varies within a rather narrow range in most complexes: -50°, -100°.

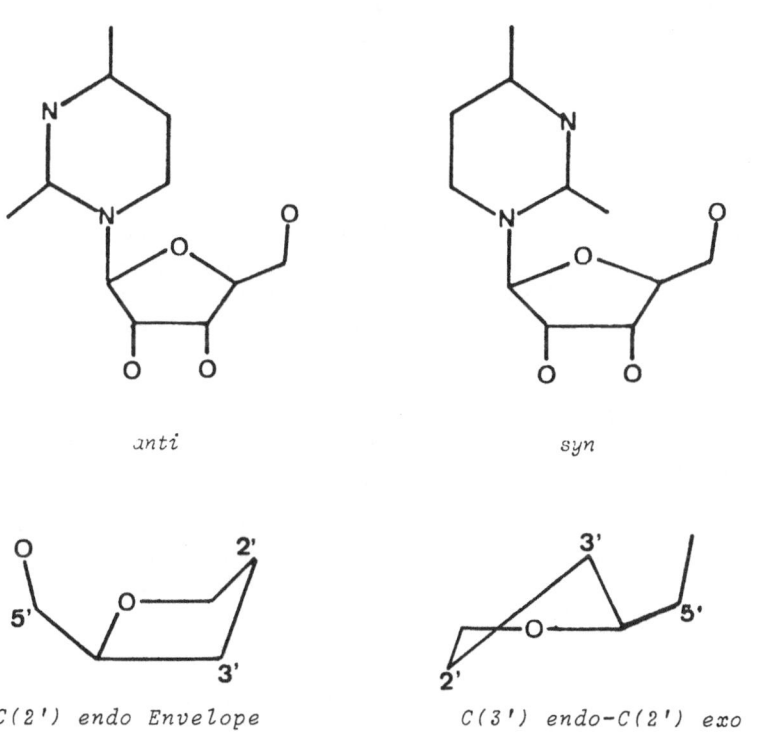

anti   *syn*

C(2') endo Envelope   C(3') endo–C(2') exo
                                  Twist

Fig. 8   Conformations of nucleoside and ribose ring.

(v) The ribose cycle exhibits two main conformations: an envelope conformation C(2')endo and a twist conformation C(2')exo (Fig. 8). The latter was chosen as the initial conformation due to its prevalence in complexes of nucleic acid constituents.

(vi) Bond angles and distances in the phosphate groups for various monophosphate nucleoside complexes were taken from the literature.

The most important feature of this study is clearly the coordination scheme.

Figure 9 shows the results obtained for a number of coordination schemes among the most specious, subsequent to solution studies. All three hypothesis (a,b,c) which correspond to a di- or tri-coordination for ATP, lead to a very poor agreement between theoretical and experimental curves. Consequently, the solution represented in Fig. 9d may be unambiguously considered, although some discrepancies remain between experimental and theoretical curves. The latter are partially due to the nonrefinement of the conformation of the ligand: long-distance interactions ($r > 7$ Å) are highly sensible to the variations of the conformation. Further, the intermolecular interactions and the interactions complex molecule - sodium ions have not been considered.

Despite the approximations, this study allowed the following conclusion; adenosine triphosphate is four-coordinated to the manganese atom via the nitrogen atom N(7) of the puric base and via one oxygen atom of each of the three phosphate groups. The octahedral environment of the metal is completed by two water molecules (Fig. 10). This confirms both the results reported by Kahn et al. [20] on the basis of a solution study and the hypothesis expressed by Sundaralingam [21] subsequent to exclusively crystallographic reflections on the structures of monophosphate nucleotides and mineral phosphates.

3. The complex (glycyl-L-leucine)(cytidine)copper (II)

As part of the same study on model compounds in bioinorganic chemistry, the interactions metal-proteine represent the most extensively investigated research field. The compilation and the utilization of hundreds of structural determinations such as amino acids, peptides and their complexes, have allowed to suggest a quite precise model of geometry for the active site of carboxy-peptidase A.

A similar approach directed the study of ternary protein-metal-nucleic acid interactions. However, the studies are less advanced so far and the amount of structural data is insufficient.

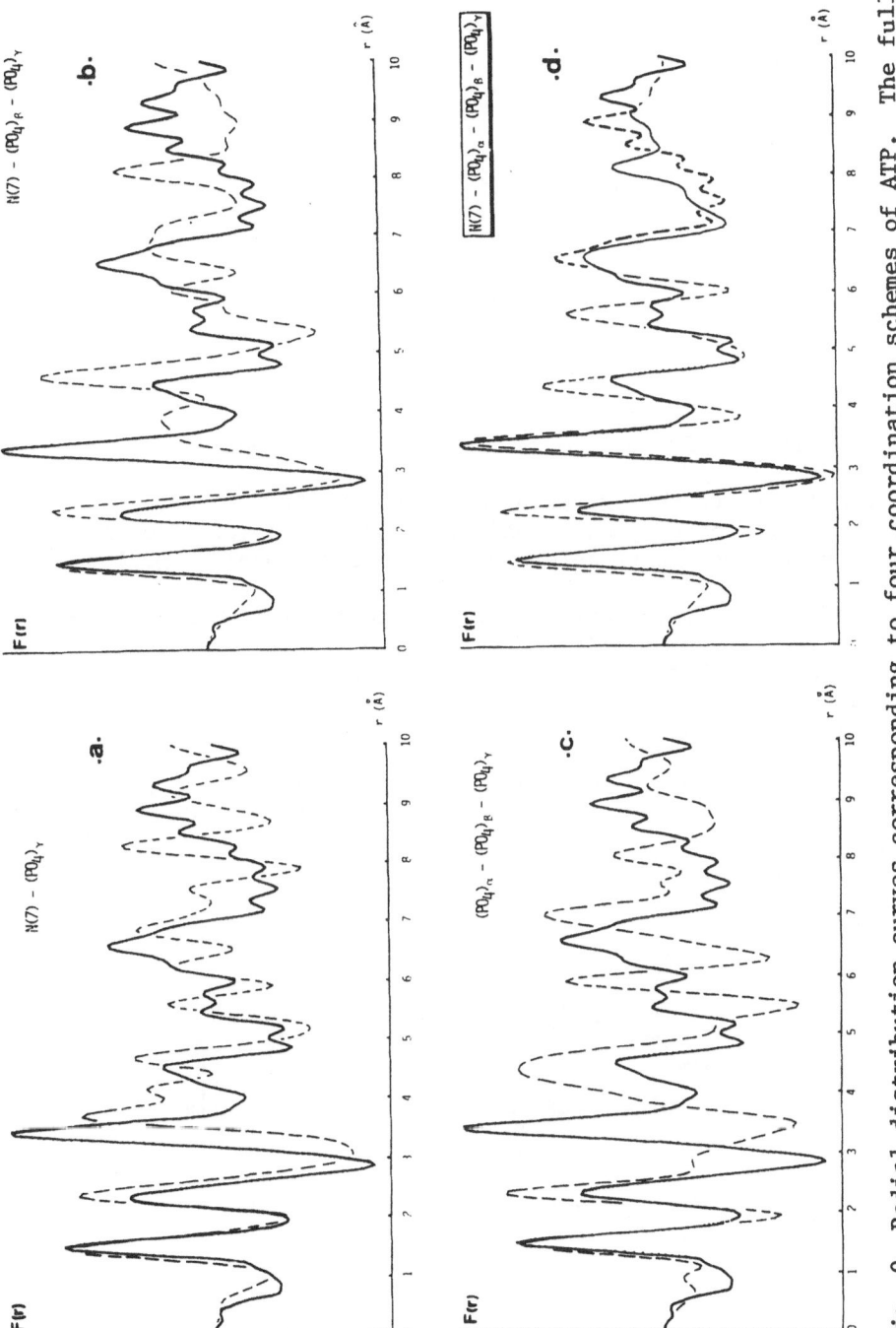

Fig. 9 Radial distribution curves corresponding to four coordination schemes of ATP. The full line represents the experimental curve and the dashed line represents the theoretical curve.

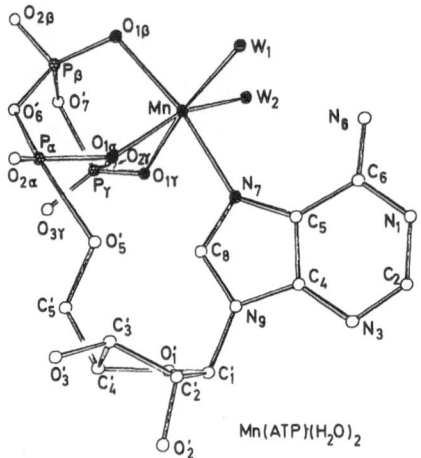

Fig. 10 Proposed model for the Mn–ATP complex.

Similarly, to the preceding complex, the lack of structural studies for parent compounds lead us to collect information from various sources so as to determine the initial model (Fig. 11).

(i) We have thus carried out the single crystal diffraction study of two copper complexes: (glycyl–L–tyrosine)copper(II), $(H_2O)_4$ [22] and (L–methionyl–glycine)copper(II) [23], so as to have a valuable geometrical model for the chelating double cycle N(5)–C(10)–C(9)–N(4)–C(8)–C(7)–O(3).

(ii) The geometry assumed for the lateral chain of L–leucine originates from the structural study of the complex copper(II)–glycyl–L–leucine–L–tyrosine tripeptide [24].

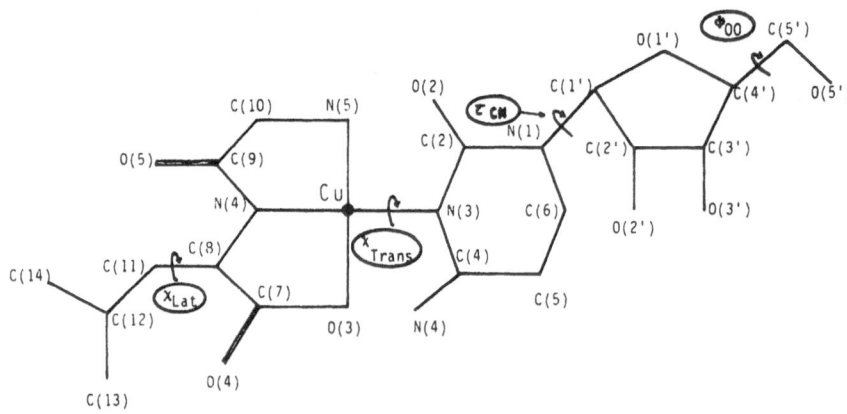

Fig. 11 Numbering and definition of dihedral angles in the (cytidine (glycyl–L–leucine) copper complex.

(iii)   The single crystal structural study of the complex
ZnCl$_3$(cytidine),N(CH$_3$)$_4$ has then been carried out to determine the
geometry of coordinated cytidine [25].

(iv)   Just like in the study of the complex manganese–ATP, a
thorough survey of literature data allowed to show a definite trend
for pyrimidine derivates to coordinate through the nitrogen atom
N(3) of the base (75% probability). Consequently, a mean geometry
of the coordination site around N(3) has been built (Fig. 12).

The initial model thus defined leads to the theoretical radial
distribution curve shown in Fig. 13a.

The theoretical and the experimental curves are in fair agreement for distances below 3 Å. This means that bond angles and
bond lengths are roughly acceptable in any part of the complex
(peptide, pyrimidine base, ribose). Unlikely, the two curves
largely differ when the distances are larger than 4 Å. Consequently,
we had (i) to check the validity of the coordination scheme chosen
(Fig. 11) and (ii) to refine the conformation of the complex.

The theoretical curve resulting from this work, which corresponds to the final conformation put forward, is shown in Fig. 13b.
The coordination of the dipeptide and of the nucleoside is that
initially chosen. This result was quite expected: tricoordination
of a peptide on a metallic center is usual and coordination of
cytidine through the only nitrogen atom N(3) of the pyrimidine
base is the most usual, as already stated. Refined dihedral angles
are listed in Table I.

The modifications brought to these dihedral angles allowed to
improve perceptibly the correlation between the theoretical and
the experimental curves. But for very long distance interactions
(> 8.8 Å) no peak remains unexplained and the divergence between
the corresponding experimental and theoretical peaks remains below
0.2 Å. Excluding experimental errors, both divergence sources
between experimental and theoretical curves may be reasonably

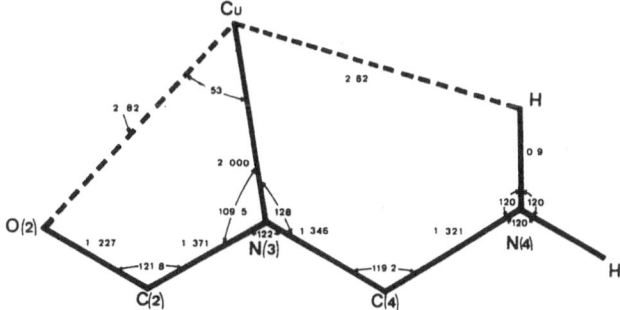

Fig. 12 Pyrimidines: average geometry for the coordination in N(3).

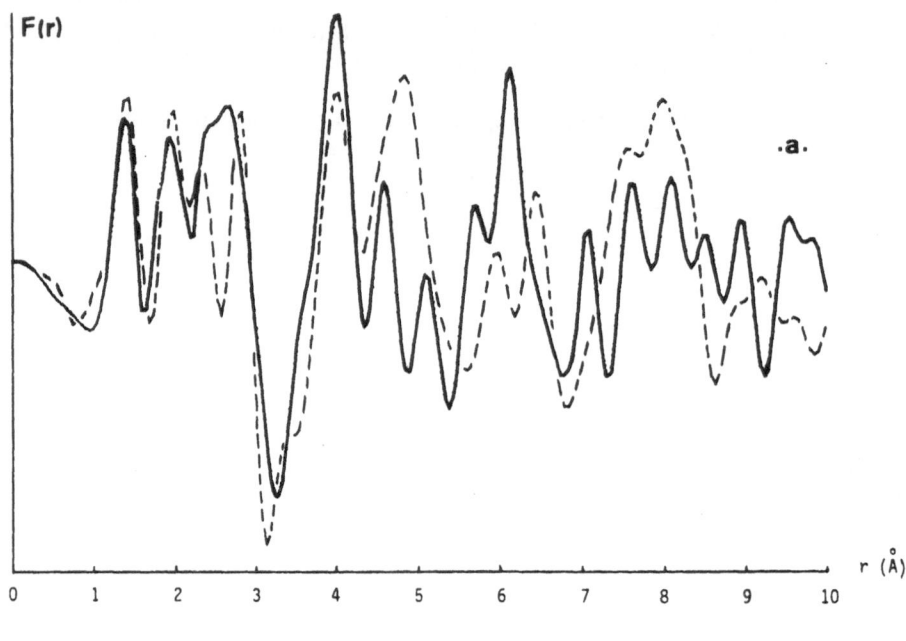

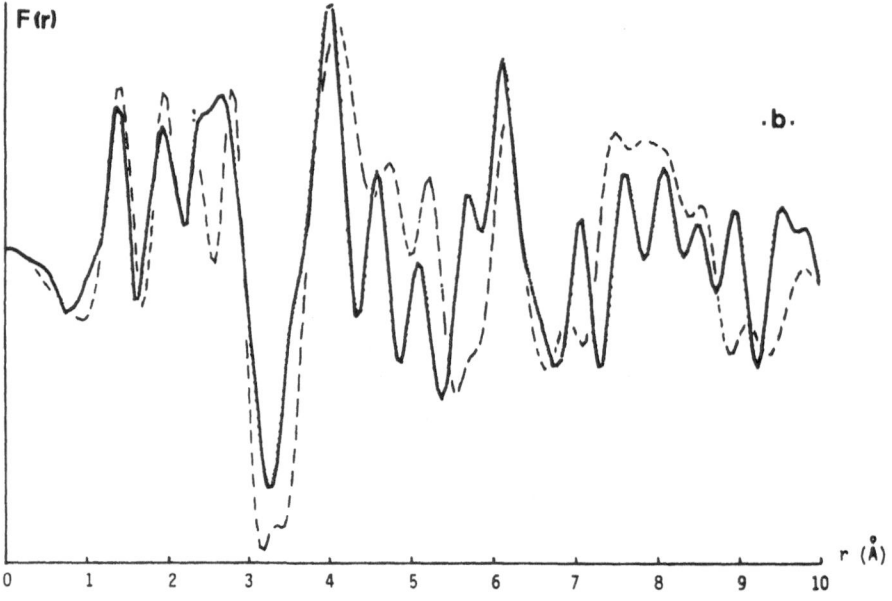

Fig. 13 Radial distribution curves correspond to two different sets
         of dihedral angles (see text). The full line represents
         the experimental curve and the dashed line represents the
         theoretical curve.

TABLE I Partial refinement of the model.

Dihedral angles	Definition	Initial	Final
τ_{CN}	C(6)-N(1)-C(1')-C(2')	$-69.7°$	$-82 \pm 5°$
χ_{trans}	O(3)-Cu-N(3)-C(2)	95.0	60 ± 10
χ_{lat}	N(4)-C(8)-C(11)-C(12)	-52.8	-80 ± 10

attributed to: (i) the lack of refinement of bond distances and bond angles; (ii) interactions between the complex and the three water molecules of the average formula which have not been taken into account.

Obviously, we cannot exclude the possibility that a quite different combination of the values of the dihedral angles leads to a theoretical curve close to that shown in Fig. 13b. However, a great number of theoretical curves have been calculated for values of the dihedral angles very distinct from the values chosen. Such a coincidence has not been encountered. Further, a number of dihedral angles are not allowed to take any value whatever. τ_{CN} usually ranges from $-50°$ to $-100°$ (cf. previous paragraph). χ_{trans} also has a limited range, and the free rotation around the bond Cu-N(3) is not possible due to the extra-cyclic functions (O(2),N(4)) grafted on the pyrimidine base.

Consequently, a high reliability factor may be reasonably attributed to the structure proposed for the complex (cytidine) (glycyl-L-leucine)copper(II) and which is described by the values of the dihedral angles listed in Table I (Fig. 14).

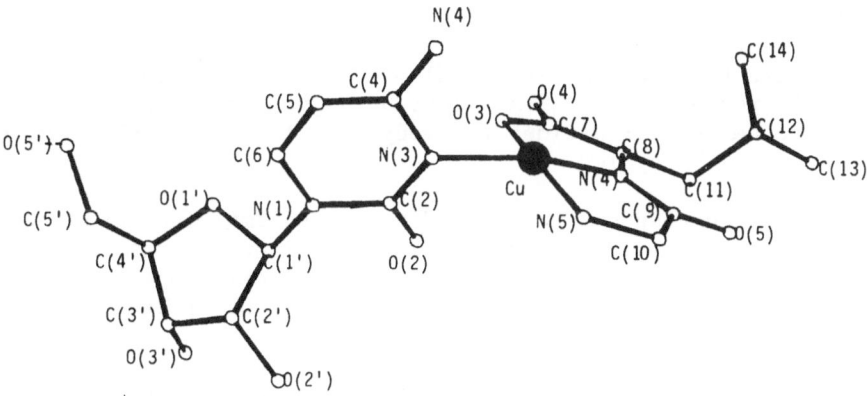

Fig. 14 Proposed structural model for the (cytidine)(glycyl-L-leucine) copper complex.

4.   Complexes of the dithiooxamide ligand

The metallic complexes of the dithiooxamide ligand, i.e., rubeanic acid, are well known as being used in various fields of chemistry, especially analytical chemistry.

But recent development in the knowledge of the magnetic, semi-conducting and catalytic properties of the complexes of rubeanic acid have aroused interest for such compounds and led to investigation of their structures. A great number of solutions or solid state studies have been carried out, using various spectroscopic methods which should allow us to know the most likely coordination schemes of such ligands. These studies which imply very different hypotheses concerning their structures have not been corroborated so far by x-ray diffraction studies, due to the extreme insolubility of the metallic complexes of rubeanic acid -- either substituted or nonsubstituted -- in most usual solvents. These complexes are then always obtained in the amorphous state or very poorly crystallized.

Once more, the complete lack of single crystal structural determinations of nonsubstituted dithiooxamide complexes compelled us to use various data sources to build our model.

(i) The coordination scheme of the ligand is the first point of importance. A careful survey of the literature data subsequent to spectroscopic studies allows to evolve six structural schemes (Fig. 15).

Model 1 corresponds to a bidentate ligand with a trans-conformation which only coordinates through its sulfur atoms. This scheme leads to the formation of a polymer with an average metal-metal distance equal to 7.8 Å.

Models 2 and 3 show a tetra-coordination of ligand on two metallic centers. The ligand still has a trans-conformation. The polymers thus formed have metal-metal distances equal to 6.3 and 5.6 Å, respectively. The situation suggested in Model 2 is assumed to be less propitious thermodynamically due to the formation of four-link cycles.

Model 4 corresponds to a mixed condition implying both scheme 1 and scheme 2. However, this model requires the cis conformation for one ligand. This model has been proposed for platinum and palladium complexes [26], but has never been observed during x-ray diffraction studies.

Last, Models 5 and 6 imply the tetracoordination of the ligand on four metallic centers. Model 5 leads to the formation of strong interactions between the metallic centers (M-M = 3.1 Å) analogous

MODEL N°   M–M

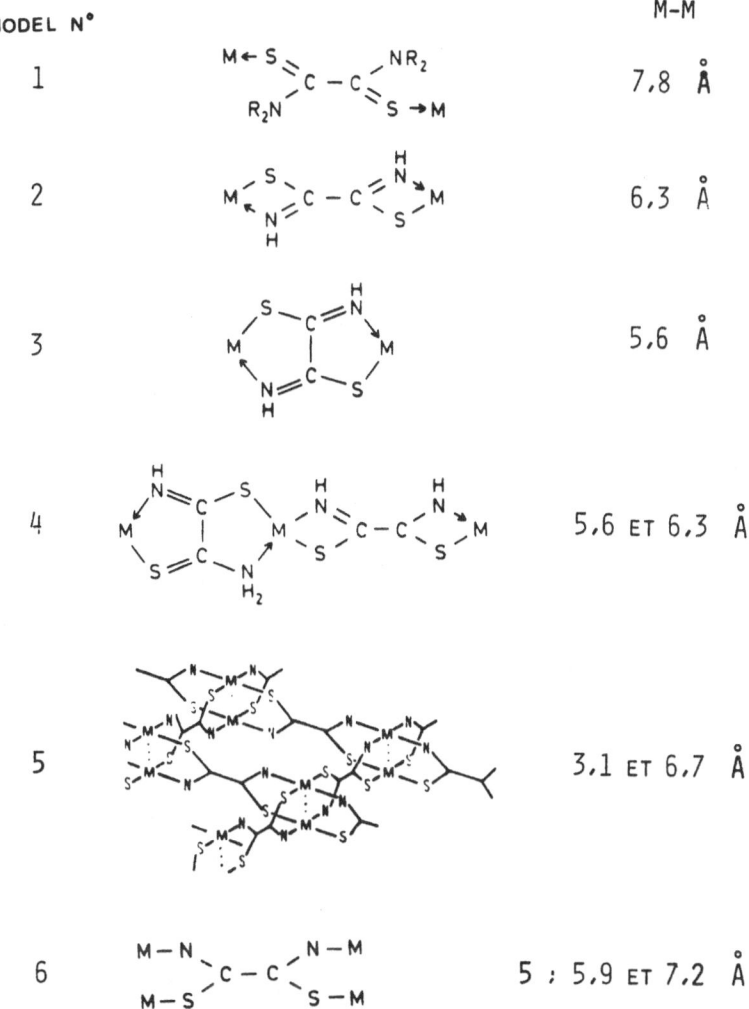

Fig. 15   Different coordination schemes proposed in the literature
          for the dithiooxamide complexes.

to that observed in copper acetate.   Model 6 implies that the ligand
has a cis-conformation.

     None of these structural models may be dismissed a priori
because of crystallographic or chemical constraints.   Such a
multiplication of models requires experiment prior to the choice
of a preliminary model.

     (ii)   A detailed crystallogenesis study allowed the first
achievement of two copper(I)-dithiooxamide complexes as single

crystals. Their structural study has been carried out to make
available the bond angles and bond lengths for the coordinated
ligand. However, due to the oxidation state of copper, no information has been obtained concerning the coordination scheme [27].

(iii) We perused the structural literature available for
metallic complexes analogous to dithiooxamide. Most studies have
been devoted to dithiooxalic acid, $HO(S)CC(S)OH$. From this survey,
it emerges that two types of assembling are usually found: polymeric
ribbons with a metal-metal distance averaging 5.6 Å and stackings
of ML_2 units or of ribbons with a distance between planes in the
range 3.5-4 Å.

The radial distribution functions of the two compounds studied
[28], $Ni_4L_5(H_2O)_{0.25}$ and $CuL(H_2O)$, have been represented in Fig. 16.
It clearly appears that the inner architectures are quite similar.
The shifts of the peaks may be explained by variations in
coordination distances. The differences in the intensity of a number of main peaks are attributed to differences in the length of
the chains, as developed later.

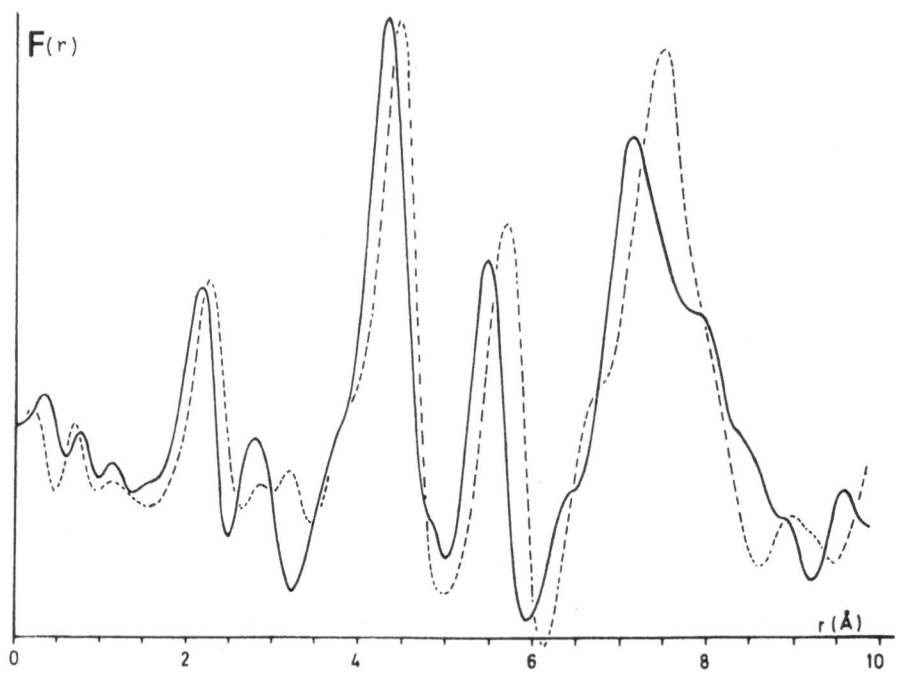

Fig. 16   Radial distribution curves for the Ni complex (full line)
          and the Cu complex (dashed line).

The first peak of the radial distribution functions is averagely centered at 2.2 Å and reflects the accumulation of coordination distances: metal-sulfur, metal-nitrogen. The three main peaks are averagely centered at 4.4, 5.6 and 7.3 Å.

Due to their high intensities, these peaks should necessarily result, either totally or partially, from the accumulation of metal-metal interactions. The distances for the various models are listed in Fig. 15, from which it appears that:

(i) The peak centered at 4.4 Å cannot be elucidated from the models. It is probably due to intermolecular or interchain interactions;

(ii) Models 2, 4, 5 and 6 may be dismissed as less likely because the corresponding metal-metal interactions have values corresponding with "holes" in the radial distribution function (6.3, 3.1, 5, and 5.9 Å, respectively). Model 1 partially allows to interpret the bulk centered at 7.3 Å but provides no information to simulate the beginning of the curve.

(iii) The peak at 5.6 Å is explained by model 3 which implies no "forbidden" distances.

These preliminary remarks allowed us to specify an initial structural model which is represented in Fig. 17 and characterized by the following criteria:

- bond angles and bond lengths are from copper(I) complexes studies;

- the rubeanic acid ligand is in a trans conformation;

- coordination occurs as in model 3 and leads to the formation of polymeric ribbons;

- the ribbons must be stacked to cause an accumulation of interatomic distances around 4.4 Å and 7.3 Å.

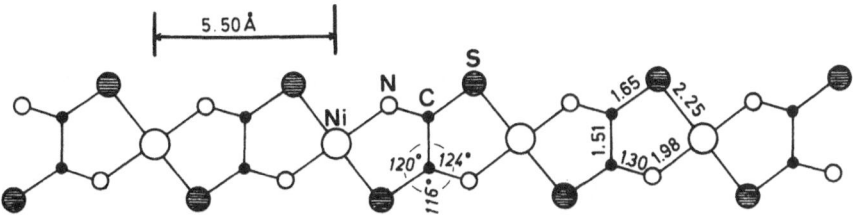

Fig. 17   The chain geometry used as initial model.

As a first step, only two chains (as described in Fig. 17) were taken into account. Their relative position was defined by fixing a number of metal–metal or metal–sulfur distances (Cf. Fig. 18): Ni(3)–Ni(4) = 4.30 Å; Ni(4)–S(7) = 5.50 Å (theoretical radial distribution in Fig. 19). The comparison with the experimental function calls for some comments.

 - The peaks at 4.4 and 5.6 Å are slightly shifted, which reflects an indefinite relative ordering of the chains.

 - The agreement between the theoretical and the experimental curve is very poor for the broad peak centered at 7.3 Å. This clearly shows the necessity of a third chain for a fair definition of the model.

 - The peaks of the theoretical curve at 4.4 and 5.6 Å are too weak and the model does not account for the shoulder at 3.7 Å. All the models tested in which the second chain follows from the first only by a translation of the reference trihedron have a similar defect.

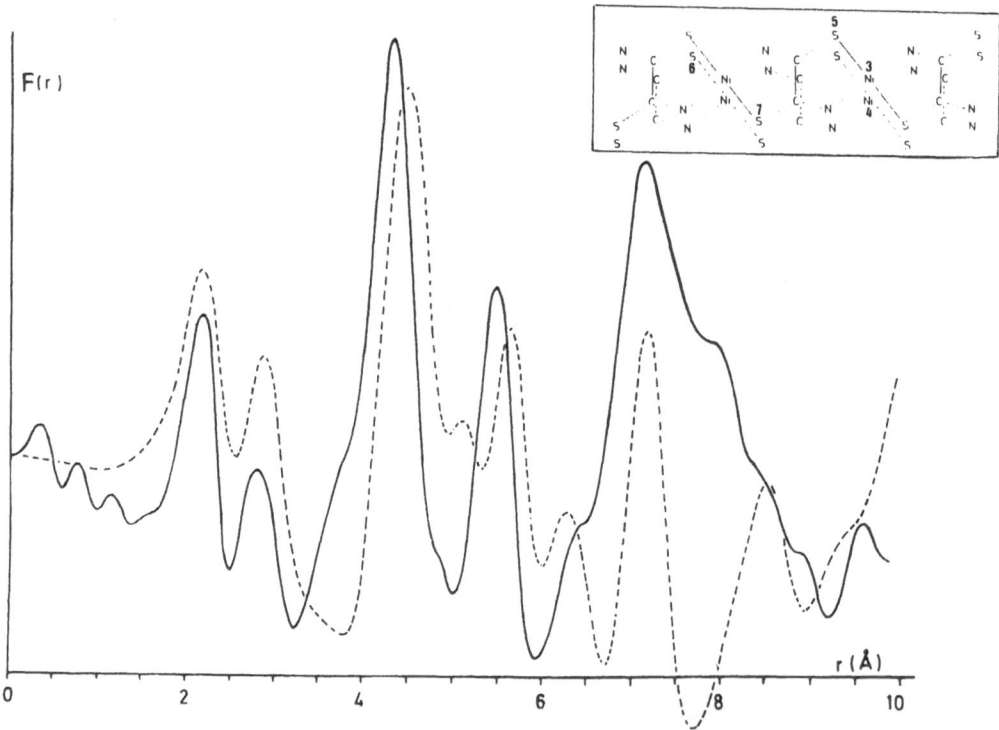

Fig. 18   Experimental (full line) and theoretical (dashed line) radial distribution curves for a model with two chains (see text).

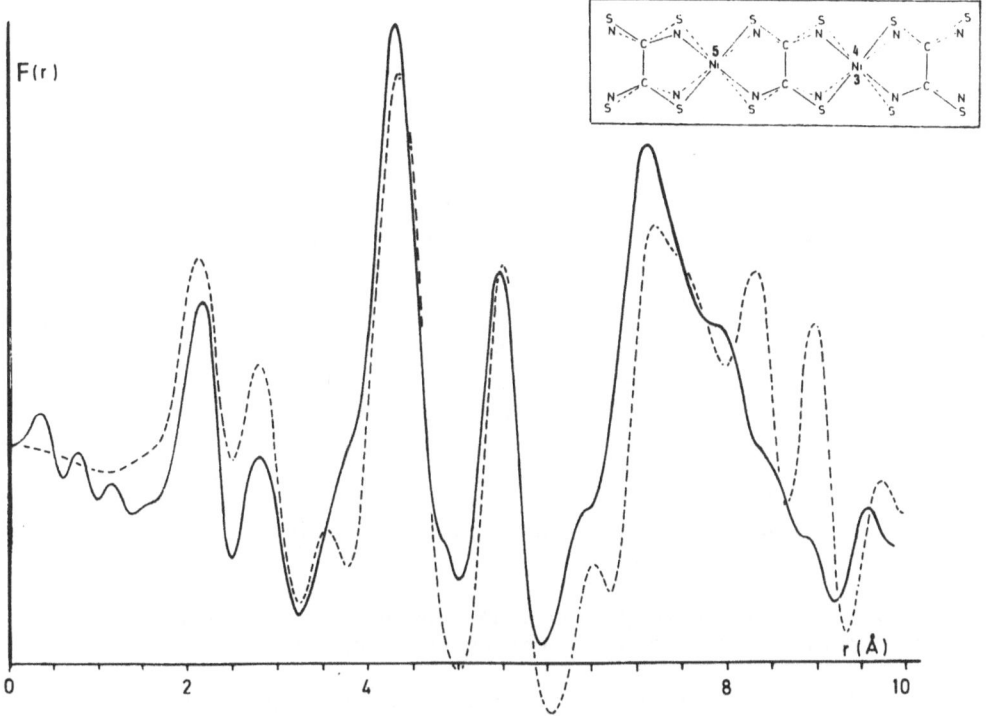

Fig. 19 Experimental (full line) and theoretical (dashed line)
radial distribution curves for the final model.

The only fair correlation was obtained by assuming the following model: the second chain follows from the first by a 3.6 Å
translation, perpendicularly to the plane of the chain followed by
a 180° rotation around the chain axis. As shown in the small
inset of Fig. 19, this operation results in an intersuperposition
of the nickel atoms and in a superposition of sulfur atoms with
nitrogen atoms. The metal—metal distances resulting from such a
disposition are Ni(3)-Ni(4) = 3.6 Å; Ni(3)-Ni(5) = 6.7 Å. This
model, extended to three chains with a 3.6 Å space between two
chains, leads to the theoretical curve represented in Fig. 19.

The agreement between experimental and radial distribution
functions is quite appropriate up to 6 Å and reasonable up to 8 Å.
Beyond this distance, considerable discrepancies appear. The
major criticisms against this structural model, which probably
explains such discrepancies, are the following :

(i) The model utilized is too symmetrical to reflect the
structural reality of the amorphous material. Small random
variations probably occur for each goemetrical characteristics:
dimensions of the dithiooxamide ligand, metal environment, chain

flatness, intervals between the chains. These variations produce
a lowering of the peaks at 2.8-3.7 Å and 6.5 Å. It is likely that
such local structural variations in the amorphous material will
result in the blunting and widening of long distance peaks, leading
to a series of shoulders which is seen in the experimental curve
between 8 Å and 9.5 Å.

(ii) Interactions between chain stackings existing in the
material have not been accounted for in the structural model. This
approximation also allows us to explain the discrepancies between
theory and experiment beyond 8 Å.

Despite these discrepancies between theoretical and experimental curves, it is assumed that the final model proposed, represented
on Fig. 20, must be affected with a high reliability factor.

In the case of the copper(II) complex, we did not attempt to
refine a specific model because of the great similarity of the
radial distribution curves (Fig. 16). The differences between the
two curves have two grounds:

(i) The mean coordination bond length, i.e., 2.16 Å for the
nickel complex and 2.25 Å for the copper complex;

(ii) The chain length: the number of metal–metal interactions
at 5.65 Å rises from 0.75 per metallic center when nickel is

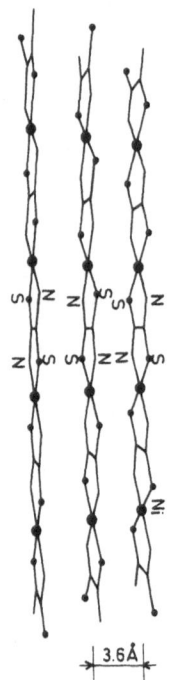

Fig. 20   Proposed structural model for
the Ni–dithiooxamide complex.

involved to 1 when copper is involved, consequently increasing
the peak intensity. Such an argument may also apply to the peak
at 7.40 Å although less easily because this peak results from the
accumulation of various contributions.

5. Complexes MM'(EDTA)(H$_2$O)$_4$,2H$_2$O

        A great number of works in the field of polymeric complexes
of transition metals have provided valuable information on the
correlations between structure and magnetic behavior, whereas
polymeric system with some structural disorders have been little
studied.

        The complexes MM'(EDTA)(H$_2$O)$_4$,2H$_2$O, where M and M' are transi-
tion metals, are highly propitious to being studied as they may be
prepared either as single crystals or in the amorphous state.

        This structural investigation was meant to elaborate a possible
structure for the amorphous variety of these complexes by confron-
ting our results with those carried out [29] on the crystallized
complex Zn$_2$(EDTA)(H$_2$O)$_4$,2H$_2$O. The structural organization consists
of infinite zig-zag strings of a kind of polymer, [MM'(EDTA)(H$_2$O)$_4$]$_\infty$,
with alternated metallic centers (Fig. 21). One site (M) exhibits
an hexacoordination through the ligand EDTA (Fig. 22), while the
other (M') is surrounded by two oxygen atoms in cis-position
belonging to two bridging acid functions of the EDTA ligand and
four water molecules.

        It clearly appears that the geometry of these chains is directly
fixed by the dimensions of the triangle MM'M.

        The first member to be studied in this [MM'] family was the
[CoNi] complex [30]. The following results were brought out:

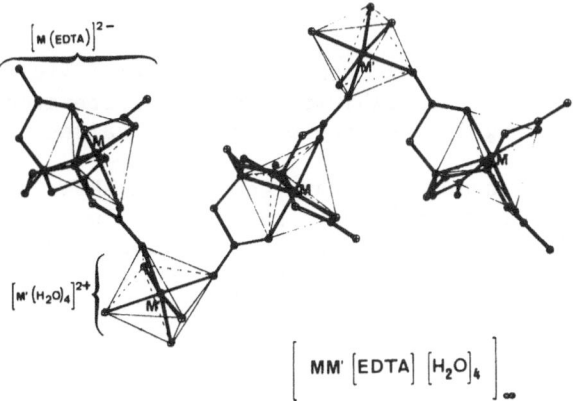

Fig. 21   Geometry of the infinite zig-zag string [MM'(EDTA)(H$_2$O)$_4$]$_\infty$.

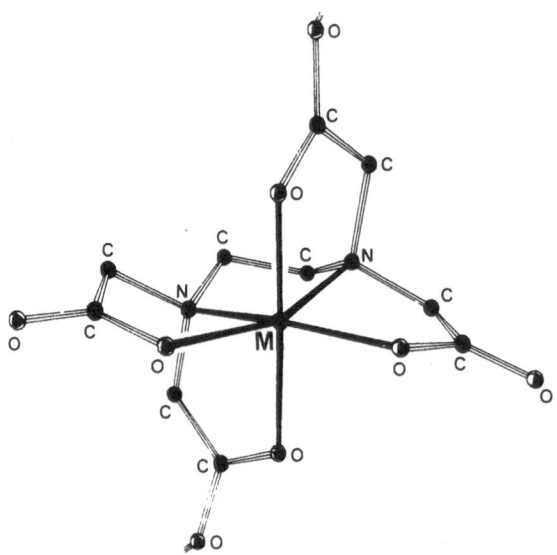

Fig. 22   Coordination scheme of the EDTA ligand.

(i) zig-zag chains are present as in the crystalline state (ii)
the EDTA coordination scheme is identical, and (iii) some enlarge-
ment of experimental peaks in the experimental radial distribution
function $F(r)_{exp}$ compared with the theoretical one $F(r)_{th}$ can be
fairly well explained by the coexistence in the amorphous state of
chains showing the MM'M angle to vary slightly around a mean value
of 79°.

     The comparison between experimental radial distribution func-
tions of both [CoNi] and [NiNi] amorphous compounds (Fig. 23) leads
to the following conclusions.

     (i) It is obvious that these complexes have similar local
structures. Both of them exhibit curves characterized by nearly
identical peaks roughly centered on the same values.

     (ii) Further, a more accurate comparison shows that the [NiNi]
curve exhibits a "slimming" and a partial resolution of the peaks
centered at 5.8 Å and 7.2 Å. Such a situation could imply that
the chain geometry distribution already mentioned for the [CoNi]
complex is less broad and that extreme conformations coexist in
the [NiNi] amorphous compound.

     These preliminary observations drive us to elucidate the
amorphous structure by considering two kinds of chain geometries
which both remain strongly related to the crystalline model.

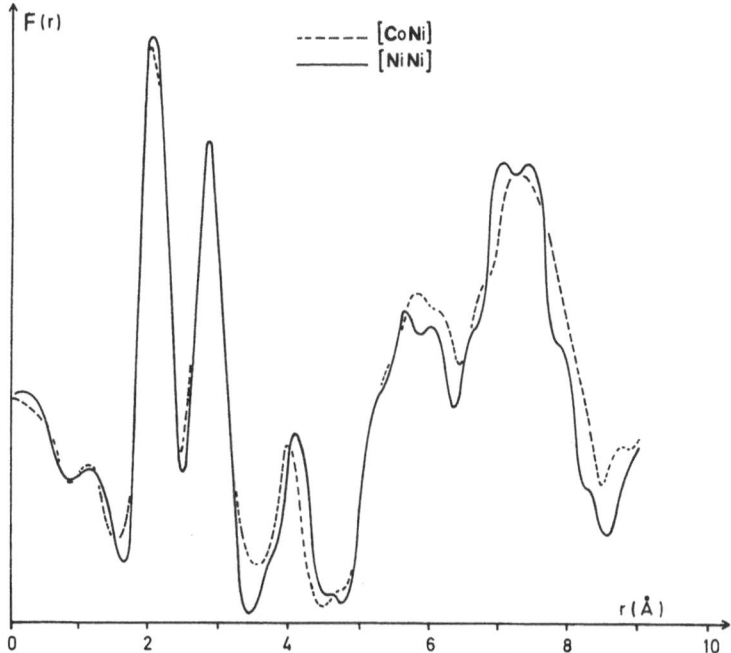

Fig. 23   Comparison of the experimental distribution curves for
CoNi (dashed line) and NiNi (full line) complexes.

The hexacoordination of the EDTA ligand implies a quasi-locked
conformation for the $[M(EDTA)]^{2-}$ moiety. The variations of the
chain geometry are thus directly connected to the acid function
bridges and to the chain angle MM'M.

In the crystalline [NiNi] complex, a "short" (5.58 Å) and a
"long" (5.99 Å) metal–metal distance are associated to a rather
distorted acid function. The third metal–metal distance is equal
to a 7.36 Å [31].

The experimental radial distribution function of the [NiNi]
complex (Fig. 24, (full line)) exhibits maxima centered around
5.8 Å and 7.2 Å which are resolved into peaks positioned at 5.67 Å
and 6.02 Å, and 7.06 Å and 7.44 Å, respectively. Such maxima
resulting from various contributions are mainly affected by the
metal–metal distances even though metal–ligand and ligand–ligand
interactions are able to affect the position of the metal–metal
peaks.

Then, a strong correlation clearly appears between the interatomic distances issued from the radial distribution and the
calculated distances, except the maximum at 7.06 Å which is not
explained; so a further more sophisticated model has to be developed.

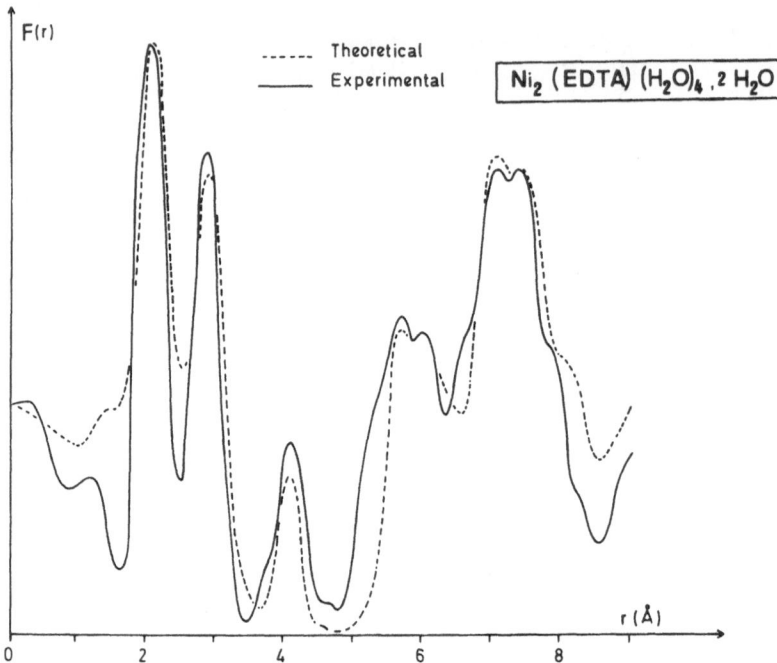

Fig. 24   Comparison of the experimental (full line) and theoretical
          (dashed line) and distribution cruves for the NiNi complex.

In order to elaborate this model, the angles ONi'O, Ni'OC, OCO,
and CONi were allowed to vary, step-by-step, within the following
limits:

   $110°-145°$ for Ni'OC, OCO, and CONi
   $85°-110°$ for ONi'O.

   After each cycle of calculations, a theoretical radial distribution curve was calculated in order to specify the influence of
the angle variations and to check the good adequation between
experimental and theoretical curves.

   The final model (Fig. 24, (dashed line)) is characterized by
two triangular MM'M units with different geometries.

   The first unit is rather close to the crystal model with
NiNi' = 5.58 Å, Ni'Ni = 5.90 Å, and NiNi = 7.45 Å; it corresponds
to a dissymetric triangle.

   The second model is symmetrical, with NiNi' = Ni'Ni = 5.58 Å
and NiNi = 7.05 Å. Such a model is obtained by ascribing the values
reported in Fig. 25 to the variable parameters. Such a scheme calls
for the following comments.

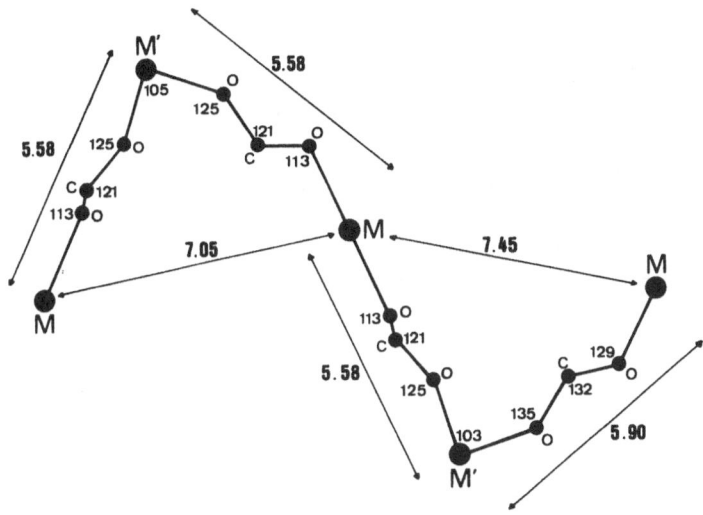

Fig. 25   Schematic drawing of the proposed structural model.

(i) The various geometries have been reported on the same drawing (Fig. 25) for convenience. In fact, these geometries are probably randomly distributed within infinite chains.

(ii) These two extreme geometries imply the possibility of a distribution of units with intermediate interatomic distances and/or angles within the amorphous solid.

(iii) Bond angles are given as indications; the method utilized does not allow a better precision than 5° for each refined angle.

Theoretical calculations, based on several models with interatomic distances or angles different from those proposed, have shown strong discrepancies with the results experimentally established by LAXS studies. It is always important to notice the non univocal answer of such calculations when establishing a structural model. Anyhow, the good agreement obtained after numerous calculations and the reasonable architectural model retained made us confident in the description of the amorphous complex $Ni_2(EDTA)(H_2O)_4, 2H_2O$ as infinite zig-zag chains built up with octahedra $(Ni^{II}(EDTA))$ and $(Ni^{II}O_2(H_2O)_4)$ sharing corners, the non-coordinated water molecules being randomly distributed in the sample [32].

CONCLUSION

The aim of this paper was to bring attention on the real possibilities for Large Angle X-ray Scattering techniques to be a useful tool in the structural study of inorganic compounds in their amorphous state.

Among the molecular compounds, very little has been done in such a field as the technique is devoted to the investigation of silica, e.g., glasses, melted or amorphous alloys or some complexes containing very heavy metals in solution.

The new tool (automatic diffusometer with a position-sensitive detector) and software, both especially designed for such studies and for an easy elaboration and checking of structural models (LASIP system), could provide routine investigation in the field of inorganic molecular or coordination chemistry.

The various examples given in the paper, covering samples issued from the solid state, i.e., V_2O_5, but essentially coming from coordination chemistry have emphasized the necessity for building structural models for the wide and deep knowledge of crystallochemistry. Through the LAXS technique, it is possible to obtain (i) the coordination scheme of the metals, (ii) the molecular conformations, and (iii) the organization of the short range order up to 7 or 8 Å and in some cases up to 10 Å (molecular chains).

LAXS studies, occasionally combined with EXAFS investigation, appear to be extremely fruitful in the field of molecular or complex chemistry of noncrystalline compounds by its ability to provide valuable information concerning bond lengths and bond angles, the core of Chemistry.

REFERENCES

1. J. Galy, A. Mosset, and P. Lecante, ANVAR French Patent No. 80.16170 and U.S. Patent No. 4.475.225 (1984).
2. P. Lecante, A. Mosset, and J. Galy, J. Appl. Cryst. 18:214-218 (1985).
3. M. Norman, Acta Cryst. 10:370-373 (1957).
4. J, Krogh-Moe, Acta Cryst. 9:951-954 (1956).
5. D. T. Cromer, and J. T. Waber, Acta Cryst. 18:104-111 (1965).
6. D. T. Cromer, J. Chem. Phys. 50:4857-4860 (1969).
7. F. Zernicke, and J. A. Prins, Z. Physik 41:184-190 (1927).
8. I. B. Patrina, and V. A. Ioffe, Soviet Phys. Solid State 6: 2581-2588 (1965).
9. A. B. Scott, J. C. McCullock, and K. M. Mar, Conduction in Low Mobility Materials (London: Taylor and Francis Ltd.):105-113.
10. J. Haemers, E. Baetens, and J. Vennick, Phys. Status Solidi(a) 20:381-387 (1973).
11. O. Kahn, J. Livage, and R. Collongues, Phys. Status Solidi(a) 26:175-180 (1974).
12. H. G. Bachmann, F. R. Ahmed, and W. H. Barnes, Z. Krist 115:110-116 (1961).
13. A. D. Wadsley, and S. Andersson, Nature 211:581-582 (1966).
14. J. C. Bouloux, and J. Galy, J. Solid State Chem. 16:385-393 (1976).

15. A. Mosset, P. Lecante, J. Galy, and J. Livage, Phil. Mag. 46: 137-149 (1982).
16. H. Morikawa, M. Miyake, S. Iwai, K. Furukawa, and A. Revcolevschi, J. Chem. Phys., Faraday Trans. I 77:361-366 (1981).
17. M. Henri, C. Sanchez, C. R'Kha, and J. Livage, J. Phys. Sect. C 14:829-837 (1973).
18. A. Mosset, P. Lecante, and J. Galy, C. R. Acad. Sc. Paris, Ser. C 290:325-328 (1980).
19. A. Mosset, Thesis, Toulouse (1979).
20. T. M. M. Kahn, and A. E. Martell, J. Phys. Chem. 66:10-18 (1962).
21. M. Sundaralingam, Biopolymers 7:821-860 (1969).
22. A. Mosset, and J.-J. Bonnet, Acta Crystallogr. Sect. B 33:2807-2812 (1977).
23. J. Dehand, I. Jordanov, F. Keck, A. Mosset, J.-J. Bonnet, and J. Galy, Inorg. Chem. 18:1543-1549 (1979).
24. W. A. Franks, and D. Van der Helm, Acta Crystallogr. B27:1299-1310 (1971).
25. A. Mosset, unpublished results (1979).
26. M. Bobtelsky, and J. Eisenstadter, Analytica Chim. Acta 17:579-587 (1957).
27. A. Mosset, M. Abboudi, and J. Galy, Z. Krist. 164:171-180 (1983a) A. Mosset, M. Abboudi, and J. Galy, Z. Krist. 164:181-188 (1983).
28. M. Abboudi, Thesis, Toulouse (1983).
29. A. J. Pozhidaev, T. N. Polynova, M. A. Porai-Koshits, and N. N. Meronova, Zh. Struk. Khim 6:1383-1394 (1973).
30. A. Mosset, E. Coronado, and J. Galy, C. R. Acad. Sc. Paris, Ser. II 296:549-552 (1983).
31. E. Coronado, M. Drillon, A. Fuertes, D. Beltran, A. Mosset and J. Galy, J. Amer. Chem. Soc. in press (1985).
32. A. Mosset, J. Galy, E. Coronado, M. Drillon, and D. Beltran, J. Amer. Chem. Soc. 106:2864-2869 (1984).

CONTRIBUTORS

R. Araujo, Research and Development Division, Corning Glass Works,
          Corning, New York 14831.

L. H. Bennet, Center for Materials Science, National Bureau of
          Standards, Gaithersburg, MD 20899

P. J. Bray, Department of Physics, Brown University, Providence, RI
          02912

W. Y. Ching, Department of Physics, University of Missouri-Kansas
          City, Kansas City, MO 64110

M. Cutler, Department of Physics, Oregon State University, Corvallis,
          OR 97330

T. Dunn, Department of Geology, University of Alberta, Edmonton,
          Alberta, Canada

A. H. Edwards, U.S. Army ET&D Laboratory, Electronic Materials
          Division, Device Physics and Analysis Branch,
          Fort Monmouth, NJ 07703

S. R. Elliott, Department of Chemistry, University of Cambridge,
          Cambridge, U.K.

G. J. Exarhos, Battelle, Pacific Northwest Laboratory, Richland, WA
          99352

W. B. Fowler, Department of Physics and Shermon Fairchild Laboratory,
          Lehigh University, Bethlehem, PA 18015

J. Galy, Laboratoire de Chimie de Coordination du CNRS associe a
          l'Universite Paul Sabatier, 205 route de
          Narbonne, 31400 Toulouse, France

S. Garofalini, Rutgers University, College of Engineering, Department
          of Ceramics, Brett and Bowser Roads, Busch
          Campus, P. O. Box 909, Piscataway, NJ 08854

A. I. Goldman, Department of Physics, Brookhaven National Laboratory,
          Upton, NY 11973

D. L. Griscom, Naval Research Laboratory, Washington, DC 20375-5000

A. Hiraki, Department of Electrical Engineering, Osaka University,
          Suita, 565 Osaka, Japan

K. Hirao, Department of Industrial Chemistry, Faculty of Engineering,
          Kyoto University, Kyoto, Japan

M. S. Hokmabadi, Department of Chemistry, Howard University,
          Washington, DC 20059

A. K. Jonscher, Chelsea Dielectric Group, Chelsea College, University
          of London SW65PR, U.K.

R. J. Kirkpatrick, Department of Geology, 1301 West Green Street,
          Urbana, IL 61801

M. Kuriyama, Center for Materials Science, National Bureau of
          Standards, Gaithersburg, MD 20899

C. Laermans, Kath. University Leuven Physics Department, VSHD 3030
          Leuven, Belguim

G. G. Long, Center for Materials Science, National Bureau of
          Standards, Gaithersburg, MD 20899

M. L. Lui, Department of Physics, Brown University, Providence, RI
          02912

R. Mosseri, Laboratoire de Physique des Solides, CNRS, 92190 Meudon
          Principal Cedex, France

A. Mossett, Laboratoire de Chimie de Coordination du CNRS associe a
          l'universite Paul Sabatier, 205 route de
          Narbonne, 31400 Toulouse, France

R. Oestrike, Department of Geology, 1301 West Green Street, Urbana,
          IL 61801

J. F. Sadoc, Laboratoire de Physique des Solides, Universite Paris-Sud
          Batiment 51091405 Orsay-Cedex, France

S. Schramm, School of Chemical Sciences, University of Illinois,
          505 S. Mathews Street, Urbana, IL 61801

J. F. Shackelford, Division of Materials Science and Engineering,
          University of California, Davis, CA 95616

K. A. Smith, School of Chemical Sciences, University of Illinois,
                505 S. Mathews Street, Urbana, IL 61801

N. Soga, Department of Industrial Chemistry, Faculty of Engineering,
                Kyoto University, Kyoto, Japan

G. Turner, School of Chemical Sciences, University of Illinois,
                505 S. Mathews Street, Urbana, IL 61801

G. E. Walrafen, Department of Chemistry, Howard University,
                Washington, DC 20059